The PHΨSICAL WORLδ

QUANTUM PHYSICS: AN INTRODUCTION

Edited by Joy Manners

IoP

Institute of Physics Publishing
Bristol and Philadelphia
in association with

The Physical World Course Team

Course Team Chair	Robert Lambourne
Academic Editors	John Bolton, Alan Durrant, Robert Lambourne, Joy Manners, Andrew Norton
Authors	David Broadhurst, Derek Capper, Dan Dubin, Tony Evans, Ian Halliday, Carole Haswell, Keith Higgins, Keith Hodgkinson, Mark Jones, Sally Jordan, Ray Mackintosh, David Martin, John Perring, Michael de Podesta, Ian Saunders, Richard Skelding, Tony Sudbery, Stan Zochowski
Consultants	Alan Cayless, Melvyn Davies, Graham Farmelo, Stuart Freake, Gloria Medina, Kerry Parker, Alice Peasgood, Graham Read, Russell Stannard, Chris Wigglesworth
Course Managers	Gillian Knight, Michael Watkins
Course Secretaries	Tracey Moore, Tracey Woodcraft
BBC	Deborah Cohen, Tessa Coombs, Steve Evanson, Lisa Hinton, Michael Peet, Jane Roberts
Editors	Gerry Bearman, Rebecca Graham, Ian Nuttall, Peter Twomey
Graphic Designers	Javid Ahmad, Mandy Anton, Steve Best, Sue Dobson, Sarah Hofton, Jennifer Nockles, Pam Owen, Andrew Whitehead
Centre for Educational Software staff	Geoff Austin, Andrew Bertie, Canan Blake, Jane Bromley, Philip Butcher, Chris Denham, Nicky Heath, Will Rawes, Jon Rosewell, Andy Sutton, Fiona Thomson, Rufus Wondre
Course Assessor	Roger Blin-Stoyle
Picture Researcher	Lydia K. Eaton

The Course Team wishes to thank the following individuals for their contributions to this book: Tony Evans, Ian Halliday, Tony Sudbery, Robert Lambourne and Ray Mackintosh. The book made use of material originally prepared for the S271 Course Team by Graham Farmelo, Joy Manners and John Walters. The multimedia packages *Electron diffraction* and *Stepping through Schrödinger's equation* were written by Joy Manners and programmed by Fiona Thomson. (*Electron diffraction* made use of material originally prepared for the S271 Course Team by John Walters and both packages made use of original prototypes programmed by Robert Hasson.)

The Open University, Walton Hall, Milton Keynes MK7 6AA

First published 2000

Written, edited, designed and typeset by the Open University.

Published by Institute of Physics Publishing, wholly owned by The Institute of Physics, London.
IoP Publishing, Dirac House, Temple Back, Bristol BS1 6BE, UK.

US Office: Institute of Physics Publishing, The Public Ledger Building, Suite 1035, 150 South Independence Mall West, Philadelphia, PA 19106, USA.

Printed and bound in the United Kingdom by the Alden Group, Oxford.

ISBN 0 7503 0720 X

Library of Congress Cataloging-in-Publication Data are available.

This text forms part of an Open University course, S207 *The Physical World*. The complete list of texts that make up this course can be found on the back cover. Details of this and other Open University courses can be obtained from the Course Reservations Centre, PO Box 724, The Open University, Milton Keynes MK7 6ZS, United Kingdom: tel. +44 (0) 1908 653231; e-mail ces-gen@open.ac.uk

Alternatively, you may visit the Open University website at http://www.open.ac.uk where you can learn more about the wide range of courses and packs offered at all levels by the Open University.

To purchase other books in the series *The Physical World*, contact IoP Publishing, Dirac House, Temple Back, Bristol BS1 6BE, UK: tel. +44 (0) 117 925 1942, fax +44 (0) 117 930 1186; website http://www.iop.org

1.1

s207book7i1.1

QUANTUM PHYSICS: AN INTRODUCTION

Introduction

Two of the most remarkable revolutions in the history of science took place in Europe at the beginning of the twentieth century. One began in 1905, when Einstein formulated the special theory of relativity. This is based on two principles, namely, that the same basic physical laws apply in all inertial reference frames and that the velocity of light in a vacuum is constant. Einstein's theory imposed modifications on the Newtonian concepts of space and time, and these modifications led to a radically new and unified interpretation of the *classical* physics of Newton and Maxwell. In this sense, the special theory of relativity can be regarded as the crowning glory of classical physics.

The same is certainly *not* true for the other great revolution, which was brought about by the advent of *quantum physics*. This was entirely new and some of its assumptions conflict with those of classical physics. This does not imply that all the physics you have learnt so far is obsolete and can be discarded. Quantum physics is required only when we try to understand phenomena on the *atomic* scale. This book will first show you how the need for quantum physics arose, and will then outline the new theory of *quantum mechanics*, as developed by Erwin Schrödinger, which aimed to explain the behaviour of particles at the atomic level.

In Chapter 1 we review a set of phenomena for which no explanation could be found within the framework of classical physics. We shall then go on to show how these difficulties were resolved, at least partially, by Planck's quantum hypothesis.

Chapter 2 introduces the revolutionary ideas of quantum mechanics and the basic principles of the theory, while in Chapter 3 these principles are applied to the understanding of atomic structure.

It is important to realize that the postulates and interpretation of quantum theory are completely different from our normal intuitions about natural phenomena. In fact, the debate about the real meaning of quantum-mechanical quantities has been fiercely argued since its inception and continues with unreduced fervour to this day. Chapter 4 is devoted to some of the aspects of the interpretation of quantum theory.

Open University students should view Video 7, *The Search for Reality*, at some stage during their study of this book. This will give you a light-hearted first look at some of these strange philosophical aspects of quantum theory. The video can be viewed at any stage, but might be most effective at the end of Chapter 1.

Chapter 1 The origins of quantum physics

1 The Compton effect — a strange dichotomy in the nature of electromagnetic radiation

In a series of experiments conducted between 1919 and 1923, Arthur Holly Compton (Figure 1.1) investigated the scattering of monochromatic (single-wavelength) X-rays from graphite targets. He found that the radiation scattered at an angle ϕ to the incident beam contained, in addition to radiation of the *same* wavelength as the incident radiation, a second component with a considerably *longer* wavelength. The existence of this second component in the scattered radiation could not be explained by the well-established theory of the scattering of electromagnetic waves by electrons. Compton considered instead the process illustrated in Figure 1.2, in which a *particle* of electromagnetic radiation, collides with a slow moving electron, which then recoils absorbing some of the X-ray particle's energy. Treating the process as a collision between *particles*, and using only the (relativistic) conservation laws of energy and momentum, Compton was able to account for the effect completely. However, at the same time, in the same experiment, Compton was using *interference* effects, depending wholly on the wave model, in order to determine the wavelength of the scattered radiation. So here was an experiment which apparently required the simultaneous use of both the wave and particle models of electromagnetic radiation for the interpretation of the results! The existence of this strange dichotomy in the nature of electromagnetic radiation was just one indication of the need for a radical revision of views regarding the physical world.

Figure 1.1 Arthur Holly Compton (1892–1962) was a native of Ohio in the United States. After receiving his doctorate at Princeton, he began his career as a research physicist at the Westinghouse Lamp Company before returning to academic circles by moving to Cambridge in 1919. He later held professorships at several American universities. An authority on X-rays, he was awarded the Nobel Prize for physics in 1927 for his discovery and interpretation of the effect that bears his name.

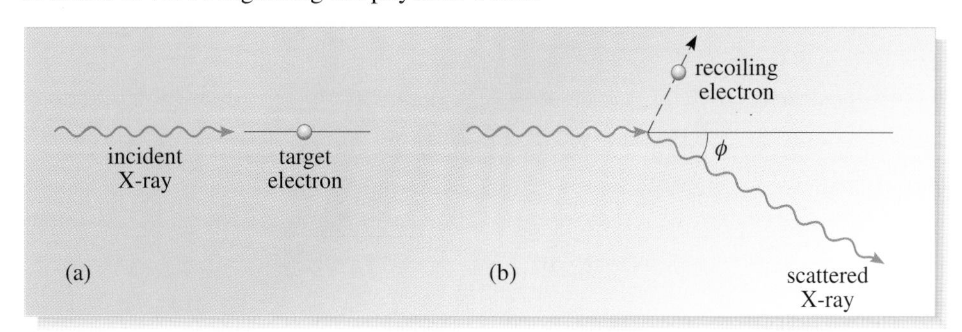

Figure 1.2 (a) The Compton effect. An X-ray or γ-ray 'particle' collides with a slow moving electron in one of the target atoms. (b) The electron recoils, absorbing energy from the X-ray particle which is scattered into a new direction and with increased wavelength.

2 Five problems for classical physics

At the beginning of the twentieth century, physicists were faced with several profound problems that could not be solved by using the classical theories of Newton and Maxwell. In this section we shall describe five of these problems and later in the chapter you will see how all were solved (at least partially) by using the ideas of quantum physics that were formulated between 1900 and 1922.

2.1 Problem 1: Understanding atoms

By the end of the nineteenth century, it was becoming accepted that matter was *not* infinitely divisible (continuous), but that it consisted of discrete parts, which were called *atoms*. Amongst the most persuasive evidence for this was the work of Dalton and Gay-Lussac on the proportions in which chemical elements combine with each other. The fact that these proportions were often in ratios of small integers indicated that the substances that were combining in these reactions were doing so in discrete amounts.

The results of some rather crude experiments had indicated that the atoms that made up the different chemical elements each had a diameter of, very roughly, 10^{-10} m (see Example 1.1). This observation prompted some scientists to ask 'What's so special about 10^{-10} m — why shouldn't atoms have diameters of, say, 10^{-6} m?' This may strike you as rather an odd question — is it reasonable to ask why atoms have a certain size? Shouldn't the typical atomic size be taken as being a fact of nature? Fortunately, the Danish physicist Niels Bohr *did* regard this as a pertinent and extremely important question for physics, and was able to give a clear answer to it.

Another important question concerned the constituents and structure of the atoms themselves. As you will see later in this chapter, important advances in this direction were made by J. J. Thomson and Ernest Rutherford.

Example 1.1

Given that the density of diamond is 3.5×10^3 kg m^{-3} and the relative atomic mass of carbon is 12.0, estimate the radius of a carbon atom.

Solution

From the information provided, we know that one mole of diamond has a mass of 0.012 kg, so that its volume is

$$\frac{\text{mass}}{\text{density}} = \frac{0.012}{3500}\,\text{m}^3 = 3.43 \times 10^{-6}\,\text{m}^3.$$

We know that the number of atoms in a mole is $N_A = 6.02 \times 10^{23}$ (Avogadro's number) and the volume occupied by one atom is therefore

$$\frac{3.43 \times 10^{-6}\,\text{m}^3}{6.02 \times 10^{23}} = 5.70 \times 10^{-30}\,\text{m}^3.$$

If this tiny volume were in the form of a cube, its sides would be of length $(5.70 \times 10^{-30})^{1/3}$ m $= 1.8 \times 10^{-10}$ m. The radius of an atom can be no larger than half of this, that is, about 10^{-10} m.

2.2 Problem 2: Understanding spectroscopy

One branch of science that had developed considerably by the end of the nineteenth century was that of **spectroscopy**, the study of the light emitted by chemical substances when they are heated in a flame.

To find the wavelengths present in, for example, the characteristic yellow light emitted by the element sodium, a beam of the light is shone on a diffraction grating (Figure 1.3a). If the light consisted of *all* wavelengths in the yellow part of the

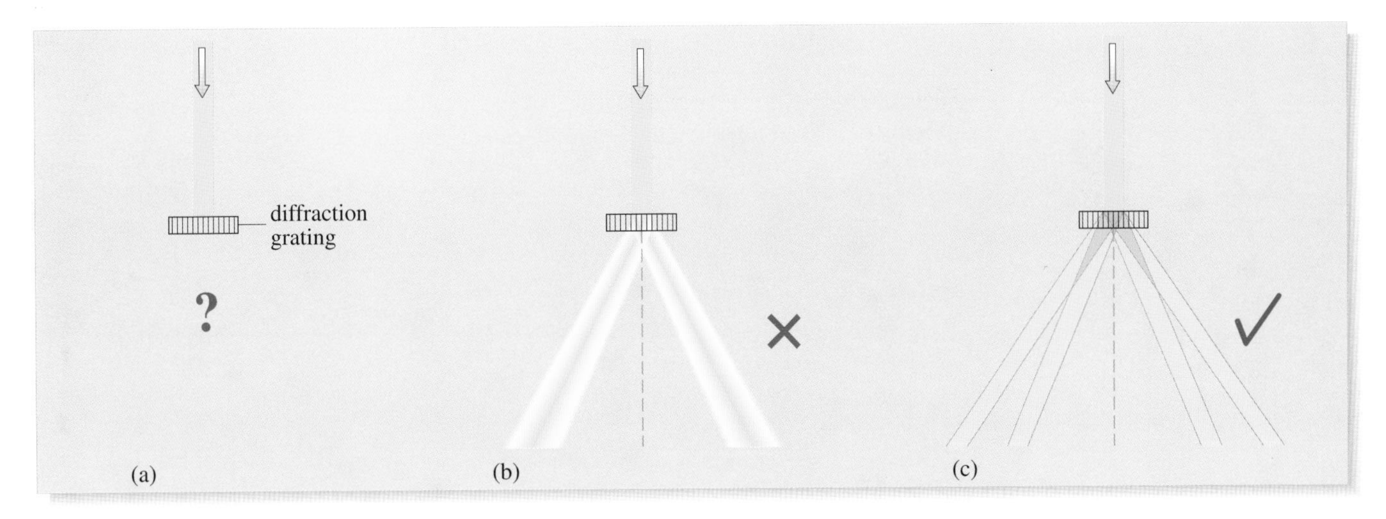

Figure 1.3 (a) What happens when a parallel beam of yellow sodium light is shone onto a diffraction grating? (b) When the light emerges from the grating, it does not fan outwards: this shows that the light does not contain all the frequencies in the range 580–600 nm corresponding to yellow light. (c) Instead, diffraction occurs only at certain definite angles, showing that sodium light consists of discrete frequencies, i.e. spectral lines. (Note that the grating spacing and the angular separation of the spectral lines are grossly exaggerated here. Only the first diffraction order is shown on each side of the straight through direction.)

visible spectrum (580–600 nm), it would simply fan out from the grating to give continuous bands on either side of the straight-through beam (Figure 1.3b). (The abbreviation nm stands for nanometre, 1 nm being equal to 10^{-9} m.) However, *this does not happen*. Instead, the light is diffracted only at certain definite angles (Figure 1.3c). This implies that the yellow sodium light consists of certain, definite wavelengths, which have come to be known as **spectral lines**. The wavelength, λ, of each line in the spectrum can be determined by the angle, θ_n, through which it is diffracted by the grating, provided that the grating spacing, d, is known. As shown in *Dynamic fields and waves*, θ_n is given by the equation $\sin \theta_n = n\lambda/d$, where n is the order of the diffraction. Each element has its own characteristic spectrum, that is, its own individual pattern of spectral lines. You can see examples of different line spectra in Figure 1.4. The fact that the elements have different spectra is the basis of an important experimental technique in chemical analysis. If the spectrum of an

Figure 1.4 The visible atomic spectra of helium, iron and neon.

element is identified in the spectrum of a chemical sample, it can be concluded that the sample contains that particular element. This method establishes the existence of particular elements in the outer layers of the Sun and other stars and even provides information about their relative abundance (Figure 1.5).

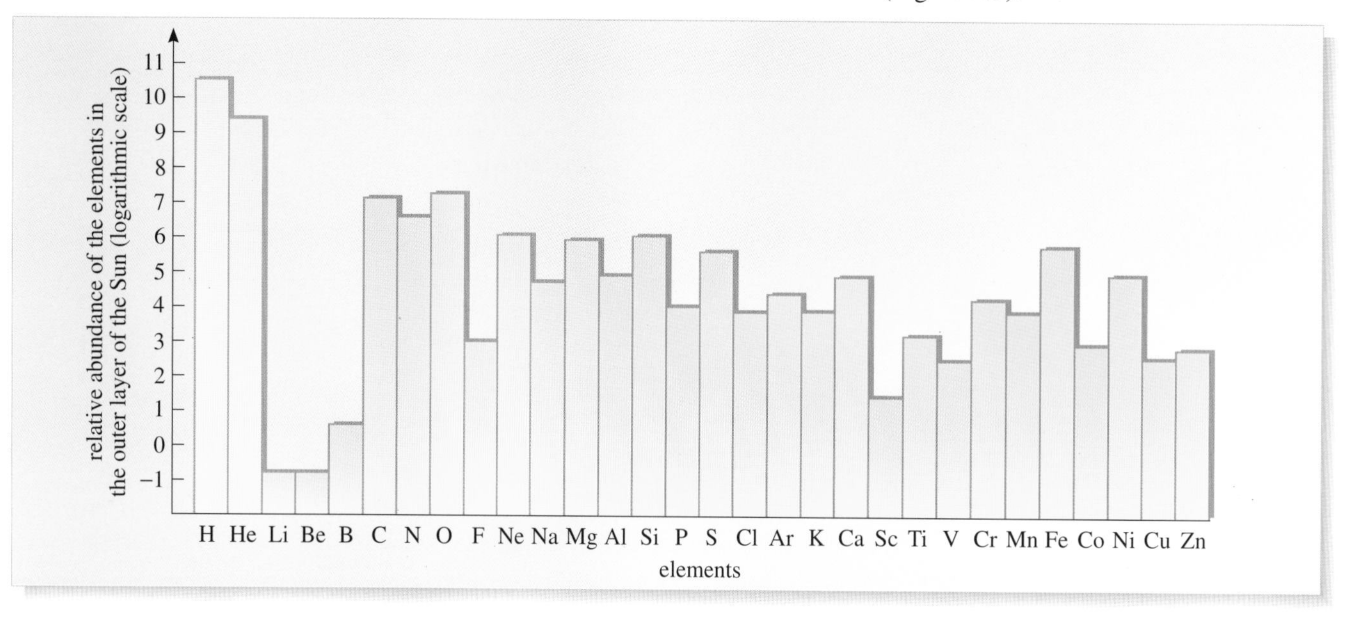

Figure 1.5 Relative abundance of the chemical elements in the solar spectrum. Note the logarithmic scale: an element with a relative abundance of 5 is ten times as abundant as one with a value of 4.

By 1860, spectroscopists had determined the spectra of many elements. That was all very well, but the really difficult problem was to *interpret* these data. What could be concluded from the patterns of spectral lines, and why *do* elements emit light of only certain specific wavelengths? It was evident that these questions were related to still deeper questions about the internal structure of atoms. The way forward was unclear. It seemed no easier for a scientist to learn anything about the structure of atoms from their spectra than for an engineer to deduce the internal construction of a piano from a performance of the 'Moonlight' Sonata.

So far, we have only drawn your attention (in Figure 1.4) to the complicated spectra of elements such as helium, iron and neon. But what about the spectrum of *hydrogen*, the element with the lightest (and simplest?) atoms of all? Perhaps it *would* be feasible to find out something about the least complicated element.

The Swiss mathematician Johann Balmer surmised in 1885 that the best way of tackling the problems of spectroscopy was to concentrate first on trying to understand the visible spectrum of hydrogen, which is shown in Figure 1.6. He made an excellent start on this problem by finding an interesting numerical pattern

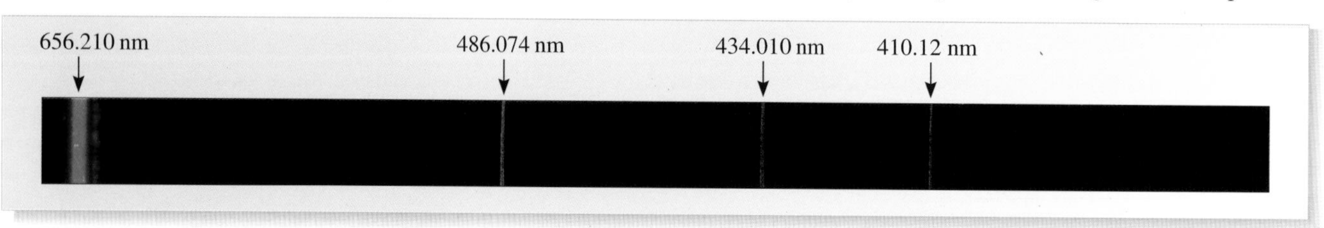

Figure 1.6 The four visible spectral lines of atomic hydrogen.

among the wavelengths of the visible hydrogen lines, which had been determined experimentally as 656.210 nm, 486.074 nm, 434.010 nm and 410.12 nm. Note that these wavelengths had been measured to five or six significant figures, testifying to a very high level of experimental precision. Balmer found that the sequence of the four wavelengths is reproduced to extraordinary accuracy by the expression

$$\lambda = 364.56\left\{\frac{n^2}{n^2-4}\right\}\,\text{nm} \tag{1.1}$$

when the integer n is set equal to 3, 4, 5 and 6. This set of spectral lines came to be known as the **Balmer series** and Equation 1.1 as **Balmer's formula**.

Balmer suggested that n in his formula might take integer (i.e. whole-number) values greater than 6 and that hydrogen might therefore have many other spectral lines which, having wavelengths outside the visible spectrum, had not been observed at that time. This prediction was later borne out by experiments in which several more spectral lines corresponding to $n = 7, 8, 9$, etc. were found in the ultraviolet part of the hydrogen spectrum.

Question 1.1 Substitute the values 3, 4, 5 and 6 for n in Balmer's formula and compare your results with the measured wavelengths of the four visible hydrogen lines. Use the formula to calculate the wavelength of the line corresponding to $n = 7$. ■

So why should the visible spectral lines of hydrogen be given by this relatively simple formula? Perhaps the most important consequence of Balmer's work on spectroscopy was that it helped to focus attention on the need to account for the spectral lines of *hydrogen* before fruitful attempts could be made to interpret the more complicated spectra of other elements. In 1885, Balmer wrote

> 'It appears to me that hydrogen … more than any other substance is destined to open new paths to the knowledge of the structure of matter and its properties. In this respect, the numerical relations among the wavelengths of the first four hydrogen spectral lines should attract our attention particularly.'

These words were to prove prophetic as you will see later in the chapter.

2.3 Problem 3: Understanding blackbody radiation

It is a familiar fact that a heated body emits electromagnetic radiation and that the intensity of the radiation increases as the temperature rises. When sufficiently hot, the body becomes incandescent and emits visible light, glowing dark red at about 600 °C, orange at about 1000 °C and white at 1400 °C. The spectrum of this *thermal radiation* is continuous (Figure 1.7), i.e. its intensity varies smoothly across all wavelengths, in sharp contrast to the discrete line spectra of gases and vapours. This difference can be understood in the following way. In very low-density material, radiation is emitted or absorbed by individual atoms and these processes are little affected by any interaction *between* atoms. In a solid, such as a lamp filament, the atoms are in very close proximity to each other and they undergo complicated coupled motions as a result of thermal agitation. Radiation generated in these random motions is *thermalized* by repeated absorption and re-emission before escaping from the surface of the solid. Experiments show that the spectra of such 'thermal' sources approximate to a theoretical ideal that depends on the temperature but not on the structure and composition of the source. This ideal thermal spectrum

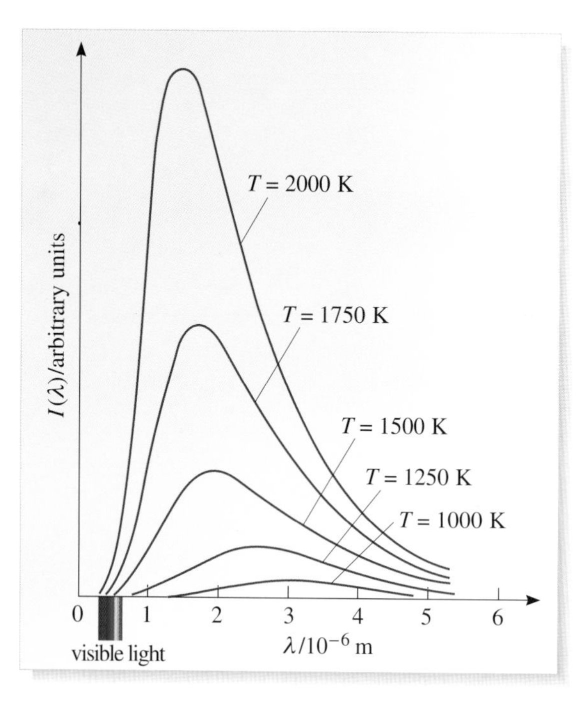

Figure 1.7 The spectrum of blackbody radiation at various temperatures. At any particular temperature, T, the curve shows the variation of intensity I with wavelength λ. (Note that the temperatures are given in kelvin.)

is known as a **blackbody spectrum**. The blackbody spectrum at a number of different temperatures is shown in Figure 1.7. Any radiation that is found to have a blackbody spectrum is said to be **blackbody radiation**.

The reason for the term 'blackbody' is that the *ideal* blackbody spectrum would be that emitted by an object that was a *perfect* emitter and absorber of radiation, and a perfect absorber is referred to as a blackbody. You may have noticed that on a sunny day, a black object in the Sun, warms up more quickly than a white one. This is because it absorbs a greater proportion of the radiation incident on it. However, blackbodies do not necessarily appear black: the Sun and stars provide rough natural approximations to blackbodies at various temperatures, but the best laboratory approximation to a blackbody is a cavity, or box, with all its walls maintained at a fixed temperature T. The cavity is filled with radiation streaming in all directions, which is constantly absorbed and re-emitted by the walls. If there were a small hole in the cavity the radiation emerging from the hole would be blackbody radiation with a spectrum appropriate to the temperature of the cavity walls.

The shape of this blackbody spectrum had already been determined experimentally by the end of the nineteenth century, and a number of attempts had been made to find an equation that would describe the form of the curve at any temperature. Some of these were relatively successful but the equations were empirical and none of them fitted the spectrum everywhere and at all temperatures.

A satisfactory theory of blackbody radiation should provide a precise mathematical expression for the blackbody spectrum. In principle, this should have been straightforward, as it does not depend on knowledge of the internal structure of atoms, but only on the properties of electromagnetic radiation, which were thought to be well understood within the framework of classical physics. However, to the astonishment of physicists at the time, classical physics was quite unable to provide a satisfactory solution.

In simplified terms the problem was this. The walls of the cavity restrict the possible modes of electromagnetic radiation that can exist within it, in much the same way as the allowed modes on a stretched string are restricted to those for which the wavelength fits with the length of the string (Figure 1.8). Classical physics states that the energy of the blackbody spectrum should be shared out equally among the possible modes of electromagnetic radiation that can fit into the cavity. This is an application of a very important theorem of classical physics: the **equipartition theorem** (see *Classical physics of matter*). This theorem states simply that, if a system in thermal equilibrium at absolute temperature T has n degrees of freedom, each of those degrees of freedom should possess an energy of $\frac{1}{2}kT$ (where k is Boltzmann's constant). Now, each mode of the electromagnetic radiation in the cavity has *two* degrees of freedom, due to the two possible polarizations, and should therefore have an energy of kT. Classical physics predicted the number, Δn, of modes available in any small wavelength range $\Delta\lambda$ between λ and $\lambda + \Delta\lambda$. The blackbody spectrum should then have been obtained by plotting ΔnkT against λ.

So, how many modes are available at different wavelengths? Well, the electromagnetic radiation is simply a set of standing waves fitted inside the cavity. This is a three-dimensional version of fitting standing waves on a one-dimensional string. The modes allowed for a string of length L are given by the formula $n\lambda = 2L$, where n is an integer. The first ten are shown in Figure 1.8. As you can probably see, as the value of n increases, the wavelengths of adjacent modes get closer and closer together: the wavelength of the $n = 9$ mode is very similar to the wavelength of the $n = 10$ mode. This means that the number of modes in a given wavelength range increases as the wavelength decreases. In three dimensions the effect is even more marked, and, in mathematical terms, the number of modes Δn in the range λ to $\lambda + \Delta\lambda$ turns out to be $8\pi V\Delta\lambda/\lambda^4$, where V is the volume of the cavity. The energy density (energy per unit volume) at wavelength λ, in the small range $\Delta\lambda$, in the blackbody spectrum should therefore be given by this expression, multiplied by kT and divided by V, that is $8\pi kT\Delta\lambda/\lambda^4$.

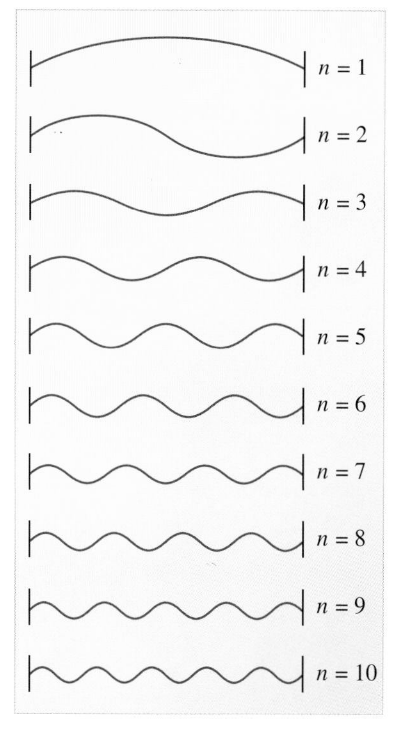

Figure 1.8 The first ten standing wave modes on a string of length L.

This formula is the classical prediction for the blackbody spectrum. At long wavelengths it fits the experimental spectrum very well (Figure 1.9). However, at short wavelengths the expression simply blows up. Classical physics predicts that all the energy in the blackbody spectrum should be at the short wavelengths. This was clearly a completely unsatisfactory prediction and became known as the '**ultraviolet catastrophe**'. In Section 3 you will see how the application of quantum physics to this problem produced a very satisfactory result.

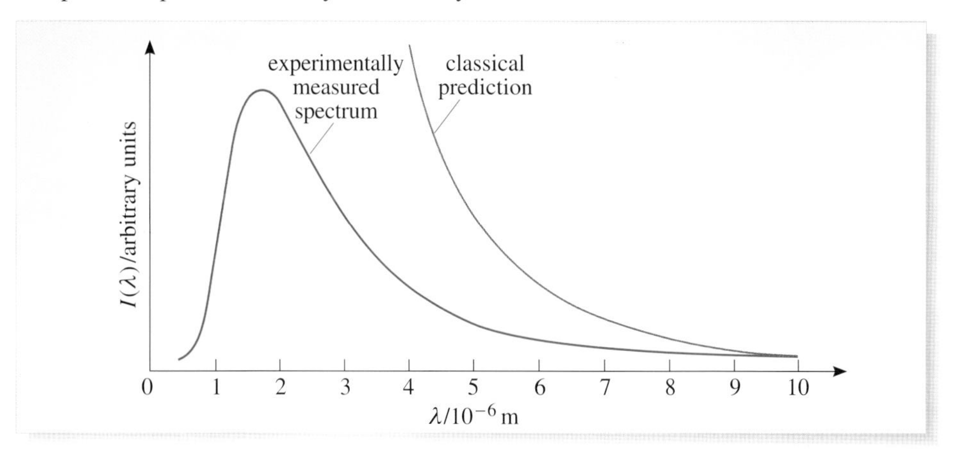

Figure 1.9 The ultraviolet catastrophe.

Question 1.2 A blackbody cavity is in the shape of a cube of side 5 cm and is at a temperature of 2000 K. Find the number of modes available in the cavity in the wavelength range 449 nm to 451 nm and hence find the total radiant energy in the cavity in this wavelength range according to the classical theory. ■

Figure 1.10 Philipp Lenard (1862–1947) did his most important and influential work around the beginning of the twentieth of the century, and he was awarded the 1905 Nobel Prize for physics for his experimental investigations into the photoelectric effect. In 1924 he became a Nazi and he expended much effort in attempting to discredit Jewish physicists, in particular, Albert Einstein.

2.4 Problem 4: Understanding the photoelectric effect

The ejection of electrons from metals which are illuminated by high frequency electromagnetic radiation was studied in a series of experiments carried out by Philipp Lenard (Figure 1.10) in 1902. This phenomenon is called the **photoelectric effect** (Figure 1.11). Lenard found that the ejected electrons did not all emerge with the same kinetic energy, indicating that some electrons are bound more tightly in the metal than others. Clearly, the *least* strongly bound electrons will emerge with the *greatest* kinetic energy.

That much was fairly easy to understand, problems arose only when it came to accounting for the details of Lenard's observations. In particular, he obtained the two results summarized below. (Recall that the frequency, f, and the wavelength, λ, of electromagnetic radiation are related by $\lambda f = c$, where c is the speed of light.)

Two key results from Lenard's observations

(i) The *maximum* kinetic energy of the ejected electrons is independent of the intensity of the incident electromagnetic radiation: it depends only on the *frequency* of the radiation.

(ii) No electrons are emitted from a metal if the frequency of the incident radiation is lower than a critical threshold frequency f_t that is characteristic of the metal.

Why were these observations so remarkable? Well, think carefully how the photoelectric effect might have been viewed by Maxwell, whose electromagnetic theory had been so successful in accounting for the behaviour of light. He would have pictured the radiation impinging on the metal as *waves*, delivering energy continuously (Figure 1.12). Since the intensity of a wave is defined as the energy it transfers per unit area per unit time, it follows that the energy delivered by these waves is determined only by their intensity and is independent of their frequency.

In this way, it can be argued that a beam of red light should eject electrons with the same maximum kinetic energy as a beam of higher frequency blue light of the same intensity. But Lenard's results showed that this was not the case. For each metal that

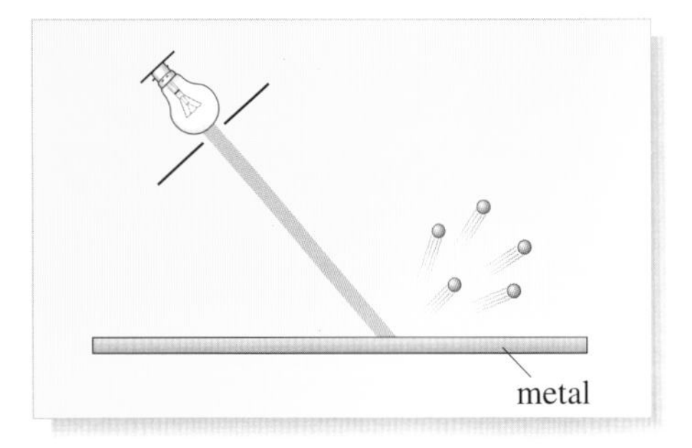

Figure 1.11 When electromagnetic radiation (of sufficiently high frequency) impinges on the surface of a metal, electrons are ejected. This is known as the photoelectric effect.

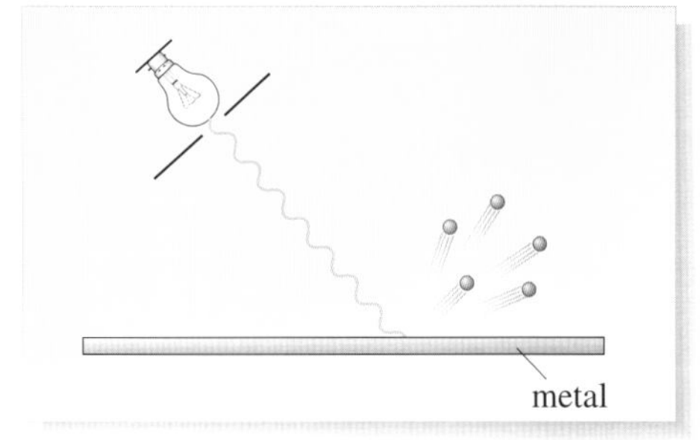

Figure 1.12 According to the classical *wave* theory of radiation, the energy of the beam of radiation is delivered continuously, like the energy of water waves. The energy of the beam should, according to this theory, depend on its intensity, *not* on its frequency. For this reason, it is difficult to understand why the *frequency* of the radiation should determine the maximum kinetic energy of the ejected electrons.

he investigated, the electrons ejected by a beam of light of higher frequency *always* had a greater maximum kinetic energy than that of electrons ejected by a beam of light of a lower frequency. He also found that if the frequency of his beam was below the threshold frequency f_t of the metal, then no electrons were ejected, no matter what the beam's intensity!

Clearly something was wrong somewhere. Maxwell's wave theory seemed to suggest that the energy of radiation should depend crucially on its *intensity*, whereas Lenard's experiments seemed to show that the energy of radiation depended on its *frequency*. This contradiction was perceived very clearly by Albert Einstein, who, in 1905, put forward a new theory. This enabled Lenard's observations to be understood and also explained another observation, namely that there was no measurable time delay between the switching on of the light source and the appearance of ejected electrons, a delay which would be expected using Maxwell's wave theory. Section 3.2 is devoted to Einstein's theory of the photoelectric effect.

2.5 Problem 5: Understanding heat capacities of solids

The molar heat capacity of a substance, denoted by C_m, is the quantity of energy which, when absorbed by a mole of the substance, will raise its temperature by 1 K. For the case of solids, the difference between the molar heat capacity at constant pressure, $C_{P,m}$, and that at constant volume, $C_{V,m}$, is very small, so we will simply refer to C_m.

We assume that the internal (thermal) energy of a solid is contained in its lattice vibrations, i.e. the coordinated vibrations of its constituent atoms. If the solid is in thermal equilibrium at some absolute temperature T, then according to the equipartition theorem, each atomic vibrator has an average energy of $3kT$ since it turns out that there are two degrees of freedom associated with *each* of its three possible directions.

The internal (thermal) energy, $U_m(T)$, of a mole of the solid will therefore be

$$U_m(T) = 3N_m kT = 3RT$$

where N_m is Avogadro's constant and

$$R = N_m k = 8.31\,\mathrm{J\,K^{-1}mol^{-1}}$$

is the molar gas constant. The molar heat capacity at constant volume of the solid is then given by the rate of change of the internal energy with temperature. Thus, using the language of differential calculus:

$$C_m = \frac{dU_m(T)}{dT} = 3R. \quad \text{(CLASSICAL THEORY)} \tag{1.2}$$

This expression for C_m is clearly independent of the temperature and it has the value $24.93\,\mathrm{J\,K^{-1}mol^{-1}}$. The question we must now ask is 'How does this classical prediction for the molar heat capacity of solids compare with experiment?'

As long ago as 1819 the French researchers Pierre Dulong and Alexis Petit reported experimental values of C_m for a number of elemental solids, including both metals and insulators, at room temperature (Table 1.1). All their samples gave a value close to $25\,\mathrm{J\,K^{-1}mol^{-1}}$. This famous result became known as the **Dulong–Petit law**. To this extent then, classical physics seems to have provided an adequate theory of the heat capacities of solids. There are, however, some serious departures from the law. A notable example is diamond, for which C_m is approximately $6\,\mathrm{J\,K^{-1}mol^{-1}}$ at room

Table 1.1 Dulong and Petit's original data, converted into SI units. These measurements were taken at constant pressure but, for solids, the difference between C_P and C_V is very small (a few per cent).

element	molar heat capacity/ $\mathrm{J\,K^{-1}\,mol^{-1}}$
bismuth	25.77
lead	25.48
gold	24.92
platinum	23.63
tin	25.35
silver	25.38
zinc	25.15
tellurium	24.74
copper	25.19
nickel	25.70
iron	25.06
cobalt	24.80
sulfur	25.50

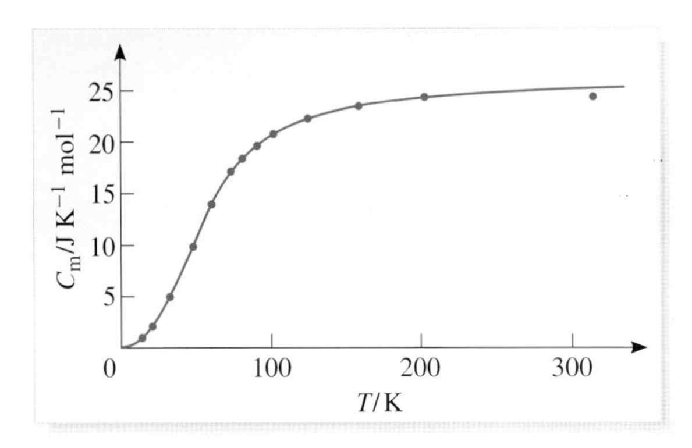

Figure 1.13 Experimentally measured values for the temperature variation of the molar heat capacity of silver.

temperature. Subsequent research has shown that the molar heats of *all* solids decrease at lower temperatures and approach zero as T approaches the absolute zero of temperature. The specific example of silver is shown in Figure 1.13.

In conclusion, the Dulong–Petit law can be regarded as only a partial success for classical physics. Although most solid elements have molar heat capacities that are close to the predicted value, there are several puzzling exceptions and the law breaks down completely at low temperatures. From the viewpoint of classical physics this is very worrying. It is hard to see how the argument that led to Equation 1.2 could ever be modified to get agreement with experiment for all elements at all temperatures. Of course, it turns out that the trouble lies with classical physics itself.

3 Physics saved by the quantum

3.1 The ultraviolet catastrophe tamed

It is appropriate to begin with blackbody radiation because it was the first of the five problems to be solved (in 1900) and because it provided the essential clue to the solution of all of the others.

As you saw in Section 2.3 classical physics predicts that the energy in the blackbody spectrum blows up at short wavelengths — the so-called ultraviolet catastrophe.

The ultraviolet catastrophe would be avoided, however, if, instead of the average energy per mode being kT for *all* modes as required by classical physics, *this average energy actually decreased rapidly with decreasing* λ. This would imply that the equipartition principle is invalid and therefore some aspect of the argument leading to it must be false. Equipartition depends on the assumption that the energy of a standing wave is continuous, in other words, it can vary by infinitesimal amounts. Max Planck (Figure 1.14) wondered what would be the consequences of *assuming* the contrary — that the energy can be changed only by adding or removing a discrete amount or **quantum** of energy? This implies that the energy of a standing wave is contained in a *whole number* of quanta — a fraction of a quantum cannot exist. Planck supposed that the magnitude of the energy quantum must be proportional to the frequency of the wave, which is a reasonable choice qualitatively: as the discrepancies between classical theory and experiment occur at the high-frequency (short-wavelength) end of the spectrum, the quantum of energy should be larger for higher frequencies. In this way Planck was led to his quantum hypothesis, which we will hereafter refer to as **Planck's law** and which may be stated as follows:

The energy E of a standing electromagnetic wave of frequency f (wavelength $\lambda = c/f$) is contained in a whole number of quanta, each of which has the value hf, or equivalently hc/λ, where h is a constant of nature. That is

$$E = hf. \quad \text{(Planck's law)} \tag{1.3}$$

Max Planck (1858–1947)

Max Planck (Figure 1.14) was born in Kiel in 1858 and studied at Munich. After appointments at Munich and Kiel he was appointed professor of physics at Berlin University in 1892. In 1937, he resigned his position as president of the Kaiser Wilhelm Institute as a protest against the Nazis treatment of Jewish scientists. He lost both his sons, one of whom was executed for plotting against Hitler. After the war Planck was reappointed as president of the Institute, which was renamed the Max Planck Institute, and moved to Göttingen. He died in 1947.

He is famous for his discovery that the key to understanding blackbody radiation is to assume that radiation can be emitted or absorbed only in the form of discrete *quanta*. He was awarded the Nobel Prize for physics in 1918 in recognition of this achievement, which is generally thought to mark the beginning of modern physics.

Figure 1.14 Max Planck.

The constant h is now known as **Planck's constant** and has the value $6.626\,18 \times 10^{-34}$ J s. Max Born, writing in 1926, said that Planck's quantum hypothesis 'marked the beginning of an entirely new conception of nature'. Its immediate impact on the problem of blackbody radiation was to provide the following new expression for the average energy of a standing wave

$$\text{average energy per mode} = \frac{hc}{\lambda}(e^{hc/\lambda kT} - 1)^{-1}, \quad \text{(QUANTUM THEORY)} \tag{1.4}$$

which obviously differs radically from its classical counterpart, namely kT.

In contrast with the constant value kT, Equation 1.4 exhibits just the sort of rapid decrease as the wavelength *shortens* which is required to tame the ultraviolet catastrophe. This result can be understood intuitively as follows. The quantum argument assumes that the energy of a standing wave cannot increase continuously under thermal agitation, but must climb a kind of ladder on which the distance apart of the rungs depends on the wavelength. At long wavelengths, the rungs are close together, many quanta will be excited and as we have seen, the classical case of continuous energy is approached. At the other extreme, when the wavelength is very short, the rungs on the ladder are far apart and the probability of even one quantum being excited is very low. In this situation, the thermodynamic average energy of a standing wave is close to zero.

● What is the significance of the factor $(e^{hc/\lambda kT} - 1)^{-1}$ in Equation 1.4?

○ It is the average number of thermally excited quanta in a standing electromagnetic wave, of wavelength λ, in a cavity at absolute temperature T. (That is, it is the thermodynamic average energy of the wave mode divided by the energy quantum (hc/λ) appropriate to that wavelength.) ■

When the quantum expression (Equation 1.4) is multiplied by the number of modes per unit volume, $\Delta n/V$, in the wavelength range λ to $\lambda + \Delta\lambda$ ($\Delta n/V = 8\pi\Delta\lambda/\lambda^4$ from Section 2.3), the resulting expression for the energy density (energy per unit volume) in blackbody radiation agrees extremely well with experiment (Figure 1.15). This expression is known as **Planck's radiation law**.

Question 1.3 Using Equation 1.4, calculate the total radiant energy in the same wavelength range for the blackbody cavity described in Question 1.2, this time according to Planck's hypothesis. Compare your result with the one you obtained using the classical theory. ■

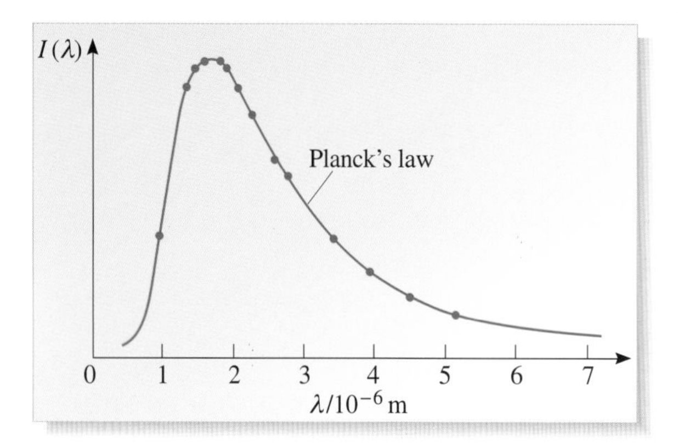

Figure 1.15 The energy density in the blackbody spectrum according to Planck's radiation law (solid line) and experimental measurements (dots). The agreement between the two is very good.

3.2 Einstein's theory of the photoelectric effect

Figure 1.16 Albert Einstein (1879–1955). By the age of 27 Albert Einstein had written three of the most famous papers in the canon of physics literature. They concerned the Brownian motion, the special theory of relativity and finally, the quantum theory of radiation, which was cited explicitly when he was awarded the 1921 Nobel Prize for physics.

In Section 3.1 we saw how, according to Planck's quantum hypothesis, electromagnetic waves of given frequency, must transfer energy in complete quanta. Albert Einstein (Figure 1.16) sharpened up this idea by arguing that *light quanta* should move through space as localized entities, rather like particles. This would explain how they can be emitted and subsequently absorbed by atoms and suggests a new picture of the radiation in a cavity as comprising a 'gas' of light quanta, which criss-cross the cavity at velocity c and bounce off the walls. The energy E of each light quantum is given by Planck's law, $E = hf$ where h is Planck's constant and f is the frequency of the radiation. Since the magnitude of Planck's constant is so small (6.63×10^{-34} J s), each quantum of light has only a minuscule amount of energy. For example, red light has a frequency of about 4.5×10^{14} Hz so a quantum of red light has an energy of (from Equation 1.3)

$$E = hf = 6.63 \times 10^{-34} \times 4.5 \times 10^{14}\ \text{J}$$
$$= 3.0 \times 10^{-19}\ \text{J}.$$

When a light is switched on, billions upon billions of light quanta are emitted. For example, an ordinary (3% efficient) 100 W light bulb emits roughly 10^{19} quanta of visible light every second! So, in almost all circumstances, the energy in a beam of light appears to be delivered in a constant, uninterrupted stream.

In 1905 Einstein applied the quantum concept to the photoelectric effect by suggesting that electrons in the metal only received energy from the incident radiation in complete quanta of energy $E = hf$ (Figure 1.17). If it is borne in mind that a certain amount of energy must be supplied to an electron just to remove it from the metal, it is straightforward to write down an energy equation for the photoelectric process:

one quantum of incident radiation − energy required to remove the electron from the metal = kinetic energy of the ejected electron.

In Section 2.4 we saw that the electrons emitted from the metal do not all have the same kinetic energy. The least tightly bound ones should emerge as the *most* energetic electrons as they require the *least* energy to remove them from the metal. This *least* energy for removal is usually called the **work function** of the metal, and is conventionally denoted by the Greek letter ϕ (phi).

You should now be able to see that, according to Einstein's theory, when electromagnetic radiation of frequency f impinges on a metal of work function ϕ, the *maximum* kinetic energy of the ejected electrons is given by the difference $hf - \phi$. Then, if m_e is the mass of an electron and v_{max} is the speed with which the most energetic electrons emerge from the metal, it follows that

$$\tfrac{1}{2} m_e v_{max}^2 = hf - \phi. \tag{1.5}$$

This result is known as **Einstein's photoelectric equation**.

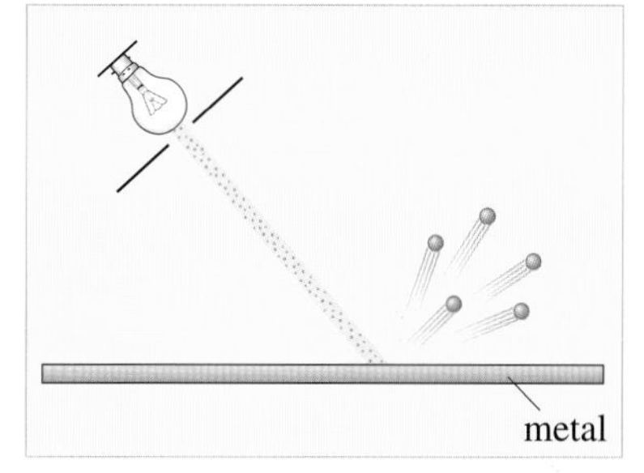

Figure 1.17 According to Einstein's theory, the explanation of the photoelectric effect is that each electron is liberated from the metal by the absorption of a single quantum of energy from the incident radiation.

Before discussing testable consequences of Equation 1.5 in detail, let us see how Einstein's theory provides an understanding of Lenard's two puzzling results, which were described in Section 2.4.

● Why does the maximum kinetic energy of the ejected electrons depend on the frequency of the incident radiation and not on its intensity?

❍ Higher frequency radiation has a higher energy quantum associated with it. Therefore, if each electron is ejected by a single quantum, the energy imparted to the electron will be greater for higher frequency radiation. Raising the intensity of the radiation increases the *number* of quanta, and hence the *number* of ejected electrons, but does not alter the maximum kinetic energy each one may have.

● Why is there a threshold frequency, f_t, below which *no* electrons are ejected from the metal?

❍ For electrons to be ejected, the energy quantum ($E = hf$) associated with the incident radiation must be sufficient to remove the least tightly bound electrons (this energy corresponds to the work function ϕ of the metal). Thus, hf must be greater than ϕ. This implies that there is a threshold value f_t for the frequency of the radiation below which electrons will not be ejected. This threshold is given by $f_t = \phi/h$. It is also possible to talk in terms of a threshold wavelength, $\lambda_t = c/f_t$, *above* which no electrons are ejected. ■

An experimental set-up for making photoelectric measurements is shown schematically in Figure 1.18a. Monochromatic light enters the evacuated tube T, and falls on the target or cathode C. The ejected electrons are collected at the anode A, and their flow may be monitored in the external circuit by means of the current meter G. By adjusting the variable resistance R_2, the electrons can be subjected to a retarding potential equal to $R_2V_0/(R_1 + R_2)$ where R_1 is a fixed resistance and V_0 is the voltage of the power supply. The procedure for checking Einstein's photoelectric equation is to adjust the retarding potential to a value, called the **stopping potential** V_{stop}, which *just* stops the flow of current through G. The maximum initial kinetic energy of the electrons as they are ejected is then eV_{stop}, where e is the charge on an electron and has a value of 1.6×10^{-19} C. A typical plot of the maximum kinetic energy of the ejected electrons (eV_{stop}) against the frequency of the incident radiation is shown in Figure 1.18b. According to Equation 1.5 this should yield a straight-line graph, and this is confirmed by the figure. The gradient or slope of the graph is equal

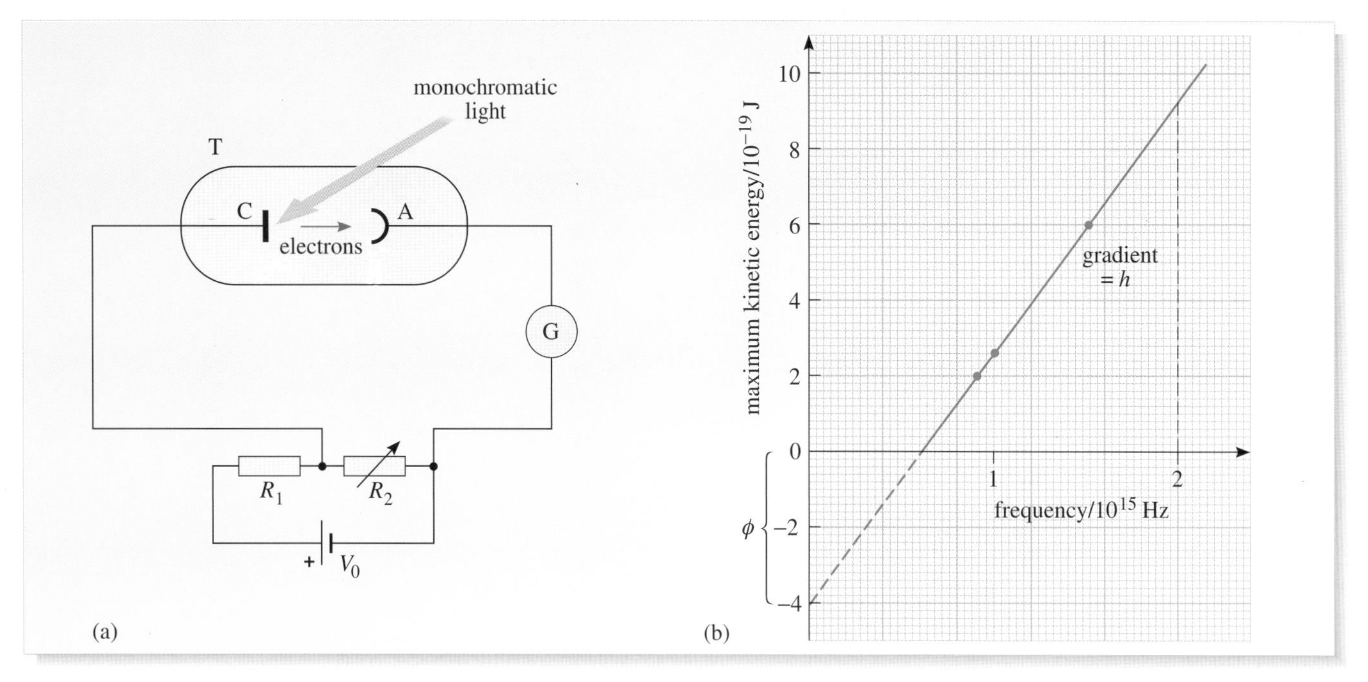

Figure 1.18 (a) An experiment for testing Einstein's theory of the photoelectric effect. For a given frequency of incident radiation, the maximum kinetic energy of the ejected electrons is given by eV_{stop}, where V_{stop} is the stopping potential. (b) The variation of the maximum kinetic energy of the ejected electrons with the frequency, f, of the incident electromagnetic radiation. The straight-line plot confirms the validity of Einstein's photoelectric equation (Equation 1.5).

to Planck's constant h and the threshold frequency f_t is the value of f at which the line crosses the frequency axis. When the line is extrapolated to lower frequencies (dashed portion), its intercept on the maximum kinetic energy axis is $-\phi$, i.e. the negative of the work function of the metal.

- How would you expect the graph in Figure 1.18b to differ for different metals?

○ If different metals are used, the graphs obtained would have the same slope h, but different intercepts, because different metals have different work functions. This reflects the fact that the amount of energy required to remove the least tightly bound electrons will vary from metal to metal. ■

Question 1.4 Use Figure 1.18b to find the value of (a) Planck's constant and (b) the work function of the metal involved. ■

Figure 1.19 Robert Millikan (1868–1953) was a native of Illinois in the United States. After obtaining his doctorate at Columbia he spent some time at the Universities of Berlin and Göttingen. In 1896 he moved to the University of Chicago where, initially, he was Michelson's assistant. He is famous for his determination of two fundamental constants of nature; e, the charge on an electron and Planck's constant h. He was awarded the Nobel Prize for physics in 1923 for his work on the photoelectric effect.

Einstein's photoelectric equation was verified in 1916 in a series of experiments carried out by the American physicist Robert Millikan (Figure 1.19). These experiments led to the award of two separate Nobel prizes: one for Millikan himself and another for Albert Einstein who, according to the citation, was awarded the prize principally for his theory of the photoelectric effect. There was no explicit reference in the citation to his more famous work on the special and general theories of relativity!

Millikan's measurements also provided yet more supporting evidence for the quantum theory of electromagnetic radiation. The time was measured between the moment the radiation impinged on the metals, and the moment that electrons were first ejected from them. It was found that the electrons were emitted immediately, that is, at the instant the metal was illuminated. (The time delay is now known to be less than 10^{-9} s.) This result is easily understood from Einstein's theory, since it is reasonable to expect that an electron in a metal might be ejected as soon as the first quantum of energy is transferred from the incident radiation. (See also Box 1.1.) Yet the simple wave theory of light predicts that about a *minute* should pass before an amount of energy sufficient to release electrons could be transferred to any individual atom in the metal!

You can verify that this long time delay is predicted by the wave theory by working through Question 1.5.

Question 1.5 A source of light emits, at a rate of 8 W, electromagnetic radiation of frequency sufficient to eject electrons from a piece of potassium that is 1 m away.

(a) Assuming that the radiation is emitted with equal intensity in all directions and that the wave theory of light is correct, show that approximately 5×10^{-21} J of energy will, in each second, be incident on an atom at the surface of the metal.
(Take the radius of a potassium atom to be 0.5×10^{-10} m.)

(b) The work function of potassium is 3.4×10^{-19} J. Assuming that a potassium atom absorbs all of the energy incident on it, estimate the time that would elapse between the instant that the waves of light impinge on the metal and the instant that the first electron is ejected.

(c) Compare your answer to part (b) with the experimentally determined upper limit on the time lag, which is about 10^{-9} s. What do you conclude?

Question 1.6 Light of wavelength 450 nm is incident on a sample of lithium. If the work function of lithium is 2.13 eV, calculate the speed of the fastest electrons ejected.

Question 1.7 The photoelectric threshold wavelength of sodium is 542 nm. Calculate the work function of sodium. ■

Box 1.1 From light quanta to photons

The idea that radiation was quantized was first introduced by Planck to explain blackbody radiation. The concept was reinforced by Einstein in his explanation of the photoelectric effect. For Einstein, electromagnetic radiation was not only emitted and absorbed in quantized amounts but could also be thought of as retaining this energy quantization as it propagated through space.

These ideas evolved gradually over the next twenty years or so into the idea of electromagnetic radiation as a stream of particles possessing both energy and *momentum*. The term **photon**, for each light particle, was introduced by the American chemist Gilbert Newton Lewis and nowadays, when physicists discuss phenomena which require the particle theory of electromagnetic radiation, they almost invariably refer to the particles as photons.

3.3 Einstein's theory of heat capacities

In Section 2.5 we discussed how classical physics predicts that the molar heat capacity of all solids at all temperatures should be equal to $3R$, where R is the molar gas constant. Although the result is valid in many cases, there are some exceptions and the prediction is completely wrong at low temperatures. In order to progress beyond the simple classical theory given in Section 2.5, it is necessary to make some assumption about the nature of the atomic vibrations within the solid. Einstein made the simplest assumption possible: that each atom of the solid oscillates at the same characteristic frequency f_v in all three directions. However, as long as the equipartition theorem is thought to be valid, these details can change nothing: the average energy of an atomic vibration will still be kT. Planck had shown that, according to his quantum hypothesis, the energy of *any* oscillator of frequency f_v should be quantized in steps of hf_v and when Einstein added this quantum condition to his model he found, as we might expect, that it changed the outcome completely. By a chain of reasoning virtually identical to that which led, in the case of blackbody radiation, to Equation 1.4, he obtained the average energy of an atomic mode of vibration as

$$\text{average energy per mode} = \frac{hf_v}{e^{hf_v/kT} - 1}. \tag{1.6}$$

To find the internal energy $U_m(T)$, of a mole of solid, we multiply Equation 1.6 by the number of independent modes of vibration in one mole, $3N_m$. Thus we have

$$U_m(T) = \frac{3N_m hf_v}{e^{hf_v/kT} - 1} \tag{1.7}$$

for the internal energy of one mole of the solid. The heat capacity is the rate of change of internal energy with respect to temperature and is obtained by differentiating $U_m(T)$ with respect to T. The result may be written, finally, (remembering that $R = N_m k$) as

$$C_m(T) = 3R\left(\frac{\theta_E}{T}\right)^2 \times \frac{e^{\theta_E/T}}{(e^{\theta_E/T} - 1)^2} \quad \text{(EINSTEIN'S THEORY)} \tag{1.8}$$

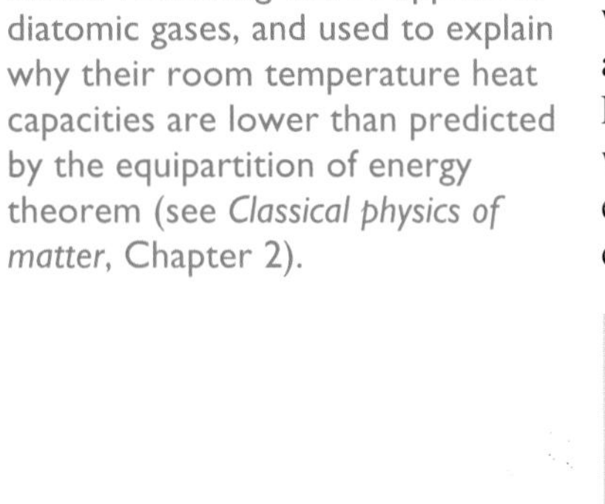

Similar reasoning can be applied to diatomic gases, and used to explain why their room temperature heat capacities are lower than predicted by the equipartition of energy theorem (see *Classical physics of matter*, Chapter 2).

where $\theta_E = hf_v/k$. The value θ_E is known as the **Einstein temperature** of the solid and is proportional to the frequency of vibration, f_v, of the atoms of the lattice. Figure 1.20 shows how Einstein's prediction (Equation 1.8) for $C_m(T)$ varies with temperature. This prediction is clearly in much better agreement with the experimentally observed heat capacity (shown for silver in Figure 1.13) than the classical prediction of a constant value of $3R$.

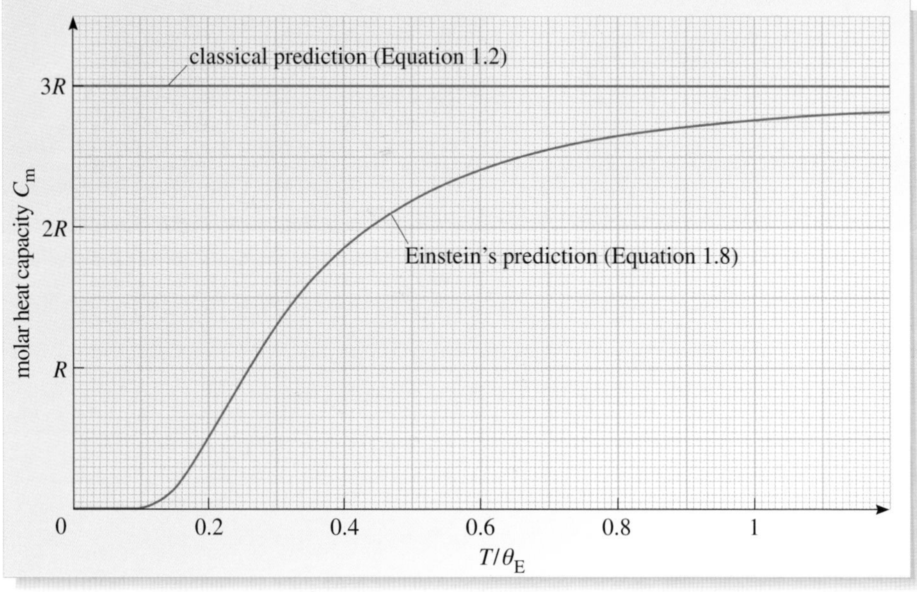

Figure 1.20 Einstein's prediction for the molar heat capacity of a solid as a function of T/θ_E, where T is the actual temperature of the solid and θ_E is its Einstein temperature.

Question 1.8 Given that for small values of x (i.e. $x \ll 1$) the function $e^x \approx 1 + x$, show that Equation 1.8 reduces to $C_m(T) \approx 3R$ in the limit of high temperature, that is, when $T \gg \theta_E$ (which is the same as saying $\theta_E/T \ll 1$). ■

3.4 Summary of Section 3

In Section 3 we have shown how, for each of the last three problems described in Section 2, the difficulty was eliminated by the introduction of a totally new idea, namely that of quantization of energy. In the cases of blackbody radiation and the photoelectric effect, the experimental results could be explained very accurately by assuming that the electromagnetic radiation was quantized so that energy of radiation at a frequency f could only be transferred in quanta of energy hf. Similarly, when it is assumed that the energy of the vibrational modes of the atoms in a solid is quantized in integer multiples of hf_v, the prediction for the molar heat capacity of the solid is a much better approximation to the experimentally measured values than the classical prediction.

4 First insights into atomic structure

4.1 Introduction

By the end of the nineteenth century it had become clear to many scientists that, to make a start on understanding spectroscopy, a theory of atomic structure was needed. Researchers in Japan, France and Germany had put forward notable atomic models, but it was two physicists working in England, Joseph John Thomson and Ernest Rutherford, whose work is now recognized as having been the most important. In this section we shall begin by describing Thomson's model before going on to say why Rutherford saw fit to formulate an entirely different model on the basis of just one piece of experimental evidence. We shall briefly compare the successes and failures of the two models, and prepare the ground for Section 5, which will describe how Niels Bohr used one of the models to form the basis of the first, partially successful, *quantum* theory of atomic structure. There you will also see how Bohr's theory accounts for Balmer's formula.

4.2 J. J. Thomson's 'plum-pudding' model

It was once believed that the substances we see around us are ultimately composed of huge numbers of tiny, *indivisible* lumps of matter, which were called *atoms*. However, the idea that atoms had no internal structure was undermined in 1897 when the existence of very light, negatively charged particles (electrons) was conclusively proved by Joseph John Thomson (Figure 1.21), a professor of physics at Cambridge University. As you know from Section 2.4, these particles can be ejected when electromagnetic radiation of sufficiently high frequency is shone on a metallic surface. This showed that metals, and possibly all matter, contain electrons. However, since it was known that matter was usually electrically neutral, it was reasonable to infer that atoms contain, in addition to *negatively* charged electrons, something with an equal and opposite *positive* charge.

Joseph John Thomson (1856–1940)

Figure 1.21 Sir Joseph John Thomson with his son George Paget Thomson.

J. J. Thomson (Figure 1.21) was one of the last great classical physicists. After obtaining a scholarship to Trinity College, Cambridge, he completed the Mathematics Tripos in 1880 (finishing second) and went on to do important work on electromagnetism, developing some of Maxwell's ideas. He was appointed Cavendish Professor of experimental physics in 1884, and although his appointment caused much surprise at the time because of his mathematical background, he inspired a generation of outstanding experimentalists. Several Nobel Prize winners spent part of their early careers in Thomson's group at the Cavendish Laboratory: Ernest Rutherford, Francis Aston, Charles Wilson, Lawrence Bragg, Owen Richardson, Edward Appleton and J.J.'s own son, George Thomson. J. J. Thomson is probably best known for the discovery of the electron and the determination of its charge to mass ratio. He was awarded the Nobel Prize for physics in 1906 for his 'investigations of the passage of electricity through gases'.

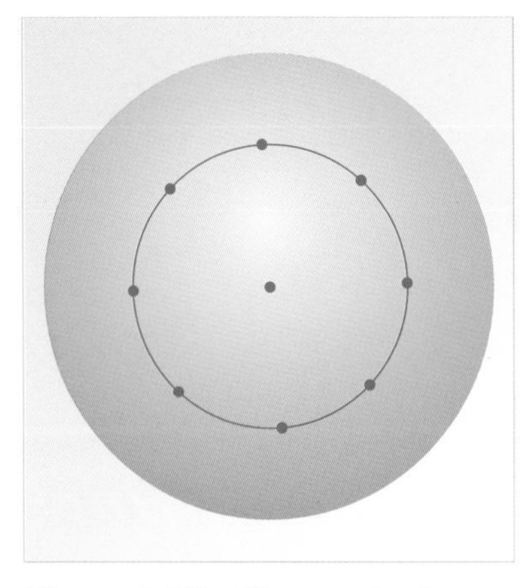

Figure 1.22 Thomson's plum-pudding model of the atom, in which electrons are contained inside a uniform sphere of positive charge.

In 1903, Thomson used these simple ideas to formulate his so-called **'plum-pudding' model of the atom**. He suggested that each atom consists of a certain number, Z, of electrons contained inside a sphere of uniform positive charge. The total positive charge would be Ze, to balance the total charge of Z electrons each carrying a charge of $-e$. He pictured the electrons as moving in concentric circular orbits (Figure 1.22), but believed that in some atoms the motion of the electrons would be stable only if one or more of them were stationary at the centre of the atom.

This formulation begs two major questions, the first of which has probably occurred to you already — what is the nature of the mysterious, positively charged sphere? The most that could be said about it was that its diameter would be roughly that of the atom, about 10^{-10} m and that, because electrons are so light, it must carry virtually all the mass of the atom. (Thomson's own rough measurement of the charge to mass ratio, e/m, of the electron had indicated that it had less than 0.1% of the mass of the hydrogen atom.)

The second major problem with the model relates to the fact that the orbiting electrons must lose energy. It follows from Newton's second law of motion that electrons in circular motion are in constant *acceleration*. Moreover, Maxwell's theory of electromagnetism implies that accelerating charged particles always continuously emit electromagnetic radiation. Proponents of this model therefore expected that atoms containing moving electrons should be *unstable*, since the energy of the electrons would gradually, but inevitably, be converted into the energy of the emitted radiation.

This was certainly a very serious problem for the model! But interestingly enough, Thomson was able to show, using Newton's laws and Maxwell's electromagnetic theory, that a given atom would be *almost* stable and might last for thousands of years, if its constituent electrons moved in certain prescribed orbits inside the positive sphere. Nevertheless, all elements should, ultimately, be unstable since atoms that contained moving electrons were doomed to collapse sooner or later. This expectation was of course, *not* borne out by observation: some rocks and fossils on Earth had been present for thousands of millions of years, whereas Thomson predicted that they should have collapsed long before this time.

It may seem to you that the plum-pudding model had too many grave shortcomings to be accorded serious consideration. But the model *was* taken seriously, and was used to make some important and productive investigations into atomic phenomena. Once a communal blind eye had been turned to its failings, the model could be used to do explicit calculations on atomic structure. For example, the model enabled Thomson to make progress in the understanding of the Periodic Table of chemical elements. He argued, correctly, that the clearly recognizable patterns among the chemical properties of certain groups of elements reflected underlying similarities in the atomic structures of these elements.

Although Thomson's model became established as a working hypothesis of atomic structure, Ernest Rutherford doubted the validity of its basic assumption, namely, that the atom is uniformly filled with positively charged matter. He was not alone in this — as early as 1903 Lenard had suggested that the atom might consist largely of empty space, having been led to this conclusion by experiments in which he bombarded thin metallic foils with electrons. Rutherford's brilliant new idea was to do the same thing with a far more penetrative probe: high-velocity **α-particles** from a radioactive source.

4.3 α-particle scattering

Rutherford had a certain proprietary interest in α-particles: he had discovered them, named them, been the first to measure their charge to mass ratio and been the first to identify them (correctly) as helium atoms that had lost both their constituent electrons, and therefore carried a charge $+2e$.

He asked his student Ernest Marsden (still an undergraduate at the time) to try to find out whether α-particles were directly reflected from thin metallic foils. This was a surprising request for Rutherford to make, as most scientists at the time would probably have been willing to wager large sums of money that Marsden would observe essentially no reflected α-particles. Such was the prediction of Thomson's model.

Why did the model predict that no α-particles would be reflected? To answer this, think how the Thomson model can be used to visualize the scattering of α-particles from a target of, say, gold or platinum foil. When the positively charged α-particles reach the surface of the metal, they encounter, according to the model, a vast array of atomic plum puddings — tiny, positively charged spheres containing electrons (Figure 1.23). The α-particles will be subject to an electrostatic force of *repulsion* from the positive charge, and to an electrostatic force of attraction to the electrons in the sphere. An α-particle is more than 7000 times more massive than an electron, so collisions with the electrons would have no discernible effect on the motion of the α-particles. (If a cannon ball and a table tennis ball collide, it is the table tennis ball that is deflected!) It turns out that the deflection suffered by an α-particle due to electrostatic repulsion from the positively charged sphere of a Thomson atom of gold or platinum is also *extremely small*. We can understand this by considering the following two situations.

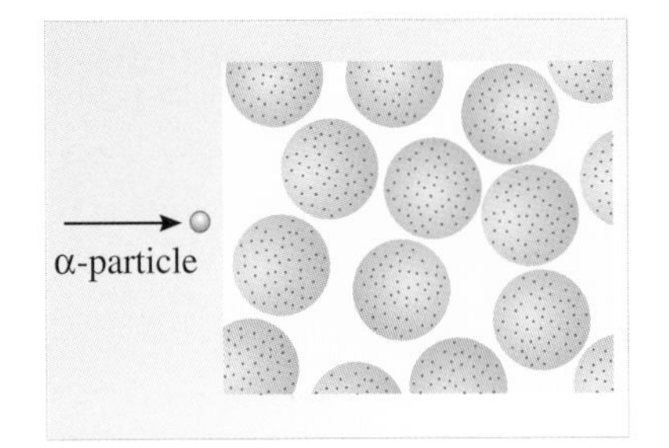

Figure 1.23 According to Thomson's model, when a high-energy α-particle reached the target of gold foil, it encountered a vast array of tiny spheres of positive charge that each contained electrons.

Let us first consider whether it is possible for the α-particle to be reflected by the electrostatic repulsion from the positive charge in a head-on collision with a Thomson atom (Figure 1.24a). Now, the α-particles are travelling at speeds of up to $2 \times 10^7\,\mathrm{m\,s^{-1}}$, corresponding to a kinetic energy of about $10^{-12}\,\mathrm{J}$. Even if it

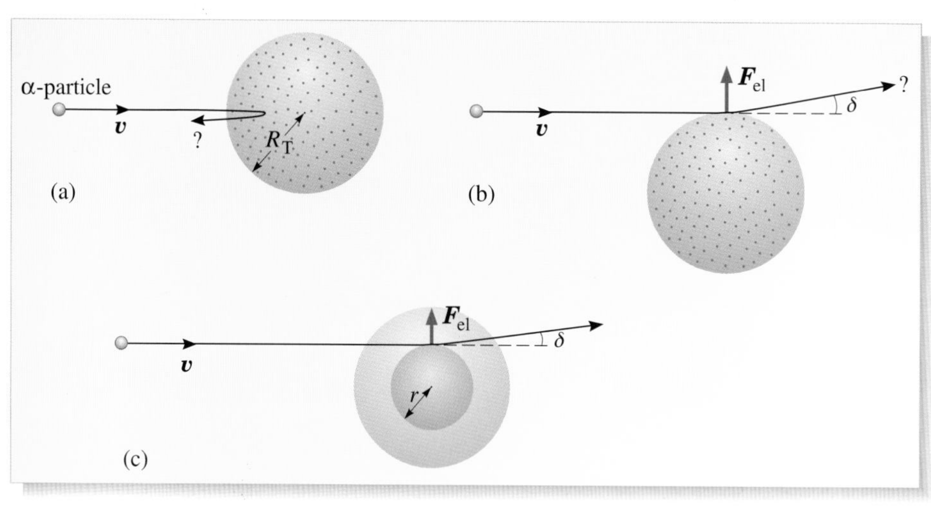

Figure 1.24 Three possible trajectories for an α-particle when it encounters a Thomson atom of radius R_T. (a) Is it possible for the α-particle to be reflected backwards from a Thomson atom? (b) Through what angle would the α-particle be deflected if it just grazed the surface of the Thomson atom? (c) If the α-particle penetrates the sphere of positive charge, it actually experiences a *smaller* repulsive force as it is only affected by the charge closer to the centre of the atom, i.e. within the radius r.

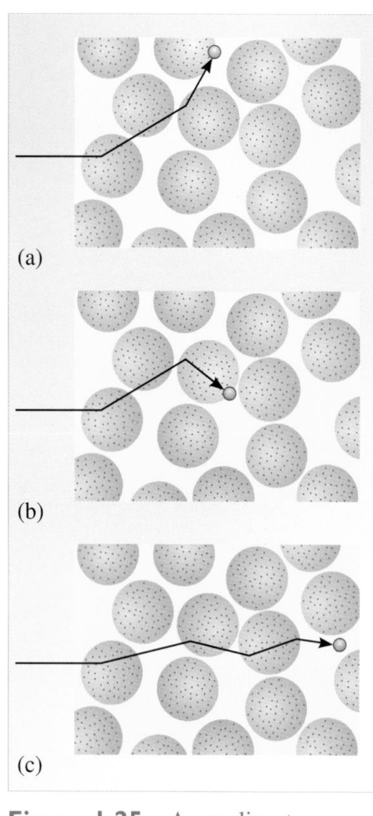

Figure 1.25 According to Thomson's model of the atom, an α-particle will be scattered many times in traversing the target. After it has been deflected once, it might (a) be deflected further in the same direction, or (b) be deflected back to its original course. The average deflection over the multiple-scattering process (c) is proportional to the square root of the total number of collisions suffered by the α-particle. Note that, for the sake of clarity, the deflection angles have been exaggerated in these diagrams.

penetrated to the centre of the atom, its electrostatic potential energy would only be about 6×10^{-16} J, nowhere near enough to bring the α-particle to a halt. So an α-particle colliding head-on with a Thomson atom will certainly not be reflected by it.

Suppose, instead, that the α-particle's trajectory is such that it would just graze the atom as in Figure 1.24b, so that the transverse electrostatic force $\boldsymbol{F}_t$ experienced by the α-particle would be the maximum possible and equal to

$$F_t = F_{el} = \frac{2Ze^2}{4\pi\varepsilon_0 R_T^2}, \tag{1.9}$$

where ε_0 is the permittivity of free space. But, the α-particle experiences this force for only a very short time and the angle δ through which the α-particle is deflected is only about 0.06°.

This is clearly a very small deflection, yet it is the *maximum* that the α-particle could expect to experience in one encounter with a Thomson atom. If the α-particle were to follow a trajectory which took it closer to the centre of the atom, i.e. *inside* the sphere of positive charge, (Figure 1.24c) it would experience a *smaller* repulsive force since it would be affected only by the fraction of charge closer to the atom centre.

An α-particle incident on a piece of gold foil 5×10^{-7} J thick might be expected to be deflected by more than one atom — it would encounter about 1000 of them in traversing the foil. The most important point to note here, however, is that the successive deflections would be essentially *random*. After the α-particle has been deflected one way, it might, on its next encounter, be deflected further in the same direction (Figure 1.25a) or, alternatively, it may be deflected back to its original course (Figure 1.25b). The *average* deflection of the α-particles is given by a statistical analysis of this **multiple-scattering process** (Figure 1.25c).

When the statistical theory of multiple scattering was applied under the assumption that the gold foil was made of Thomson atoms, it turned out that the average deflection was expected to be only about 0.6°. More astonishingly, the theory predicted that the chance of an α-particle being reflected backwards was less than 1 in 10^{1000}. We may conclude therefore that the Thomson model leads to the following two predictions:

(a) the vast majority of the α-particles fired at the target should emerge having hardly deviated at all from their original paths;

(b) no α-particles with large deflections should be seen.

Marsden performed the experiment in collaboration with the more experienced Hans Geiger, who later became famous for his counter. Figure 1.26 is a schematic diagram of the apparatus that they used. A specimen of the radioactive element radon provided the source of high-energy α-particles, which were directed towards the gold foil target. The deflected α-particles were detected by monitoring the flashes of light which were emitted when they struck the zinc sulfide screen. In order to find the relative number of α-particles that were deflected at different angles, Geiger and Marsden counted the number of flashes occurring in different areas of the screen in a given time. This entailed spending many hours in their darkened laboratory, peering down a microscope focused on the screen. All this effort was amply rewarded however, because they discovered that *prediction (b) based on Thomson's model is incorrect.* Although 98% of the incident α-particles suffered deflections of less than 5°, they observed significant numbers deflected through angles ranging from 15° to 150°. In fact, one α-particle in 10^4 was reflected backwards from the gold foil.

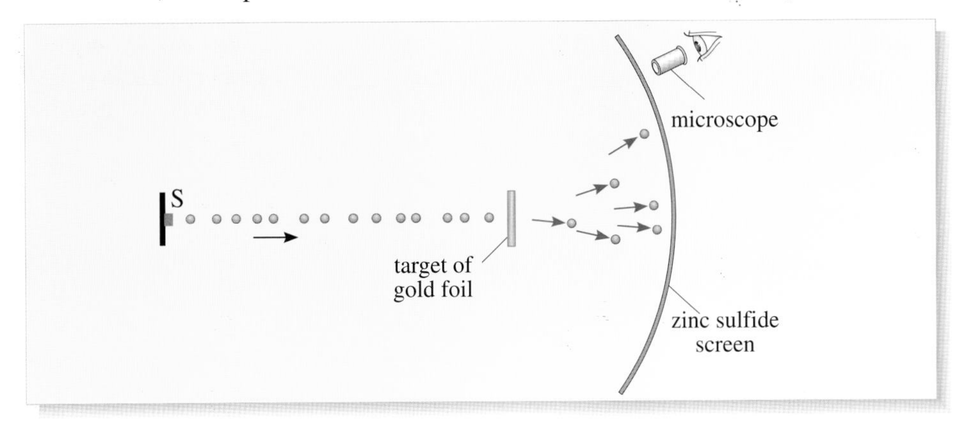

Figure 1.26 A schematic diagram of Geiger and Marsden's apparatus. The α-particles emerged from the radioactive source S and impinged on the target of gold foil before being detected as tiny flashes of light on the zinc sulfide screen. Would they be scattered *backwards*? In the interests of clarity, the diagram shows only small-angle deflections, but in the actual experiment Geiger and Marsden detected α-particles deflected through as much as 150°.

Rutherford's reaction to this extraordinary result is well captured by his often quoted remark:

> 'It was quite the most incredible event that has ever happened to me in my life. It was almost as if you fired a fifteen inch shell at a piece of tissue paper and it came back and hit you.'

You might expect that when atomic physicists in 1910 were confronted with Geiger and Marsden's α-particle scattering data, they immediately downed tools, tore up their work on Thomson's model and sat down assiduously to formulate a better model of the atom. *But that did not happen.* Many physicists paid little attention to Geiger and Marsden's data and continued to work on Thomson's model almost as if

Figure 1.27 Ernest Rutherford (1871–1937) was brought up in New Zealand and came to England in 1895 with a scholarship that had been established with some profits of the 1851 Exhibition in London. He worked with J. J. Thomson, who soon recognized his extraordinary ability. Later at McGill University in Canada, Rutherford did much important work on radioactivity for which he was awarded the Nobel Prize for *chemistry* in 1908. Yet his best work had still to be done. In 1911, he formulated his 'nuclear' model of the atom, which has since formed the basis of all atomic models. Rutherford ended his career as professor of physics at the Cavendish Laboratory, Cambridge, a public figure and recognized to be probably the greatest experimental physicist since Faraday.

the data did not exist! They argued that perhaps the model was missing a single ingredient that, once incorporated, would allow the α-particle scattering data to be understood. Alternatively, there might be something wrong with Geiger and Marsden's experiment.

Rutherford himself (Figure 1.27) was in no doubt that the experiment *had* sounded the Thomson model's death knell, and he set about finding a new model that *could* explain Geiger and Marsden's data.

4.4 Rutherford's classical model of the atom

In December 1910 Rutherford wrote, in a letter to the chemist B. B. Boltwood:

> 'I think I can devise an atom much superior to J.J.'s... It will account for the reflected alpha-particles observed by Geiger, and, generally, I think will make a fine working hypothesis. Altogether, I am confident that we are going to get more information from scattering about the nature of the atom than from any other method of attack.'

The **Rutherford model of the atom** proposed in March 1911, was radically different from Thomson's and provided a simple explanation of the results of Geiger and Marsden's α-particle scattering experiment. Rutherford's idea was that the positively charged matter in the atom, instead of taking the form of an extended sphere, as in Thomson's model, is concentrated into a very tiny central core — the **nucleus**. The size of the atom is then determined by the electrons, which orbit in the otherwise empty space outside the nucleus (Figure 1.28). According to this model, the magnitude of the electrostatic force repelling the α-particles is given by Coulomb's law *all the way to the nuclear surface*. An α-particle which approached such a nucleus to a distance of 10^{-13} m would experience a repulsive force about a million times stronger than the maximum force encountered in Thomson's model (Equation 1.9) and, in consequence, will be deflected through a large angle. This is, in essence, Rutherford's explanation of the large-angle α-particle scatterings observed by Geiger and Marsden. Notice that, in this theory, the predicted large-angle scattering is caused by a *single* collision with the nucleus of a target atom.

Figure 1.28 Rutherford's model of the atom, in which negatively charged electrons orbit a tiny, positively charged nucleus.

Rutherford used Coulomb's law and the laws of classical (Newtonian) mechanics to calculate the angle through which an α-particle (of any energy) is expected to be scattered by *any* given nucleus. It turned out that the angle of scattering depended on three factors.

The first of these is the *charge* on the target nucleus, as you might expect. The greater the nuclear charge, the stronger the repulsive electrostatic force and the greater the scattering angle.

The second factor is the *energy* of the incident α-particle. It is intuitively reasonable that the higher the energy (i.e. the higher the speed) of the incoming α-particle, the more likely it is to pass by the nucleus relatively unperturbed.

The third and perhaps the most important factor that influences the angle through which the α-particle is scattered, is the degree to which the α-particle is aimed *directly* at the target nucleus (see Figure 1.29). It should be clear that the *further* the trajectory of the α-particle is from being *directly towards* the nucleus (i.e. a head-on collision) the less is the effect of the nuclear charge on the α-particle. The distance of the trajectory of the α-particle from the line of direct aim is called the **impact parameter** b and this quantity is shown for each collision illustrated in Figure 1.29.

Figure 1.29 According to Rutherford's scattering theory, the *smaller* the impact parameter b of the incident α-particle, the *greater* the angle through which it is deflected. When the α-particle is aimed directly at the nucleus (b = 0), it stops momentarily when its distance from the nucleus is r_{cl}, the 'distance of closest approach', and returns along its original track.

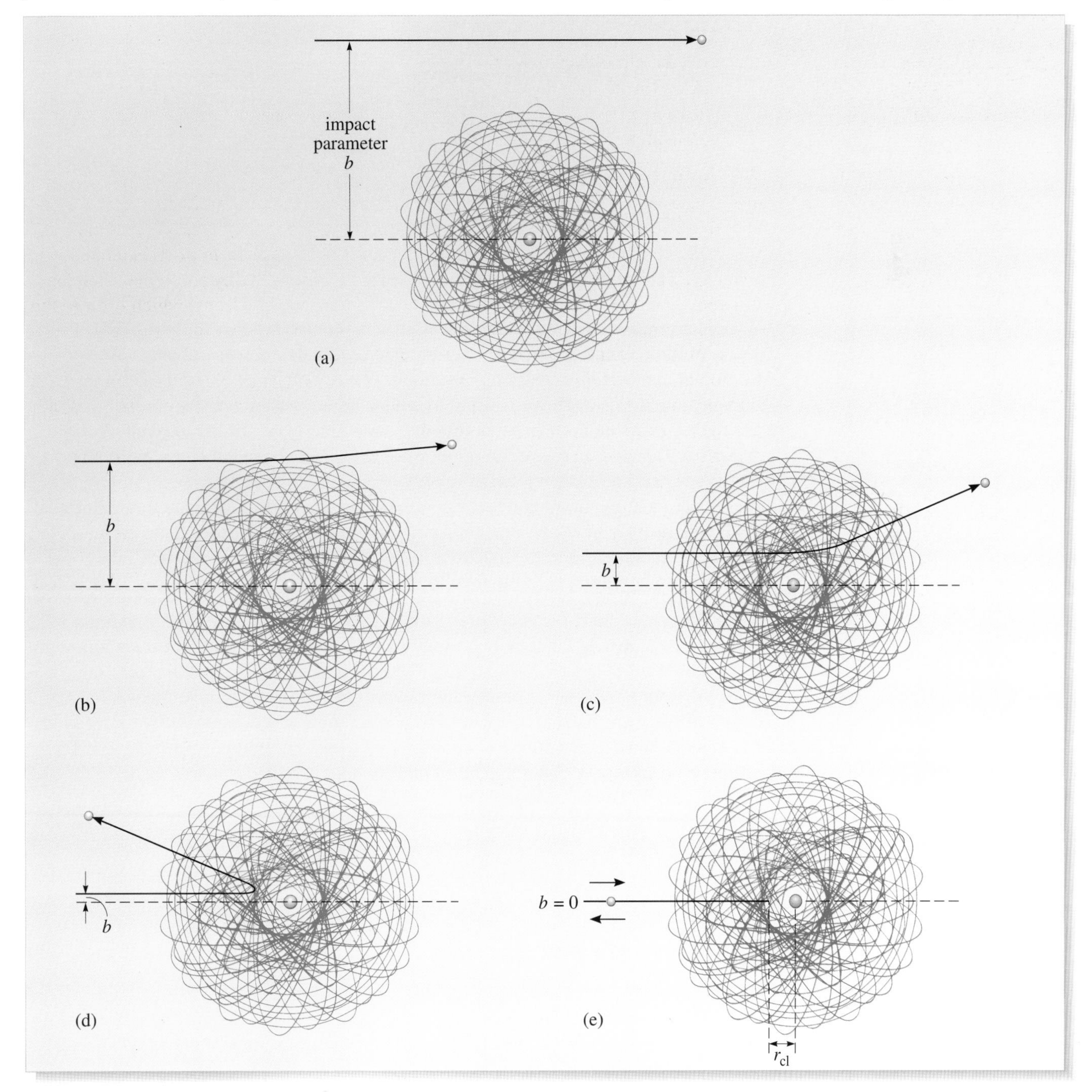

The five drawings in Figure 1.29 show the predictions of Rutherford's scattering theory for the trajectories or paths of α-particles having the same initial kinetic energy, but different impact parameters. As the impact parameter *decreases*, the angle through which it is scattered *increases*. An especially interesting case is that in which the α-particle collides head-on with the nucleus ($b = 0$), and is reflected directly backwards ($\delta = 180°$). The so-called **distance of closest approach**, labelled r_{cl} on Figure 1.29e, can easily be calculated. To do this, think about the energy of the α-particle as it approaches the target nucleus: its kinetic energy is converted into electrostatic potential energy and it gradually slows down. It stops momentarily when it is at its closest to the nucleus, and then moves back along its original track as its electrostatic potential energy is converted back into kinetic energy. You can find a formula for r_{cl} for yourself by working through the next question.

Question 1.9 (a) Write down an expression giving the electrostatic potential energy of an α-particle due to the positive charge on an atomic nucleus a distance r from it. Assume that the nucleus has a charge of Ze.

(b) Use your result from part (a) to show that r_{cl}, the distance of closest approach for an α-particle having kinetic energy E_{kin} when it is far from the nucleus, is given by the formula

$$r_{cl} = \frac{Ze^2}{2\pi\varepsilon_0 E_{kin}}.$$ ■

The confirmation, by Geiger and Marsden in a second experiment performed in 1913, of the detailed predictions of Rutherford's model was one of the most important experimental results of twentieth century physics. Not only did it demonstrate the correctness of Rutherford's nuclear model of the atom, but, since the radius of the nucleus must be less than the value of r_{cl} for the *most energetic* α-particles used, it also gave an upper limit to the size of the nucleus. Rutherford's theory also shows that the scattering power of a target material is proportional to the square of Z, the number of charges on its atomic nuclei. This parameter, usually called the **atomic number**, was obtained from α-particle scattering data for a number of metals by James Chadwick in 1920. His results were in excellent agreement with values of Z obtained by other means.

So much for the successes of Rutherford's model — now let us briefly review its shortcomings, of which Rutherford was well aware. First of all, the Rutherford atom is unstable for the same reason that the Thomson atom is unstable: the orbiting electrons should emit electromagnetic radiation continuously and collapse into the centre of the atom. In fact, the Rutherford atom would last less than 10^{-5} s. Second, the model contributed nothing to contemporary understanding of the Periodic Table of chemical elements. Finally, Rutherford's model is quite unable to account for the experimentally determined sizes of atoms. Nevertheless, Rutherford's model represented a giant step in the right direction, and became firmly established when Niels Bohr used it as the basis of his primitive quantum theory of the hydrogen atom. This new theory ultimately shed light on each of the three key problems cited above as failures of the Rutherford model.

4.5 Summary of Section 4

Although we have given some of the historical background to the story of the development of the atomic models of J. J. Thomson and Rutherford, it is the basic physics of the two models which you should try to understand and recall. The main points to remember are as follows:

1. According to J. J. Thomson's plum-pudding model, the neutral atom is thought to consist of a number of negatively charged electrons inside a uniform sphere of positive charge. Although the nature of the positive sphere is rather obscure, it provided some understanding of the structure of the Periodic Table of chemical elements.
2. Geiger and Marsden showed that when a gold foil target is bombarded with high-speed α-particles, significant numbers are reflected backwards. This result cannot be accounted for by the Thomson model.
3. The model formulated by Ernest Rutherford could explain the surprising α-particle scattering data of Geiger and Marsden. In this model, the atom is pictured as a very small, positively charged nucleus surrounded by a number of negatively charged electrons (Figure 1.27). Because of the high concentration of positive charge at the centre of the atom, an α-particle would be deflected through a large angle if its impact parameter were sufficiently small. Unlike the Thomson model, the Rutherford model sheds no light on the structure of the Periodic Table of chemical elements.
4. Both models predict that atoms are unstable, and neither is able to account for the fact that atomic radii are approximately 10^{-10} m.
5. Rutherford's model ultimately led to major advances in understanding atomic structure because it provided the basis for later quantum theories of the structure of hydrogen and heavier atoms.

Question 1.10 When an α-particle is fired at a speed of $1 \times 10^7\,\mathrm{m\,s^{-1}}$ directly towards a tin nucleus ($Z = 50$), its distance of closest approach is 6.8×10^{-14} m. Use this information and the expression for r_{cl} in Question 1.9 to calculate the distance of closest approach of:

(a) an α-particle with speed $2 \times 10^7\,\mathrm{m\,s^{-1}}$ fired directly towards the same tin nucleus;

(b) an α-particle with speed $1 \times 10^7\,\mathrm{m\,s^{-1}}$ fired directly towards a nucleus whose charge is double that of the tin nucleus;

(c) an α-particle with speed $1 \times 10^7\,\mathrm{m\,s^{-1}}$ fired directly towards a nucleus whose charge is the same as that of the tin nucleus but whose *mass* is 2% greater. ■

5 Bohr's semi-classical atomic model

5.1 Bohr's model of the hydrogen atom

In 1913 Niels Bohr (Figure 1.30) opened up new vistas for atomic research by formulating a new and powerful model of the lightest and simplest atom of all, that of the element hydrogen. By bringing together Rutherford's atomic model and the quantum ideas of Planck, Bohr achieved a spectacular breakthrough. He showed that it was possible not only to understand why atomic diameters are the size they are but also (at last!) to explain Balmer's mysterious formula and to generalize it so as to predict the wavelengths of hitherto unknown lines in the hydrogen spectrum!

Figure 1.30 Niels Bohr.

Niels Bohr (1885–1962)

Niels Bohr (Figure 1.30) was one of the most brilliant physicists of the twentieth century. His quantum theory of the hydrogen atom earned him the Nobel Prize for physics in 1922, and as the quantum theory developed he became one of its leading exponents. He alone was able to refute the ingenious arguments advanced by Einstein against the *indeterminism* inherent in the quantum theory. He was born in Copenhagen in 1885 into an academic family — his father was a professor of physiology and his own son Aage (1922–) shared the 1975 Nobel Prize for physics. He studied physics in his home town university where he obtained his doctorate in 1911. He continued his education at Cambridge and Manchester where he worked under Rutherford and J. J. Thomson before returning to Copenhagen as professor of physics in 1916 where the Institute of Theoretical Physics was built specially for him. By this time he had already proposed the atomic model that bears his name.

Later in his career Bohr continued to put forward radical new ideas. His most notable proposals include the liquid drop model of the nucleus and the 'principle of complementarity', i.e. the notion that systems can be considered in mutually incompatible ways depending on the context.

From the early 1930s, Bohr's career was blighted by the rise of Nazism. In Copenhagen, he worked to protect Jewish physicists before, in 1943, himself fleeing first to Sweden and then, via England, to the United States where he worked at Los Alamos on the atom bomb. Before leaving Denmark he dissolved the Nobel medals of Franck and Laue in acid in order to avoid the material being used by the Nazis. (On his return to Copenhagen after the war he precipitated the gold out of solution and recast the medals.)

In his later life Bohr was at the forefront of campaigns to question the development of nuclear weapons and to emphasize peaceful uses of nuclear energy. He died in his home town, Copenhagen in 1962.

According to Rutherford's model, there is just one electron orbiting the nucleus of hydrogen. The nucleus is a single **proton**, possessing a mass 1833 times that of the electron and carrying a charge e. Rutherford's successful analysis of the motion of a charged particle (i.e. an α-particle) in the electrostatic field of a heavy nucleus was based on classical mechanics and showed that Coulomb's law is valid down to distances of the order of 10^{-13} m. It is therefore natural to apply the same basic principles to the case of atomic hydrogen. The electron is attracted to the nucleus, which may be regarded as stationary, and describes an orbit around it in the manner of a planet orbiting the Sun. Let us assume that the orbit is circular, of radius r, and that the electron moves with (constant) speed v. Newton's second law requires that the centripetal acceleration of the electron, v^2/r, times its mass, m_e, should be equal to the magnitude of the electrostatic force pulling it towards the nucleus. This is expressed by the equation

$$\frac{m_e v^2}{r} = \frac{e^2}{4\pi\varepsilon_0 r^2}$$

and multiplying both sides by r gives

$$m_e v^2 = \frac{e^2}{4\pi\varepsilon_0 r}. \tag{1.10}$$

(Equation 1.10 will come in very handy shortly when we come to consider the consequences of Bohr's postulates.)

Now, you will remember that one of the serious problems with Rutherford's model of the atom (and of Thomson's as well) was that classical electromagnetic theory predicts that the orbit is unstable because the electron, by virtue of its charge and acceleration, will constantly lose energy by emitting electromagnetic radiation. It can be shown that, from a typical atomic orbit with $r \approx 10^{-10}$ m, the electron *should* spiral into the nucleus in a few microseconds! Bohr countered this difficulty by applying Planck's quantum hypothesis through the following postulate.

Postulate 1

The electron is confined to move only in orbits in which its angular momentum is an integer multiple of $\hbar$ (= $h/2\pi$). In such an allowed orbit (known as a **Bohr orbit**), the electron emits no radiation and can remain in the orbit indefinitely.

$\hbar$ is read as h-bar or h-cross and is equal to 1.05×10^{-34} J s.

Apart from solving the problem of instability at a stroke, let us see what other consequences arise from this postulate. One consequence of assuming that the electron's angular momentum is quantized is that the radius, r, of the electron's orbit can also take on only certain discrete values. Let us now determine what these values are. The magnitude, L, of the electron's angular momentum in its circular orbit is given by $L = m_e vr$ so that, according to Bohr's postulate,

$$m_e vr = n\hbar$$

giving

$$r = \frac{n\hbar}{m_e v} \tag{1.11}$$

where n = 1, 2, 3, etc. Now, Equation 1.10 provides us with an expression for v, namely

$$v = \sqrt{\frac{e^2}{4\pi\varepsilon_0 m_e r}}.$$

Substituting this into Equation 1.11 we obtain an expression for r which can be written as $r = n^2 a_0$, where

$$a_0 = \frac{4\pi\varepsilon_0 \hbar^2}{m_e e^2} = 0.53 \times 10^{-10}\text{ m}. \tag{1.12}$$

Question 1.11 Confirm for yourself the expression for the allowed values of r by substituting the expression for v into Equation 1.11. (*Hint*: It's easier if you square Equation 1.11 first and substitute for v^2.) ■

The parameter a_0 is called the **Bohr radius**. It is the radius of the *smallest* Bohr orbit in atomic hydrogen, that is, the orbit for which $n = 1$. The other allowed orbits are found by multiplying a_0 by n^2 and are thus 4, 9, 16, etc. times larger as illustrated in

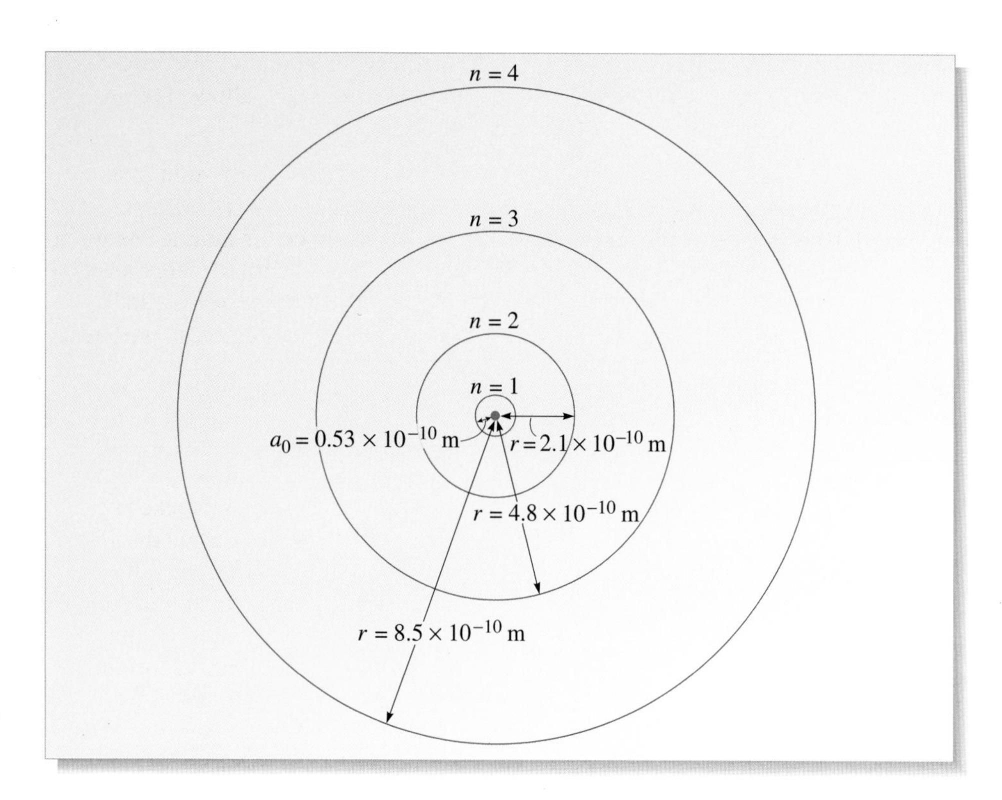

Figure 1.31 The first four Bohr orbits in atomic hydrogen.

Figure 1.31. Perhaps more importantly, a_0 is a combination of fundamental constants which sets the length scale for atomic phenomena at around 10^{-10} m. This effectively answers the question posed in Section 2.1 about the size of atomic diameters. You might like to convince yourself that the units of a_0 do indeed turn out to be those of length.

Question 1.12 By substituting the expression for r (Equation 1.11) into Equation 1.10, obtain a formula for the speed, v_n, of the electron when it is in the nth orbit in the hydrogen atom. Calculate the speed, v_1, of the electron when it is in the first Bohr orbit. Do you think the use of classical, non-relativistic mechanics is justified in the Bohr model? ■

There remains the problem of explaining the observed spectral lines of hydrogen — the Balmer series. Recall that, in Section 2.2 we discussed how Balmer had discovered a numerical expression which gave the wavelengths of the visible lines in the hydrogen spectrum with great accuracy, namely

$$\lambda = 364.56\left\{\frac{n^2}{n^2 - 4}\right\} \text{nm} \qquad \text{(Eqn 1.1)}$$

where n is an integer which can take the values 3, 4, 5 and 6. Can Bohr's model shed any light on the origin of this formula? Bohr tackled the question of the emission of radiation with another postulate.

> *Postulate 2*
>
> A quantum of radiation is emitted when the electron makes a transition, or jump, from an allowed orbit of higher energy to one of lower energy. If the energy change in the transition is ΔE, the frequency f of the radiation is given by $f = \Delta E/h$.

It follows from this that, in order to predict the frequencies that might be emitted by an atom, we need to know the energy differences between all the allowed orbits. This of course requires us to know the energy of each orbit.

The total energy of the electron, in an allowed orbit of radius r, is the sum of its kinetic energy, $\frac{1}{2}m_ev^2$, and its electrostatic potential energy, $-e^2/4\pi\varepsilon_0 r$. A glance at Equation 1.10 reveals that the potential energy is exactly twice the kinetic energy in magnitude (though, of course, it is opposite in sign), so that the total energy must be just half the potential energy, that is, $-e^2/8\pi\varepsilon_0 r$. Thus, substituting $r = n^2a_0$ and denoting by E_n the total energy (E_{tot}) of the electron in the nth Bohr orbit, we have

$$E_n = -\frac{e^2}{8\pi\varepsilon_0 a_0 n^2} = -\frac{|E_1|}{n^2} \tag{1.13}$$

where $E_1 = -e^2/8\pi\varepsilon_0 a_0 = -2.18 \times 10^{-18}\,\text{J} = -13.6\,\text{eV}$ is the energy in the lowest Bohr orbit. It is clear from this that the electron in the hydrogen atom cannot take any arbitrary value of the energy. The electron's energy is restricted to certain discrete values, namely, those given by Equation 1.13. We refer to these different values of energy as **energy levels**. The set of energy levels available to the electron in atomic hydrogen is shown in Figure 1.32. You will see that the energies get closer and closer together as they approach zero from below and, in fact, there is an infinite number of bound (i.e. negative energy) **excited states** above the **ground state**, which corresponds to the lowest Bohr orbit. In addition, there is a *continuum* of positive energy states corresponding to ionized hydrogen, in which the electron is detached from the proton and is thus completely free.

● What is the minimum amount of energy that must be supplied to the electron in atomic hydrogen in its ground state to liberate it from the nucleus?

❍ The energy required is $2.18 \times 10^{-18}\,\text{J}$ or $13.6\,\text{eV}$. ■

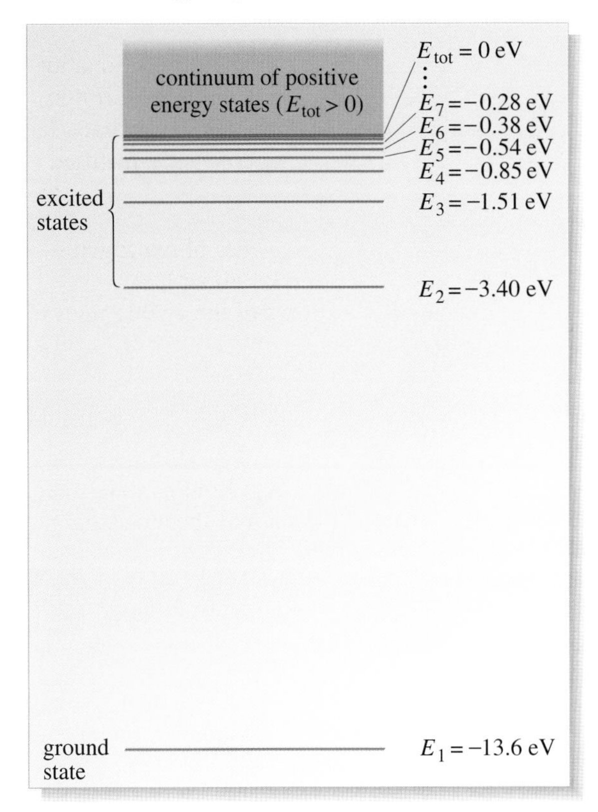

Figure 1.32 The possible energy states of the electron in the hydrogen atom. When the electron is *bound* in the atom, its energy is *negative* and *quantized* — it can take only certain discrete values. When the electron is unbound, its energy may take *any positive* value.

Question 1.13 (a) Is it correct to say that the total energy of an unbound electron is quantized?

(b) What would happen if 4.0 eV of energy were transferred to an electron in the $n = 2$ energy level of a hydrogen atom? ■

It is clear that an orbit with larger radius has higher (i.e. less negative) energy. It follows that the emission of radiation takes place when the electron makes an *inward* transition from one of the outer orbits. The process is illustrated in Figure 1.33. If an electron in the nth orbit makes a transition to the qth orbit, where $q < n$, then from Equation 1.13,

$$\Delta E = E_n - E_q = -\frac{|E_1|}{n^2} + \frac{|E_1|}{q^2} = |E_1|\left\{\frac{1}{q^2} - \frac{1}{n^2}\right\}$$

and the frequency of the emitted quantum is, according to Postulate 2,

$$f_{n\to q} = \frac{|E_1|}{h}\left\{\frac{1}{q^2} - \frac{1}{n^2}\right\} = \frac{|E_1|}{h}\left\{\frac{n^2 - q^2}{q^2 n^2}\right\}.$$

Rewriting this as a wavelength, using the relation $\lambda f = c$, we obtain

$$\lambda_{n\to q} = \left|\frac{hc}{E_1}\right|\left\{\frac{q^2 n^2}{n^2 - q^2}\right\} = 91.127\left\{\frac{q^2 n^2}{n^2 - q^2}\right\}\text{nm}. \tag{1.14}$$

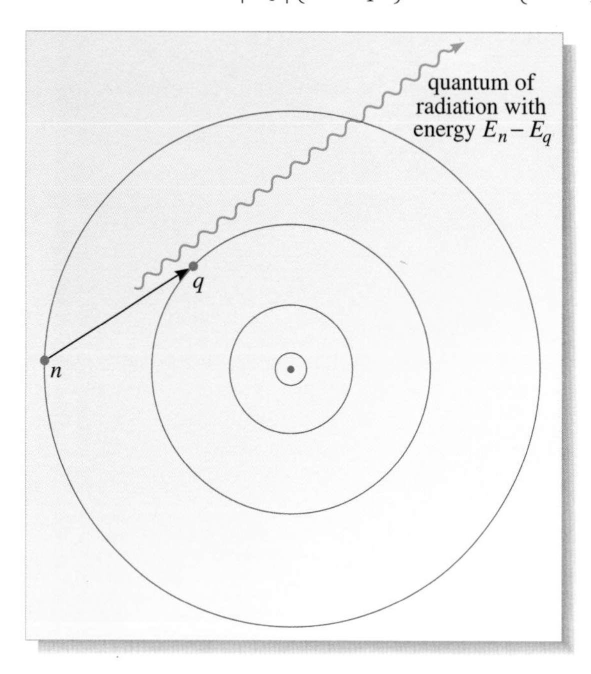

Figure 1.33 When the electron in a hydrogen atom makes a transition from the energy level E_n to the energy level E_q a quantum of light with energy $E_n - E_q$ is emitted. In the example shown here $n = 4$ and $q = 3$.

Equation 1.14 is known as **Bohr's equation**. When we put $q = 2$ into Equation 1.14, it becomes almost identical to Equation 1.1 — Balmer's formula! The numerical factor outside the brackets here is 364.51 nm compared to Balmer's value of 364.56 nm, a difference of about 1 part in 7000. Bohr's theory thus gives a clear interpretation of the four lines of the Balmer series: they arise from transitions *to* the second Bohr orbit *from* the next four orbits.

The *experimentally determined* number outside the brackets in Equation 1.14 is 91.1763×10^{-9} m, and is equal to the reciprocal of a frequently-used quantity in spectroscopy known as the **Rydberg constant** R for the hydrogen atom: $R = 1.096\,78 \times 10^7\,\text{m}^{-1}$.

It is clear from Equation 1.14 that Bohr's theory predicts the existence of other series of lines in the hydrogen spectrum arising from transitions terminating at the orbits with $q = 1, 3, 4, 5$, etc. These series have been found experimentally, and are named after their discoverers: Lyman, Paschen, Brackett and Pfund. The relevant transitions are shown, together with those of the familiar Balmer series, on the energy-level diagram in Figure 1.34.

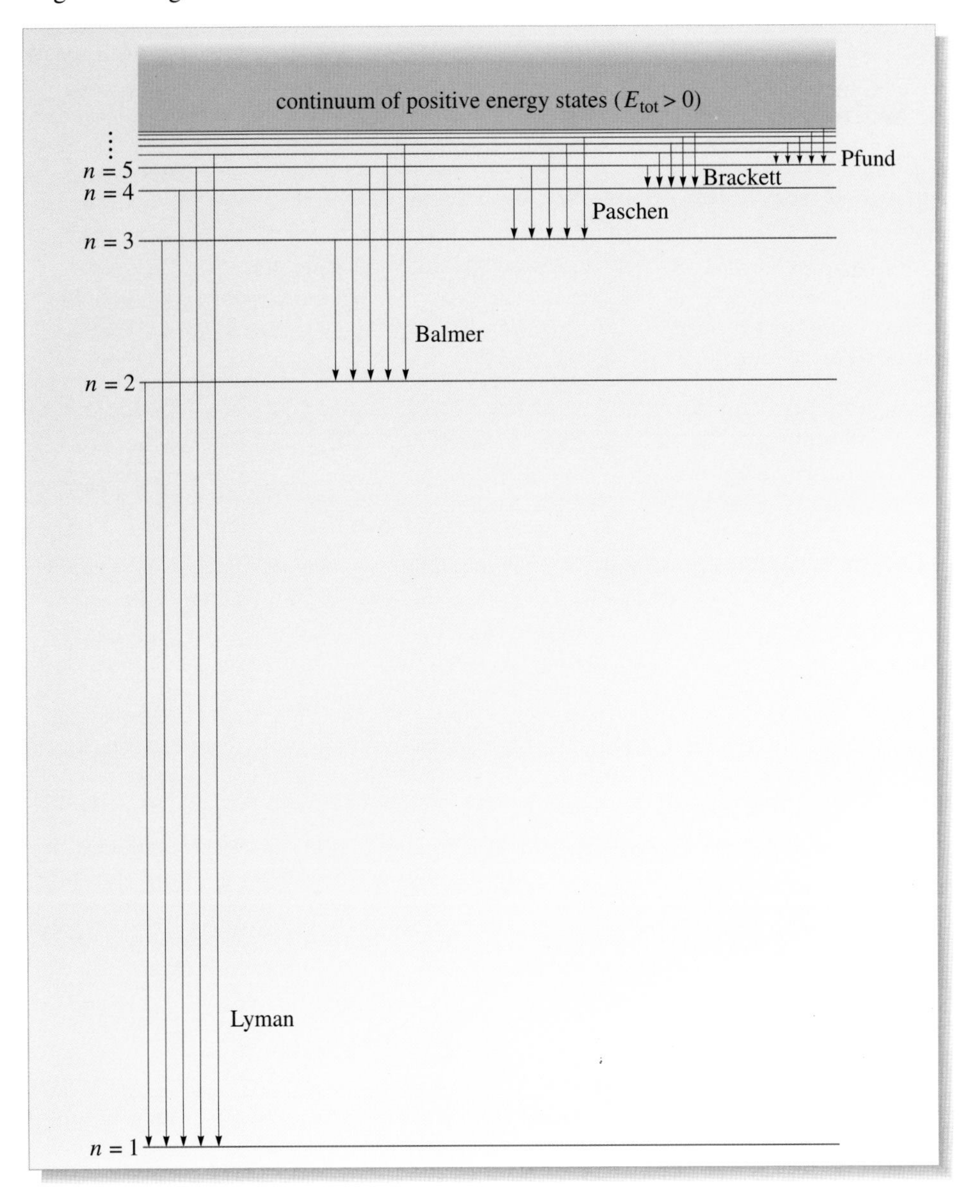

Figure 1.34 The transitions of the electron in the hydrogen atom that give rise to the Lyman, Balmer, Paschen, Brackett and Pfund series of spectral lines. (Only the first five lines of each series are shown here.)

Question 1.14 Show that, according to the Bohr model, the wavelengths of the Brackett series of spectral lines are given by

$$\lambda = 1458\left\{\frac{n^2}{n^2 - 16}\right\}\text{nm} \quad (n = 5, 6, 7, \ldots).$$

Which transition in this series is responsible for the line at wavelength 1736 nm? ■

We will end this section on Bohr's famous model of the hydrogen atom with a few comments on the two postulates on which it is based. Although they might seem rather ad hoc, they are based on principles that had been established by Planck as he refined and generalized the quantum hypothesis during the decade following his success in deriving the blackbody spectrum. Planck described the absorption and emission of radiation by the walls of an isothermal cavity in terms of electrical oscillators located in the cavity walls. In speculating about the role of the quantum in these processes, he concluded that they are associated with transitions of the oscillators between allowed energy levels and that the frequency of the radiation they emit obeys the rule $f = \Delta E/h$, where ΔE is the energy difference between two allowed energy levels.

5.2 Bohr's model extended to heavy atoms

Bohr's model of the hydrogen atom was very successful, so there was a strong incentive to develop a similar model for heavy atoms, which we define to be atoms containing more than one electron. The task of developing such a model proved to be very difficult. Bohr and many other distinguished physicists, in particular, Arnold Sommerfeld, put a great deal of effort into this. Bohr struggled for nearly ten years to refine his first model of heavy atoms, formulated in 1913. In this section, we shall briefly describe a mature version of his model.

Bohr proposed that each heavy atom contained electrons moving around the nucleus in well-defined orbits, with definite energy values. He also suggested that the electrons occupy *shells*, which are now customarily labelled by letters of the Roman alphabet. The shell nearest the nucleus is called the K-shell and the next nearest shells are labelled (in alphabetical order) the L-, M-, N-shells, and so on. He assumed that, for each shell, there is a certain *maximum* number of electrons which it can contain and these are 2, 8, 18 and 32 in the K-, L-, M- and N-shells, respectively. (The general formula for this number is $2n^2$ for the nth shell.) This implies that once a shell is full, any further electrons must go into the next higher shell.

An example may serve to clarify these ideas. Figure 1.35 is a schematic diagram of Bohr's model of the neon atom, which contains ten electrons. In this particular atom, the K-shell and the L-shell are both full, so that all the electrons have their lowest possible energies. The atom in this configuration is said to be in its ground state. If the atom is in an excited state in which one of the electrons originally in the L-shell is excited up to one of the normally empty shells, radiation of a characteristic wavelength is emitted when it returns to the L-shell.

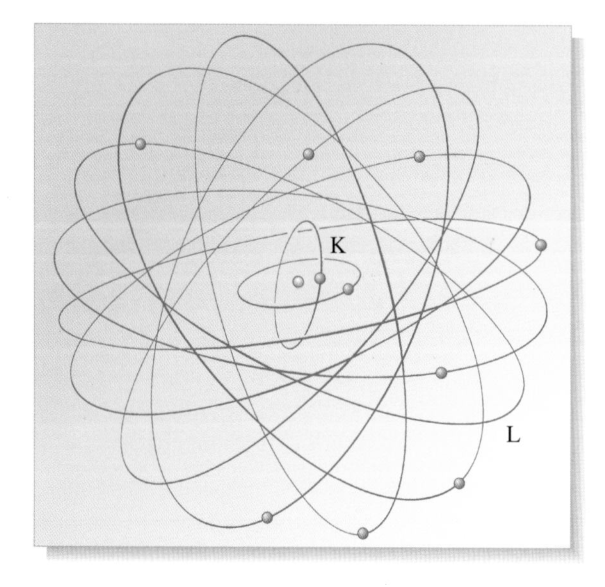

Figure 1.35 According to Bohr's model of heavy atoms, electrons normally move around the nucleus in specified orbits with fixed energy values. This figure represents the neon atom with its ten constituent electrons in their lowest possible orbits, which belong to the K- and L-shells.

The idea that electrons in atoms occupy shells allowed Bohr, and others, to obtain considerable insight into the structure of the Periodic Table of chemical elements. The model can be used to argue that the similarities among the properties of certain groups of elements reflected underlying similarities in the atomic structures of these elements. He argued, for example, that the reason sodium and potassium have similar chemical properties is that each has a single electron in its outer shell.

Despite some successes, the extension of Bohr's model to heavy atoms ultimately failed, and by the early 1920s it was clear that a new model of atomic structure was needed. The main shortcomings of the Bohr model are listed below.

1. The formula $2n^2$ for the maximum number of electrons which could be contained in the nth shell could not be understood.
2. No method existed for calculating the contribution to the energy of the atom resulting from the electrostatic repulsion between the electrons, even in the case of the helium atom where there are only two of them. This meant that the wavelengths of the spectral lines emitted by atoms other than hydrogen could not be computed theoretically.
3. No explanation could be given for the difference in brightness (intensity) between one spectral line and another. (This was, in fact, a shortcoming of the Bohr model even in the case of the hydrogen atom.)

6 Towards a quantum model of the atom

6.1 Introduction

The failure of Bohr's model in heavy atoms plunged physics into a new crisis. The rather unhappy marriage between classical mechanics and Planck's quantum hypothesis had clearly broken down and a completely new approach was needed. Bohr himself saw this clearly:

> 'In atomic physics all previous concepts have proved inadequate; we know from the stability of matter that Newtonian physics does not apply to the interior of the atom … it follows that there can be no descriptive account of the structure of the atom; all such accounts must be based on classical laws which no longer apply … anyone trying to develop such a theory is trying the impossible …'

In due course, a new theory *was* developed, mainly by European physicists, during the decade 1923–1933. This period saw some of the greatest intellectual achievements in the entire history of science, with the creation of a new and revolutionary quantum theory of matter. The new theory and its applications form the subjects of Chapter 2.

6.2 Wave–particle duality

We will finish this chapter with a description of the **wave–particle duality** which signposted the way towards a new understanding of matter on the atomic scale. Let us ask an apparently simple question. *What is light*? Newton's answer to this question was that it is a stream of tiny particles or corpuscles. In the seventeenth century, the known phenomena involving light could usually be explained quite naturally in corpuscular terms.

However, even in Newton's time, a wave theory of light was put forward by the Dutch scientist Christiaan Huygens. At the beginning of the nineteenth century,

These effects are discussed in *Dynamic fields and waves*.

Thomas Young demonstrated that sunlight, passing through two pinholes, produces interference effects. He concluded that light must be a form of wave motion. This view was not widely accepted at first, but became generally established after confirmation of Young's experimental results by the French scientist Augustin Fresnel. Further impetus was given to the wave theory by the discovery of the Doppler effect in 1842. This famous phenomenon: the apparent change in frequency of a light signal received by an observer in motion relative to the source, is most naturally explained in terms of wave motion. Indeed, the effect was first demonstrated for *sound*. The triumph of the wave theory of light seemed complete when the second great pillar supporting the edifice of classical physics (the first being Newtonian mechanics) was erected by Maxwell: the theory of electromagnetism. According to this theory, light is a transverse electromagnetic wave. Effects associated with the polarization of light, for which no satisfactory explanation had existed, now came within the compass of the theory.

It was against this background of a pre-eminent wave theory of light that Einstein put forward his 'light quantum' ideas in 1905. He saw that certain phenomena, notably the photoelectric effect, could not be explained unless particle-like behaviour was attributed to light. To paraphrase Mark Twain: 'reports of the demise of the corpuscular theory of light have been greatly exaggerated.'

We must conclude that electromagnetic radiation can, under different circumstances, exhibit behaviour characteristic of either a wave (in, for example, diffraction or polarization) or a particle (in, for example, the photoelectric effect). Obviously, under such circumstances, it would be inappropriate to say that light 'is' either a wave or a particle, and, indeed, there is no reason why a phenomenon such as light *should* necessarily be explicable in terms of the simple classical concepts of waves and particles. Rather, we should accept that both models of light have some merit and we should ask if physics can provide some deeper reason for this so-called wave–particle duality. As you will see, that is, indeed, what quantum physics does.

6.3 The de Broglie formula and its verification

If electromagnetic radiation exhibits a wave–particle duality, could the same be true of ordinary matter? Perhaps the matter that we normally call 'particles' must sometimes be modelled as a wave? These extraordinary questions occurred to a young French postgraduate student, Louis de Broglie (Figure 1.36) in 1922. He hypothesized that this was indeed the case and included it in his doctoral thesis.

De Broglie showed that if wave-like properties are associated with matter particles, then the wavelength of the wave must be given by

$$\lambda_{\text{dB}} = \frac{h}{p} \qquad (1.15)$$

where p is the magnitude of the particle's momentum. This important equation is called the **de Broglie formula**, and the wavelength λ_{dB} is called the **de Broglie wavelength** of the particle. This relationship between the wavelength and the magnitude of the momentum of a supposed 'particle' such as an electron is also true for radiation. To see this, recall that, according to special relativity, the relationship between the energy, E, and momentum, p, of a particle of mass m is $E^2 = p^2c^2 + m^2c^4$. The light quantum may be thought of as a particle of zero mass, in which case this relationship simplifies to $E = pc$, or, rearranging $p = E/c$. Now, according to Planck, $E = hf = hc/\lambda$ so that $p = h/\lambda$ or $\lambda = h/p$.

Louis Victor Pierre Raymond de Broglie (1892–1987)

Figure 1.36 Louis de Broglie.

Louis de Broglie (Figure 1.36) was born into an aristocratic French family in Dieppe in 1892. He gained a degree in history at the Sorbonne before following the traditional route for his family into the Army. There he became interested in radio communications and science. After the First World War he concentrated on science and, in 1924, he gained his doctorate. He was the younger brother of Maurice, the sixth Duc de Broglie (1875–1960), also a distinguished physicist and whom Louis assisted at his private laboratory at the family home. Louis de Broglie was professor at the Henri Poincaré Institute from 1932 to 1962. He made one outstanding contribution to the development of quantum theory — the de Broglie hypothesis. Guided by the relationships between mass and energy (established by Einstein) and frequency and energy (established by Planck) de Broglie suggested a corollary to the Planck energy quantum for radiation, namely, that matter could behave as waves and he set out the linking equation $\lambda_{dB} = h/p$. For this he was awarded the Nobel Prize for physics in 1929. His idea was used by Schrödinger and others in the new 'wave-mechanics'. De Broglie died in 1987 at the age of 95.

Question 1.15 Calculate the de Broglie wavelength of the following particles, using the non-relativistic definition of momentum:

(a) a billiard ball of mass 0.1 kg travelling across a billiard table at $2\,\mathrm{m\,s^{-1}}$;

(b) a dust particle of mass 10^{-6} kg falling to the ground at $0.1\,\mathrm{m\,s^{-1}}$;

(c) an electron with a kinetic energy of 10 eV. ■

You saw in your answer to Question 1.15 that even the most minute everyday object has an extremely short de Broglie wavelength — many orders of magnitude smaller even than the typical interatomic spacing in a crystal which is about 0.5 nm. Little wonder then, that we do not observe any wave-like properties for macroscopic particles. The de Broglie wavelength of the 10 eV electron, however, *is* similar to the interatomic spacing in a crystal. This suggests that it might be possible to observe the *diffraction of electrons* using a crystal as a diffraction grating.

It was clear from the extensive and accelerating output of experimental work in atomic physics in the early 1920s, that it would not be long before de Broglie's idea was put to the test. In April 1925, at the Bell Telephone Laboratories in New York, Clinton J. Davisson and Lester H. Germer were doing some fairly routine measurements of the angular distributions of electrons scattered by a nickel target. Their apparatus is shown schematically in Figure 1.37. The double-walled collector box had an adjustable retarding potential difference applied between the outer and inner walls.

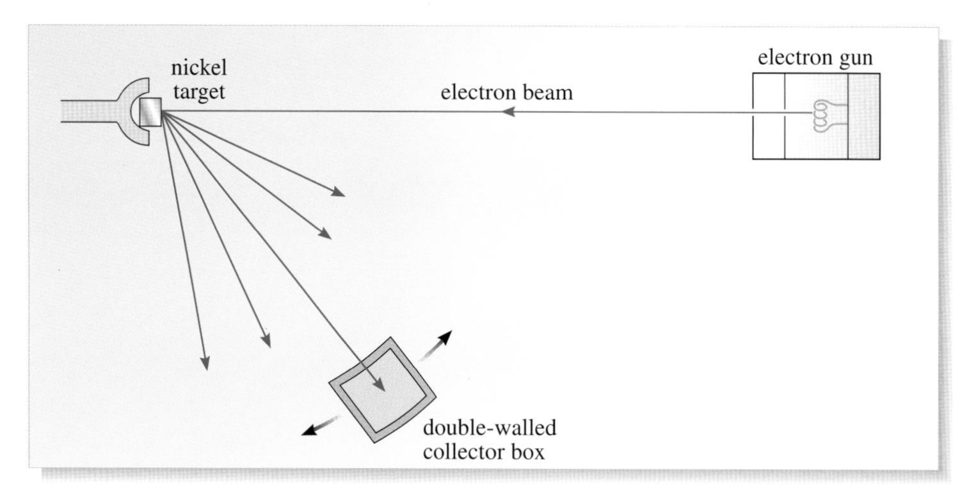

Figure 1.37 The apparatus used by Davisson and Germer. A narrow beam of electrons with a well-defined energy was scattered from a nickel target. The intensity of electrons scattered at different angles was determined using the double-walled collector box, which could be rotated about an axis perpendicular to the plane of the page. The orientation of the target with respect to the incident beam could also be adjusted.

By varying this retarding potential difference, Davisson and Germer ensured that they detected only electrons that had undergone elastic collisions, i.e. had suffered no loss of kinetic energy. The target was polycrystalline nickel — it was composed of many randomly oriented crystals. During the experiment, an accident took place in the laboratory causing Davisson and Germer to obtain a remarkable set of results. As a consequence of the explosion of a liquid-air bottle, the surface of the target became oxidized. While the oxide was being removed, by baking in a high-temperature oven, the target *recrystallized* into a few large crystals. The striking difference between the observations *before* and *after* the accident is shown in Figure 1.38.

You can see that the effect of the accident was to produce distinct variations in the number of electrons scattered at different angles, reminiscent of the maxima and minima in the diffraction pattern produced when light waves impinge on a diffraction grating. In fact, the analogy is a good one. Davisson and Germer went on to interpret their results quantitatively in terms of the scattering of *electron waves* by regularly spaced atoms in a crystal. Moreover, they found that the electron wavelength determined by their measurements was given by de Broglie's formula.

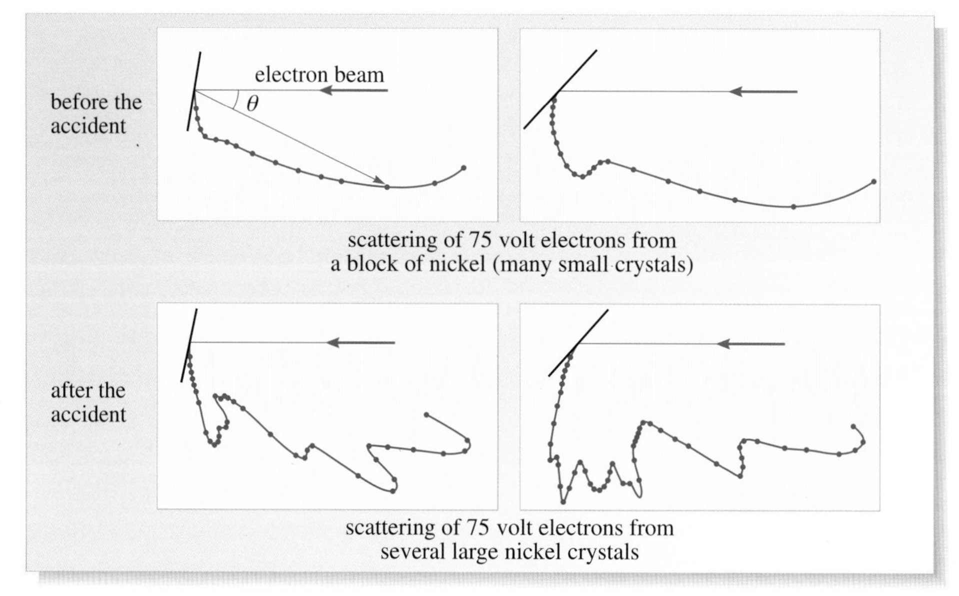

Figure 1.38 Some of Davisson and Germer's data on the scattering of electrons by nickel targets. The results are shown for two different target orientations. In each diagram, the scattered intensity at each angle θ is proportional to the distance of the relevant point on the curve from the point of scattering. Before the accident, the intensity of the scattered electrons varies quite smoothly with θ, but after the accident, diffraction maxima and minima can be seen clearly in the intensity distribution.

The original result of Davisson and Germer was soon corroborated by other measurements. George Paget Thomson, (the son of J. J. Thomson,) together with O. Oldenburg, carried out a set of experiments in which electrons were passed through thin metallic foils. They used electrons with de Broglie wavelengths of 0.01 nm, about ten times smaller than those used by Davisson and Germer, and observed beautiful diffraction patterns (Figure 1.39). (It is interesting to reflect that thirty years after the father showed that the electron is a particle, the son showed that it is also a wave.)

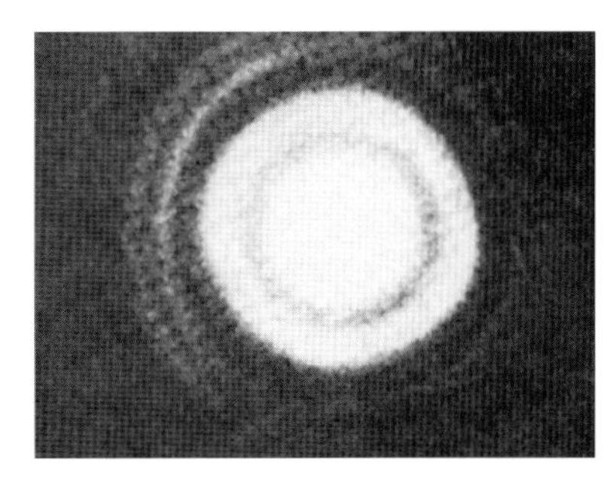

Figure 1.39 This diffraction pattern was recorded by G. P. Thomson and O. Oldenburg when they fired electrons through a thin gold foil.

A large number of experiments have amply demonstrated that other particles, such as neutrons, helium atoms and much heavier atoms all show wave-like properties. They all exhibit diffraction and interference effects, under the right conditions, consistent with a wavelength given by de Broglie's formula. But what is a de Broglie wave? Can it be pictured as a real, physical oscillation, and if so, what is oscillating? These questions will be explored in Chapter 2.

7 Closing items

7.1 Chapter summary

1. Major problems confronted physicists in the early 1900s. In particular classical physics could not explain the experimental observations in the following five areas:
   (i) atomic structure and atomic diameters;
   (ii) atomic spectral lines, especially the Balmer series in hydrogen;
   (iii) the shape of the blackbody spectrum;
   (iv) the photoelectric effect;
   (v) the specific heats of solids.
2. Planck solved the problem of blackbody radiation by introducing the idea of a quantum of energy for electromagnetic radiation.
3. Einstein used the idea of quanta to explain the photoelectric effect.
4. Einstein formulated a quantum theory of the specific heats of solids which accounted for the observed decrease in the specific heat with decreasing temperature.
5. The plum-pudding model of atomic structure was put forward by J. J. Thomson. This assumed that the atom was filled with positively charged matter and electrons. After the α-particle scattering experiment of Geiger and Marsden, Rutherford proposed a rival model in which the positive matter is concentrated into a tiny nucleus with the electrons orbiting outside it.
6. Bohr used Rutherford's nuclear model of the atom together with Planck's quantum hypothesis to formulate a semi-classical atomic model.
7. Bohr's model was extremely successful when applied to hydrogen, accounting for both the size of the hydrogen atom and the Balmer series. It predicted the existence of other spectral lines subsequently observed experimentally.
8. In contrast to the case of hydrogen, Bohr's theory achieved only modest success when applied to heavy atoms. Its failure indicated the need for a more radical theory.
9. The quantum theory requires light to have particle-like properties. As the wave properties of light are well established, it is said to exhibit a wave–particle duality.

10 It was suggested by de Broglie that particles, such as electrons, might show wave properties. This hypothesis was confirmed experimentally by Davisson and Germer. Thus, the wave–particle duality is a feature of both radiation and matter at the atomic level.

7.2 Achievements

After studying this chapter you should be able to:

A1 Explain the meaning of all the newly defined (emboldened) terms introduced in this chapter.

A2 Describe the basics of atomic spectroscopy, blackbody radiation, the photoelectric effect and the molar heat capacities of solids.

A3 Describe in detail the failure of classical physics to account for the experimental results on all four of these topics.

A4 Explain Planck's quantum hypothesis and how it was used to account for the key facts of the blackbody spectrum, the photoelectric effect and the temperature variation of the molar heat capacities of solids.

A5 Recall and apply Planck's law for the energy of a quantum of radiation $E = hf$, and Einstein's photoelectric equation, $\frac{1}{2}m_{\text{e}}v_{\text{max}}^2 = hf - \phi$.

A6 Describe the main features of the atomic models of J. J. Thomson and Ernest Rutherford.

A7 Describe the Geiger–Marsden α-particle scattering experiment, discuss its implications for the atomic models of Thomson and Rutherford and derive a formula for the distance of closest approach r_{cl} of an α-particle to a nucleus during a collision.

A8 Give an account of Bohr's semi-classical atomic model, its successes and failures.

A9 Use Bohr's equation

$$\lambda_{n \to q} = 91.127 \left\{ \frac{q^2 n^2}{n^2 - q^2} \right\} \text{nm} \qquad (1.14)$$

to derive wavelengths of different series of lines in the hydrogen spectrum.

A10 Interpret the energy-level diagram of the electron in the hydrogen atom.

A11 Explain what is meant by wave–particle duality and discuss its experimental basis.

A12 Recall and apply the de Broglie formula $\lambda_{\text{dB}} = h/p$.

7.3 End-of-chapter questions

Question 1.16 The peak of intensity in the spectrum of light emitted from a certain furnace occurs at a wavelength of 2.3 μm. Use Figure 1.7 to estimate the temperature of the furnace assuming that it radiates as a blackbody.

Question 1.17 The photoelectric threshold wavelength of sodium is 542 nm. Calculate the work function of sodium and the maximum kinetic energy of electrons ejected by incident electromagnetic radiation of wavelength 425 nm. Give energy values in eV.

Question 1.18 The Einstein temperature of diamond is 1300 K. Calculate the frequency of vibration of the carbon atoms in a diamond crystal according to Einstein's theory.

Question 1.19 Summarize very briefly (one short sentence for each) the experimental facts about (i) atomic spectroscopy, (ii) blackbody radiation, (iii) the photoelectic effect and (iv) the molar heat capacities of solids, which classical physics is unable to explain.

Question 1.20 Under ideal conditions, the human eye can detect as little as 10^{-18} J of electromagnetic energy. Estimate roughly how many photons this represents in the middle of the visible spectrum.

Question 1.21 Consider a large number of oscillators of natural frequency f in thermal equilibrium at absolute temperature T. Write down expressions for the thermodynamic average energy of an oscillator according to (a) classical physics (b) Planck's quantum hypothesis.

Question 1.22 Imagine yourself to be a guest at a certain Knightsbridge soirée, held during the winter of 1911. The guests are all physicists and among them are two gentlemen who have strong views on the merits of the current atomic models. The first, a Mr Smith from Cambridge, is a passionate advocate of Thomson's. The second, a Mr Jones from Manchester, is enthusiastically in favour of Rutherford's model and is generally scathing about the rival model. You decide to be awkward by challenging the views of both men.

(a) Give to Mr Smith, the advocate of Thomson's model, *one* reason why you think his favourite model is intrinsically unsatisfactory and one piece of experimental evidence that *his* model cannot possibly explain but that *is* explained by Rutherford's new model.

(b) Give to Mr Jones, the advocate of Rutherford's model, *one* reason why you think his favourite model is intrinsically unsatisfactory and one piece of experimental evidence that cannot be understood by *his* model but that *can* successfully be investigated by using Thomson's model.

(c) Messrs Smith and Jones are both offended and angered by your incisive attacks on their prejudices. To placate them, show that really you are equally sceptical of both models by citing *two* pieces of experimental evidence that cannot be explained by *either* model.

Question 1.23 The most energetic α-particles available from radioactive sources have kinetic energies of about 1.3×10^{-12} J. Calculate the distance of closest approach, r_{cl}, of an α-particle of this energy to a platinum nucleus carrying a charge of $78e$. What is the significance of your result for atomic physics?

Question 1.24 What is the de Broglie wavelength of a tennis ball of mass 0.06 kg served at a speed of $60\,\text{m}\,\text{s}^{-1}$? ■

Chapter 2 Schrödinger's wave mechanics

1 Quantum mechanics — a new approach to describing atomic matter

If experimental evidence conflicts with an accepted scientific theory, adjustments to the theory can usually be made, but only up to a point. As mounting experimental evidence becomes overwhelming, scientists sometimes have to revolutionize their ideas with a new theory capable of accommodating experimental observations comfortably. Perhaps the most significant scientific revolution since the time of Newton, began at around the turn of the twentieth century and peaked in the mid-1920s, with the advent of Schrödinger's formulation of quantum mechanics. The need for a revolutionary theory of microscopic nature began to be recognized as the results of experimental atomic physics accreted into a body of evidence which, scientists eventually realized, undermined many of the basic ideas of classical mechanics. (Figure 2.1.)

The challenge to classical mechanics could hardly have been more formidable, for the world of classical mechanics is **deterministic**. Classical mechanics supposes that, while the mathematical difficulties are often insurmountable, it is *in principle* possible, given sufficient information about the present, to predict the future exactly, that is, to predict the progress of any experiment with absolute certainty. The evidence from atomic physics implied that this was not true of microscopic nature. In microscopic nature, *identical conditions do not produce identical results.*

The new theory that embraces the *indeterminism* of the microscopic world is called **quantum mechanics** and is the main subject of this chapter. It provides an intellectually satisfying and coherent framework for solving problems in atomic physics, and hence for understanding the behaviour of matter. The new theory answers most of the questions left unanswered by Bohr's old quantum theory, as well as revealing many new, exciting and unforeseen effects.

The empirical quantum ideas of Bohr and others are usually said to belong to the *old quantum theory* to distinguish it from those of the fully developed *quantum mechanics* that followed.

In the previous chapter, you saw that electromagnetic radiation *and* matter exhibit what is often termed *wave–particle duality*: in different circumstances, either a wave

Figure 2.1 The old and the new. (The participants at the 1933 Solvay Council.) Niels Bohr (seated, third from the left), a pivotal figure in the old quantum theory, together with Werner Heisenberg (standing, fourth from the left) and Erwin Schrödinger (seated, first on the left) — the pioneers of the new quantum mechanics.

model or a particle model may be required to account for some observed aspect of their behaviour. In classical physics these two models are contradictory. In quantum physics they must somehow be brought together to provide a single self-consistent description. This is what quantum mechanics achieves.

You will realize as you read this chapter that the study of quantum mechanics draws on many ideas and principles from other parts of physics. In particular, we shall make use of concepts from Newtonian mechanics, electricity and magnetism, and the study of waves. You should note that the major proportion of the material in this chapter is contained in Section 4 and you should reserve at least half of your study time for it.

It is very important to note at the outset that the laws of quantum mechanics are *completely* different from the laws of classical mechanics. And, whereas classical mechanics is based on intuitively reasonable ideas, quantum-mechanical ideas are very unintuitive and may therefore appear somewhat harder to grasp. Nevertheless, it is essential to come to grips with these new ideas in order to understand the behaviour of matter at the atomic level.

2 Towards quantum mechanics

In this section, we start by studying some idealized experiments, the results of which motivate the need for a quantum-mechanical description of matter. In Section 2.2, we shall very briefly discuss the scope of quantum mechanics, and we shall contrast some aspects of this theory with the more familiar ideas of classical mechanics. Section 2 ends with a brief description of the kind of waves that are required to give a quantum-mechanical description of matter.

2.1 Electron diffraction experiments

In Section 6 of the last chapter we saw how Louis de Broglie suggested a model of nature which assigned wave-like attributes to what our experience of reality had formerly designated a *particle* (say, an electron). Certain observations can be understood by associating with the particle a de Broglie wavelength given by

$$\lambda_{\mathrm{dB}} = h/p \qquad \text{(Eqn 1.15)}$$

where p is the magnitude of the particle's momentum and h (= 6.626×10^{-34} J s) is Planck's constant. Although we can't go into details, it is worth noting that a large number of experiments have amply demonstrated that particles such as neutrons, helium atoms and much heavier atoms, and molecules, all show wave-like properties. They all exhibit diffraction and interference effects under the right conditions and the de Broglie formula gives the correct value for the associated wavelength.

We shall now explore a well known thought experiment, which demonstrates the essence of quantum behaviour. The experiment involves firing a beam of electrons of known energy at a narrow slit or double slit and observing the pattern of electrons as they impinge on a screen on the other side of the slit (Figure 2.2).

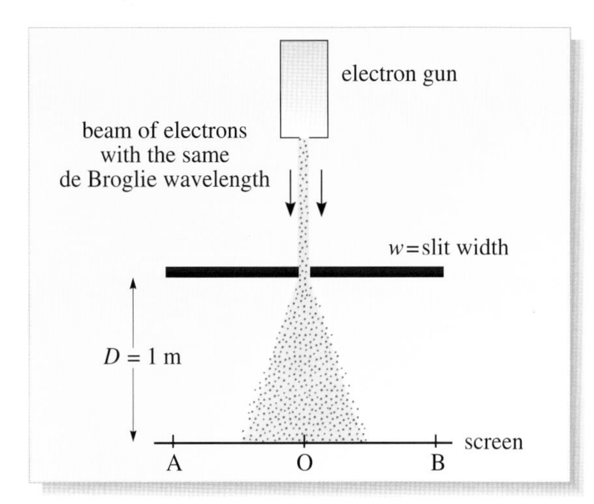

Figure 2.2 The set-up for the electron diffraction thought experiment.

The thought experiment, as we describe it, represents a cohesive account of the results of related experiments that have actually been performed, and it contains the essence of what those experiments tell us of quantum nature.

Open University students should leave the text at this point and study the multimedia package *Electron diffraction* that accompanies this book. When you have completed the package (which should take about 2 hours), you should return to this text.

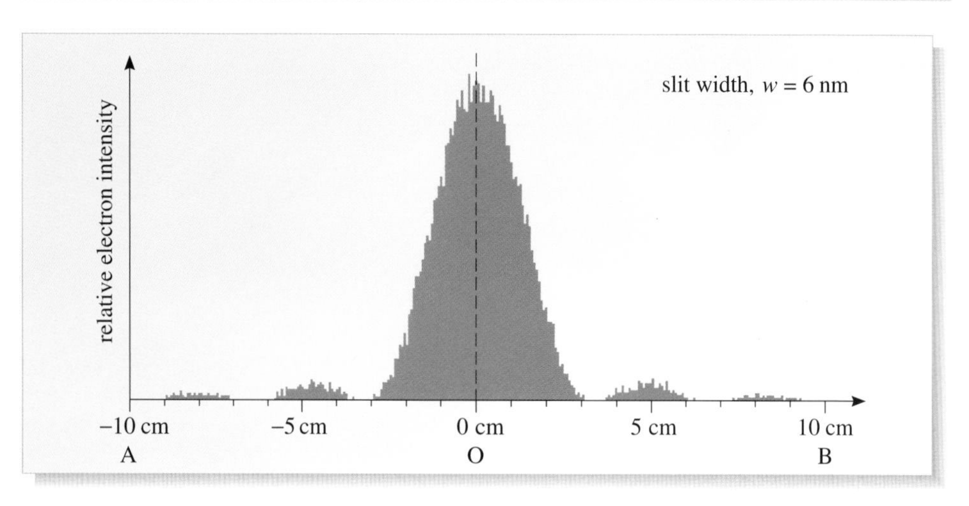

Figure 2.3 The intensity distribution of electrons arriving at a screen after passing through a single slit of width w. The first minima occur at an angle $\theta = \lambda_{\text{dB}}/w$ either side of the straight through direction ($\theta = 0$).

When a beam of electrons of known kinetic energy E_{kin}, and therefore known de Broglie wavelength, is directed at a narrow slit of width w, as shown in Figure 2.2, then the intensity distribution of the electrons arriving at a screen placed behind the slit is that of a single slit diffraction pattern (Figure 2.3). It is the same intensity distribution as that obtained by shining electromagnetic radiation of the same wavelength at a slit of the same width. This intensity distribution for a very large number of electrons is predictable and reproducible. However, if the beam intensity is reduced to such a level that there is effectively only one electron passing through the apparatus at a time, then there is no way of knowing where any particular electron will impinge. All that can be said is that it is most likely to arrive at a point where the distribution shows a maximum and has low probability of impinging near the positions of the intensity minima (zero actually *at* the minima). In order to explain the intensity distribution at the screen we need to call upon the wave-like aspects of the electron. Each individual electron, however, is detected at a highly localized position and thus demonstrates what we would normally consider to be particle-like behaviour.

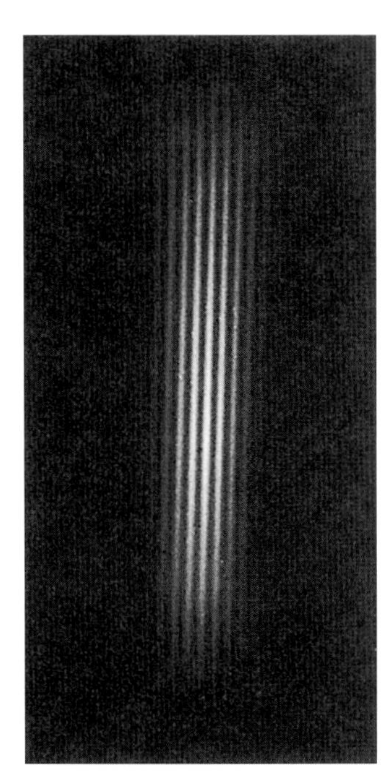

Electron two-slit interference pattern.

If we replace the single slit by a double slit, with the distance between the slits equal to d, then the intensity distribution (Figure 2.4) at the screen is a series of interference fringes in which the maxima occur at angles θ, given by the relation $n\lambda = d \sin\theta$, where n is an integer. Once again, if the beam intensity is reduced to such a low level that only one electron appears at the screen at a time, it is impossible to predict where each of these electrons will arrive, but the pattern will still build up to show the series of interference fringes as shown in Figure 2.4. This implies that *each* electron somehow interacts with *both* slits, and it is not possible to know *which* slit the electron passed through. It is not possible, for instance, to reproduce the double-slit diffraction pattern by closing one of the slits for the first half of the experiment and the other for the second half. The resulting intensity distribution in this case is simply the sum of two single-slit patterns (Figure 2.5).

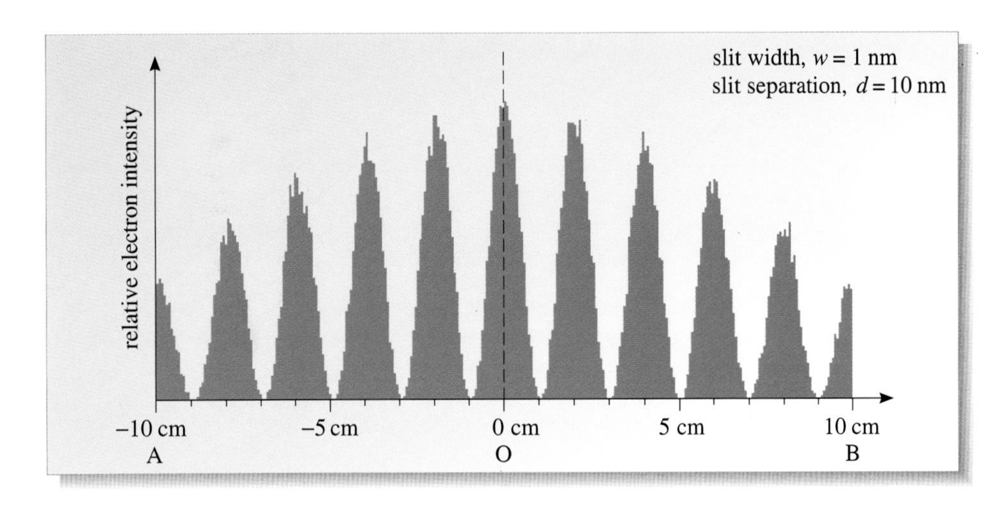

Figure 2.4 The intensity distribution of electrons arriving at the screen after passing through a double slit (slit separation d). The intensity maxima occur at angles given by $n\lambda = d \sin\theta$, where n is an integer.

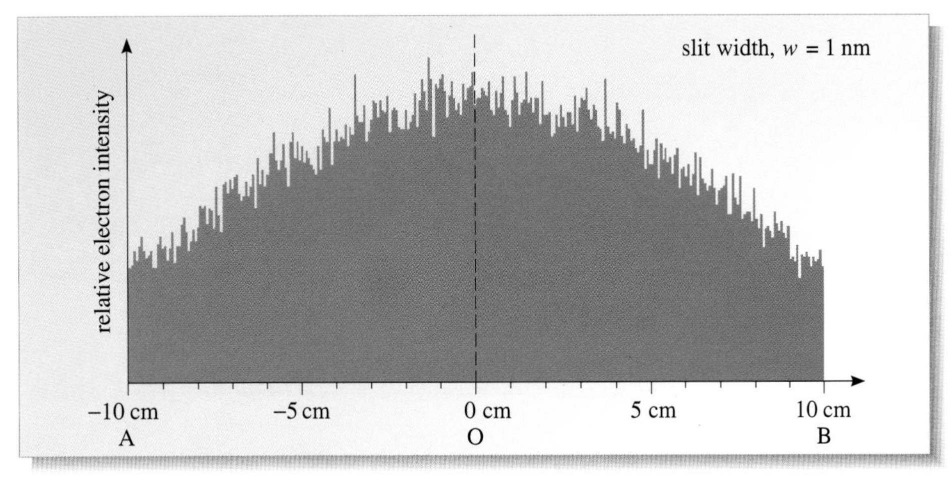

Figure 2.5 The intensity distribution if each slit is open for half of the experimental time.

We might try to determine which slit each electron passed through by, for example, placing a light source close to the slit (Figure 2.6). The electron would then scatter light and betray its position with a flash. However, under these circumstances, when we examine the intensity distribution that builds up on the screen we would again observe the sum of the two single-slit patterns. When the light is removed, the original two-slit interference pattern would be restored.

These are the peculiar facts of the electron diffraction experiment. Can quantum mechanics help us to understand how they come about?

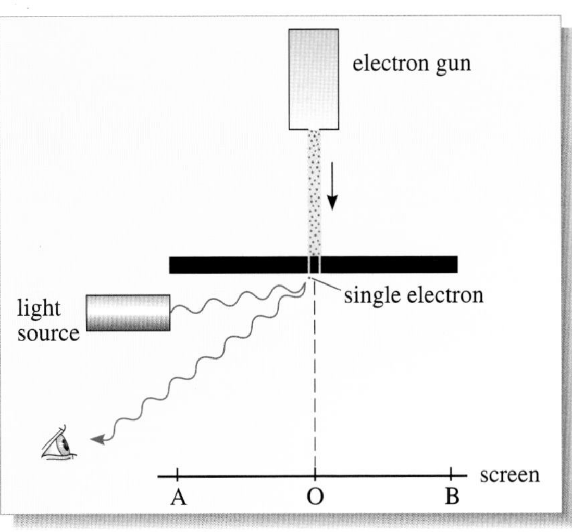

Figure 2.6 Can we observe which slit an electron passed through? The answer is 'yes' but only at the expense of destroying the two-slit interference pattern.

2.2 The scope of quantum mechanics

The results of the two-slit electron diffraction experiment can be simply interpreted provided we use *both* the wave *and* particle models of the electron. However, we use the two models to account for different aspects of the experiment — the wave model is used to predict the distribution of detected electrons, i.e. where they are most likely to arrive on the screen, whereas the particle model is used to explain why the electrons are detected at the screen one at a time.

Can these two apparently contradictory models be reconciled in some way, so that they both arise as different aspects of a single model? The answer is that they *can* be, but only when our classical preconceptions have been abandoned. According to ideas based on classical physics, it should be possible to block one of the two slits without altering the behaviour of electrons that pass through the other. It should always be possible, *in principle*, to predict where an electron will strike the screen. However, you know from the previous subsection that these expectations are *incorrect*. It is simply *not* possible to monitor the path of an electron without disturbing the interference pattern, nor is it possible to predict where a particular electron will strike the screen. It is possible only to determine the probability that the electron will strike a given region of the screen. This is an indication that deterministic classical physics *cannot* be used to understand the results of the two-slit experiment.

However, quantum mechanics *can* be successfully applied to this problem. We have not yet discussed this theory in detail, so we obviously cannot describe how it is applied. Nevertheless, it is possible to describe the *scope* of the quantum-mechanical analysis.

Quantum mechanics enables us to predict the possible results of a measurement made during an experiment on a physical system. It also enables us to predict the *probabilities* with which these possible results will occur. However, the theory says nothing about the *unobserved* details of the experiment.

Notice that quantum mechanics is, in a sense, a less ambitious theory than classical mechanics. Quantum mechanics deals only with *observable* quantities and with the *probabilities* of an experiment yielding certain results, whereas, according to classical mechanics, it should be possible (in principle) to predict *with certainty* the details of an experiment, whether they are observed or not.

In the case of the two-slit diffraction experiment, the scope of a quantum-mechanical analysis is such that it enables us to predict that electrons may strike the screen, and to predict the probability that an electron will strike any given region of the screen. However, the analysis says nothing about unobserved details such as the exact path of a particular electron. Moreover, the probability distribution will be dramatically altered if attempts are made to determine which slit each electron went through.

Now, having briefly discussed the scope of quantum mechanics, we can go on to discuss *how* waves can be used to describe the behaviour of particles of matter.

2.3 Probability waves

Let us start by summarizing the wave aspects of electron behaviour that have emerged from the diffraction experiments discussed earlier.

1. The wavelength of the wave associated with an electron is given by the de Broglie formula $\lambda_{\mathrm{dB}} = h/p$ (Equation 1.15).

2 The waves obey the *principle of superposition*: the resultant wave at any point is the sum of contributions from the individual waves reaching that point, and this sum can lead to constructive or destructive interference.

3 The intensity of the interference pattern at the screen is proportional to the relative number of electrons arriving in any region and therefore indicates the *probability* that an electron will arrive in any particular region of the screen. This suggests that the de Broglie wave is connected in some way with the probability of the electron's detection.

Now, in a two-slit experiment with light, the intensity of *light* is a measure of the number of *photons* arriving at different positions, and therefore tells us the probability that an individual photon will be detected at different positions. But in *Dynamic fields and waves*, Chapter 2, it was established that the intensity of a wave at a point is proportional to the square of the amplitude of the wave at that point. In the case of light, which is an electromagnetic wave, the intensity is proportional to the square of the amplitude of the electric field (Figure 2.7a), that is

$$I = \mathscr{E}^2_{\text{max}}. \tag{2.1}$$

It therefore follows that the relative probability of detecting a photon at different positions is proportional to $\mathscr{E}^2_{\text{max}}$, the square of the amplitude of the electric field at those positions.

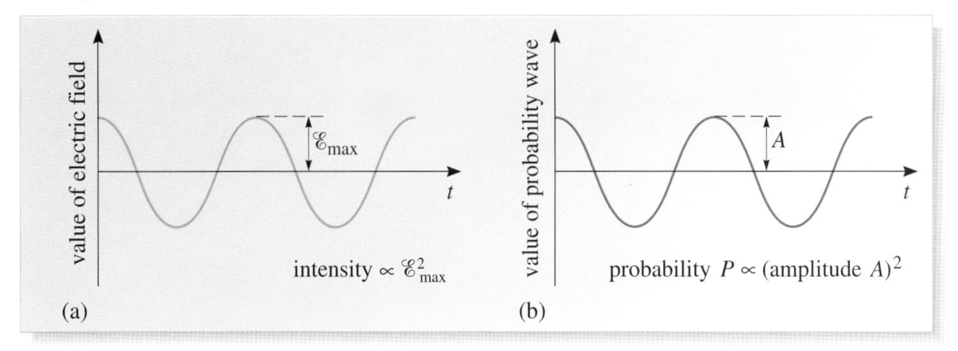

Figure 2.7 (a) The value of the electric field at a point in the path of a simple electromagnetic wave varies sinusoidally with time. The intensity is proportional to the square of the amplitude. (b) The value of a simple probability wave (for an electron) at a point may also vary sinusoidally with time. In this case, the probability of detecting an electron is also proportional to the square of the amplitude.

A similar meaning may be attached to the amplitude of the de Broglie wave for electrons. That is, at a particular point (Figure 2.7b), the square of the amplitude of the de Broglie wave for a beam of electrons is proportional to the probability of detecting an electron, i.e.

probability of detecting an electron at a certain position ∝ square of the amplitude of the de Broglie wave at that position

$$\text{probability } P \propto (\text{de Broglie amplitude } A)^2. \tag{2.2}$$

This connection between de Broglie waves and probability is actually a simplified form of an important general quantum-mechanical relationship to which we shall return later. Because of this relationship, de Broglie waves are often referred to as **probability waves**.

In this interpretation, the phenomenon of interference of de Broglie waves can be understood in exactly the same way as for any other type of wave. If two de Broglie waves interfere constructively at some point, then their amplitudes add (if the waves are exactly in phase) and the probability of detecting the particles at that point will be proportional to the square of the resultant amplitude. If the waves are exactly out of phase, they interfere destructively resulting in a much reduced probability of detection — zero if the two interfering waves have the same amplitude.

The important point to note is that it is the *square* of the wave's amplitude that is related to the probability of detecting an electron. But the question that arises is 'What is varying in a de Broglie wave?' For light, which is an electromagnetic wave, it is the electric and magnetic fields that are varying with time and position, and Figure 2.7a shows the variation of the electric field with time at a certain position for a simple light wave. Is there anything in a probability wave that's analogous to $\mathscr{E}$ (and $\boldsymbol{B}$) in a light wave? The answer is 'no', because *the wave itself does not correspond to a measurable physical quantity.* In Figure 2.7b, you can see the variation with time of a simple sinusoidal probability wave at a certain position. But there is no experiment that can *measure* the value of the wave at a particular instant and at a particular position. What *can* be measured is the *intensity*, or the *probability* of detecting an electron, and this is proportional to the square of the amplitude A of the wave (Figure 2.7b). Since quantum mechanics is concerned with quantities that can be measured, the lack of a direct physical meaning for the wave is not a great concern and will not prevent us from using a generalized form of probability wave when we discuss quantum mechanics in Section 4.

The idea of a probability wave is a difficult one to grasp. It is hard to accept that we cannot attach any direct physical meaning to the quantity that is varying in the wave. However, if you bear in mind that the square of the amplitude of the wave at a certain position is proportional to the probability of detecting a particle at that position, then you should be able to interpret the various waves that you will see in the later sections of this chapter.

2.4 Summary of Section 2

You saw in Section 2.1 that both wave and particle models of the electron are required in order to understand the two-slit diffraction experiment. Contrary to ideas based on classical mechanics, it is not possible to predict the point on the screen that an electron will strike. One can predict only the *probability* that an electron will strike the screen in a certain region, and this probability is altered if attempts are made to determine which slit the electron went through. A description of these experimental results *is* possible within the scope of a quantum-mechanical analysis, though no such analysis has been given here — we have merely asserted that it can be done.

In Section 2.3, you saw that a de Broglie wave is really a probability wave. The probability P of detecting an electron at a particular place is proportional to the square of the amplitude A of the electron's de Broglie wave at that place:

$$P \propto A^2. \qquad \text{(Eqn 2.2)}$$

The idea that the behaviour of particles can be described in terms of the probabilities of possible measurement outcomes, and that such probabilities can be calculated using waves, will both emerge as important aspects of quantum mechanics in Section 4.

Question 2.1 The French mathematician and astronomer Pierre Laplace (1749–1827) once said 'Give me the initial data on all particles and I will predict the future of the Universe'. He was, in effect, asserting the power of classical (Newtonian) mechanics, of which he was a great exponent. Explain why Laplace would probably not have made this statement if he had known the results of the electron diffraction experiments that you met in Section 2.1. ■

3 Heisenberg's uncertainty principle

In this section, we shall pursue the consequences of describing the behaviour of matter in terms of probability waves. As you will see, a description in terms of simple de Broglie waves would fail to do justice to the behaviour of matter, so we shall be mainly concerned with developing a more comprehensive description in terms of *wave packets*. Detailed considerations concerning wave packets will lead us to one of the most far reaching principles of quantum mechanics, *Heisenberg's uncertainty principle*. This principle, which has no analogue in classical mechanics, imposes fundamental limitations on what we can expect quantum mechanics to be capable of predicting.

3.1 A wave packet description of the electron

Let us now return to the wave–particle duality of electrons. According to quantum mechanics, it is not meaningful to ask whether an electron is a particle or a wave: quantum mechanics deals with the observable behaviour of matter (such as electrons), so it allows us to ask different and more pragmatic questions such as:

> Given a certain type of measurement, what is the appropriate model that describes the electron's behaviour? Is it a particle model, or is it a wave model?

It might be possible to take the argument a little further. We could ask the more interesting question:

> Is there some way in which we can combine some particle-like and some wave-like features in one composite model?

This will be the approach we will pursue in this section, and we shall begin by examining more carefully the characteristics of waves and of their motion.

The simplest possible probability wave is the de Broglie wave that describes a free particle, which has no forces acting on it and which therefore has a constant momentum of magnitude p. This type of wave (Figure 2.8a) is really an infinitely long travelling wave with a fixed amplitude and a wavelength λ_{dB} given by the de Broglie formula $\lambda_{\text{dB}} = h/p$ (Equation 1.15). But an *infinitely long* de Broglie wave can't be used to describe an electron that is more likely to be located in one place rather than another. As you saw in Section 2.3, the probability of detecting an electron at a certain place is proportional to the square of the amplitude of its probability wave at that place (Figure 2.7b). Since the amplitude of the wave shown in Figure 2.8a *doesn't* depend on position x, it is equally likely that the electron it describes will be found at any position along the x-axis.

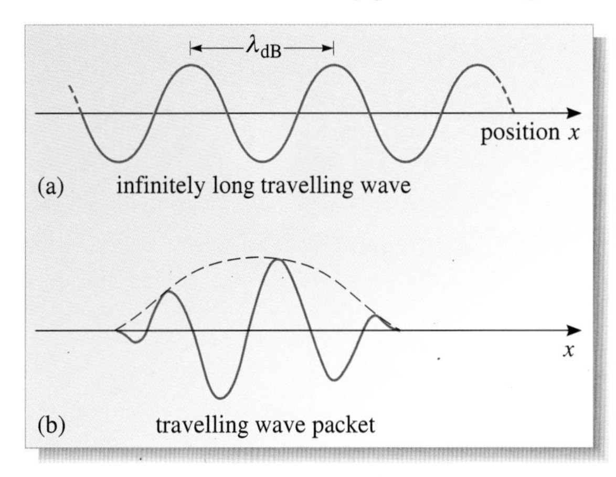

Figure 2.8 (a) A snapshot of an infinitely long travelling probability wave of fixed amplitude and with de Broglie wavelength λ_{dB}. When a particle is described by this probability wave, we know the magnitude p of the particle's momentum from the de Broglie formula $\lambda_{\text{dB}} = h/p$, but we know nothing about the particle's position coordinate x. (b) A snapshot of a travelling *wave packet*. Such a wave packet can be used to specify the approximate position of a particle, but does it have a definite wavelength? What information about the particle's momentum can be provided by this wave packet?

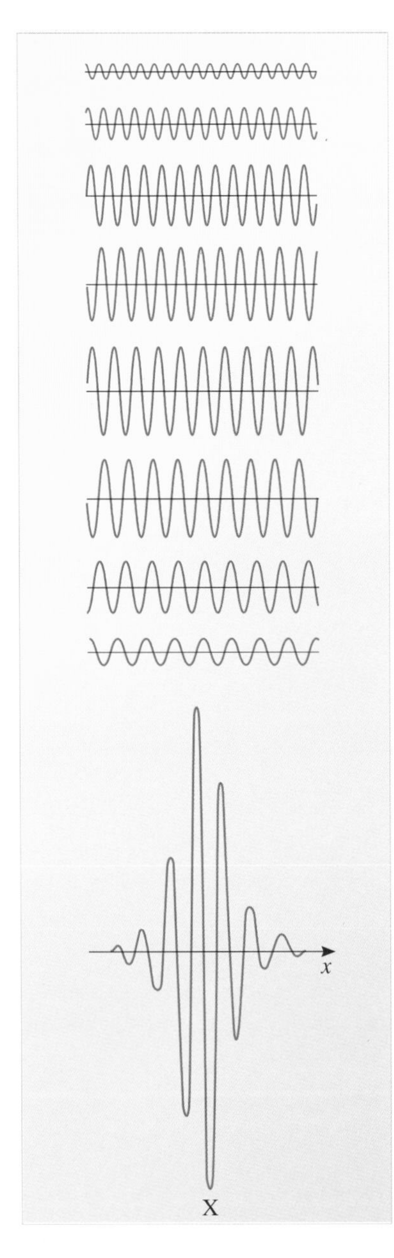

It is obviously desirable to find a wave-based description of the free electron that would allow us to specify (to some degree of accuracy) the position of the electron at a particular time. Such a description of the electron is given in Figure 2.8b. This time the electron is described by a travelling *wave packet* — a short 'burst' of waves that moves along the x-axis. The amplitude is large only in the region of the wave packet, and it falls off rapidly on either side. Since the probability of detecting the electron is proportional to the square of the amplitude, the electron is likely to be found only in the region of the wave packet. This means that the position of the electron described by the wave packet shown in Figure 2.8b is *localized* in the region of the wave packet itself.

Let us look more closely into this wave packet description of the electron. The first important point to note is that a wave packet can be 'built up' by adding together infinitely long wavetrains whose wavelengths lie within a finite *range*. Figure 2.9 is an attempt to illustrate this approach. It shows snapshots of eight infinitely long travelling wavetrains from such a range. The result when these waves are all added together is a travelling wave packet centred on the point X.

The process shown in Figure 2.9 can be refined by adding more wave trains of the appropriate amplitude and wavelength to produce a single wave packet of any desired shape. The mathematical ideas that underlie this technique of 'building up' a wave packet from other, simpler waves were developed by the French mathematician Jean-Baptiste Fourier at the beginning of the nineteenth century. The technique is known as *Fourier synthesis*.

The wave packet model of a localized *free* electron is important because it helps us to answer the question posed at the beginning of this section: 'Is there some way in which we can combine some particle-like and some wave-like features in one composite model?' The wave packet model obviously allows us to take into account the wave-like properties of the electron. However, the important point to notice is that because the model allows us to give a *localized* description of the electron, it helps us to take into account its *particle-like* features as well.

Figure 2.9 The bunching produced when a range of waves with different amplitudes and wavelengths are added together using the principle of superposition. A sample of eight waves from a narrow range of wavelengths is illustrated at the top of the diagram. The resulting wave packet is shown at the bottom. (It is actually necessary to have a *continuous* spread of wavelengths over this range: a few discrete wavelengths taken from this range, when added together, will not produce a *single* localized wave packet.)

3.2 Some more properties of wave packets

It is worth exploring a little more closely the make-up of wave packets, as this will give us some deep insights into the behaviour of matter. First, it is useful to recall the quantity known as the angular wavenumber (introduced in *Dynamic fields and waves*, Chapter 2) where it was defined by

$$k = 2\pi/\lambda. \tag{2.3}$$

You may see k ($= 2\pi/\lambda$) referred to as 'wavenumber' in other texts. However, in *The Physical World* we use the term 'angular wavenumber' to distinguish it from 'wavenumber' ($\sigma = 1/\lambda$).

As you will see shortly, the quantity k is useful because it provides a convenient way of expressing the spread of wavelengths needed to produce a wave packet. If you look at Figure 2.10a, you can see that a broad wave packet is produced if the wave trains that make it up have a fairly restricted range of angular wavenumbers.

By introducing *more* wave trains with a *wider spread* of angular wavenumbers, the width of the packet can be greatly *reduced*, as you can see by comparing Figures 2.10a and b.

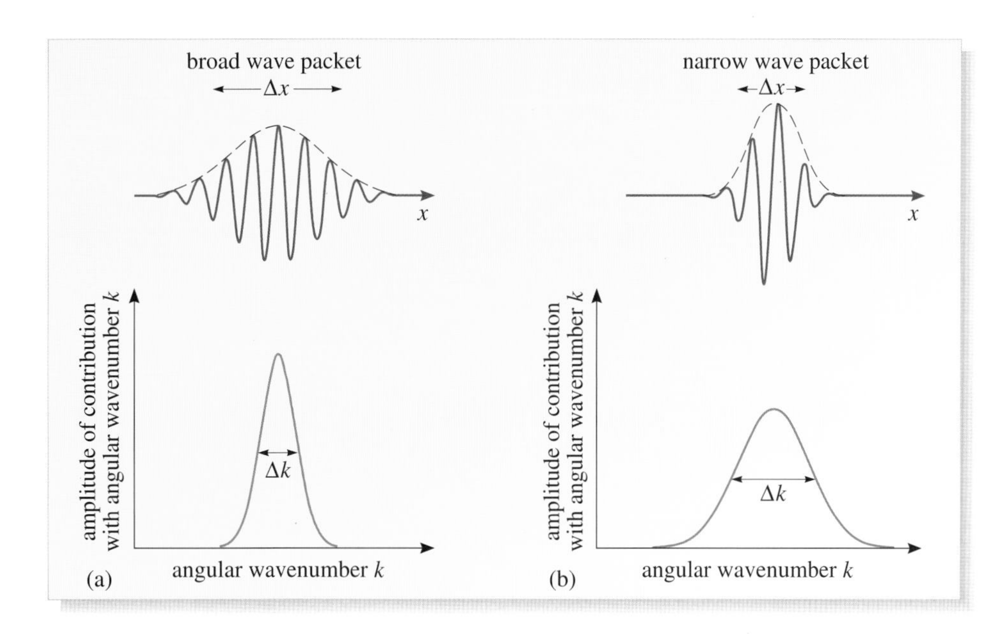

Figure 2.10 (a) A *broad* wave packet can be built up from waves with a narrow range of angular wavenumbers. However, a narrow wave packet is built up from waves with a broader range of angular wavenumbers (b).

The greater the spread Δk of angular wavenumbers, the narrower the width Δx of the corresponding wave packet.

By using the techniques of Fourier synthesis, it can be shown that the product of Δx, the width of the wave packet, and Δk, the range of angular wavenumbers required to produce it, is typically given by:

$$\Delta x\,\Delta k \approx 1.$$

The numerical value, one, on the right-hand side of this equation is approximately correct for *all* types of wave packet (including probability waves), and it varies very little with the shape of the wave packet.

Under some circumstances, the product $\Delta x\,\Delta k$ is greater than 1, so it is more realistic to write

$$\Delta x\,\Delta k \gtrsim 1, \tag{2.4a}$$

where the symbol $\gtrsim$ means 'is greater than a value approximately equal to'.

We can rearrange Equation 2.4a to obtain

$$\Delta x \gtrsim \frac{1}{\Delta k}. \tag{2.4b}$$

This indicates that Δx is inversely proportional to Δk — a narrow wave packet can be built up from waves with a large range of angular wavenumbers — thus confirming the dependence shown qualitatively in Figure 2.10. We shall make use of Equations 2.4a and 2.4b in the next subsection.

3.3 Wave packets and Heisenberg's uncertainty principle

In Section 3.1, you saw that a free electron whose momentum has magnitude p is described by a de Broglie wave that is *infinitely long* (Figure 2.8a). It is impossible, therefore, to state *where* the electron is at a given time. So although we know the particle's momentum, we have no idea of its position.

You have also seen that it is possible to describe the electron in terms of a wave packet (Figure 2.8b). In this case, the particle is localized (its position is fairly well defined), but the wave packet is built up from waves having a range of different wavelengths so the de Broglie formula ($\lambda_{dB} = h/p$) implies that the *momentum* of the particle is poorly defined. Hence, when we know the particle's position reasonably well (Figure 2.8b), we have a poor knowledge of its momentum.

Let us try to express these subtle ideas in a mathematical form. You saw in Section 3.2 that

$$\Delta x \, \Delta k \gtrsim 1, \qquad \text{(Eqn 2.4a)}$$

where Δx is the width of a wave packet and where Δk is the spread of the angular wavenumbers of the waves that build up the wave packet.

We also have the de Broglie relationship between the wavelength and momentum of a particle:

$$\lambda_{dB} = h/p_x.$$

The subscript x on p_x indicates that it is the component of momentum in the x-direction that is involved here. Substituting this equation into the definition of the angular wavenumber k (with $k = 2\pi/\lambda$, Equation 2.3), we obtain

$$k = \frac{2\pi p_x}{h}. \qquad (2.5)$$

This simple proportional relationship between the angular wavenumber k and the component p_x of the momentum implies that any spread or uncertainty Δk in the angular wavenumber k will result in a proportional spread or uncertainty Δp_x in the momentum p_x. Hence,

$$\Delta k = \frac{2\pi \Delta p_x}{h}.$$

Substituting this expression for Δk into Equation 2.4a we find,

$$\Delta x \, \Delta p_x \gtrsim \frac{h}{2\pi}. \qquad (2.6)$$

As we mentioned in Chapter 1, the quantity $h/2\pi$ occurs so often in quantum mechanics that it is given a special symbol $\hbar$ (said as h-bar or h-cross) which we shall use from now on. Thus we have

$$\Delta x \, \Delta p_x \gtrsim \hbar.$$

In fact, a rigorous derivation shows that the minimum possible value for the product $\Delta x \, \Delta p_x$ is actually $\hbar/2$. So we have

$$\Delta x \, \Delta p_x \geq \hbar/2. \qquad (2.7a)$$

Now, Δx and Δp_x, as used in the above formulae, are the theoretical minimum values for the uncertainties in x and p_x. They are uncertainties that are inherent in the quantum-mechanical description of the world, and Equation 2.7a is an expression of the fact that it is not possible to know the exact position *and* the exact momentum of a particle *simultaneously*. These uncertainties have nothing to do with any uncertainties associated with particular measuring apparatus.

Equation 2.7a expresses one of the most profoundly important principles in quantum physics — **Heisenberg's uncertainty principle**. It was first written down by the German physicist Werner Heisenberg (Figure 2.11) in 1927. The basic idea underlying the principle can be stated informally as follows:

Heisenberg's uncertainty principle for the position and momentum of a particle:

There is a fundamental limit to the precision with which the position x and the momentum component p_x of a particle can be simultaneously known.

Equations analogous to Equation 2.7a hold for the other momentum components and the corresponding position coordinates:

$$\Delta y\, \Delta p_y \geq \hbar/2 \tag{2.7b}$$

$$\Delta z\, \Delta p_z \geq \hbar/2. \tag{2.7c}$$

Werner Heisenberg (1901–1976)

Born in Duisburg, Germany, Werner Heisenberg (Figure 2.11) worked under Max Born in Göttingen and Niels Bohr in Copenhagen and became professor of theoretical physics at Leipzig in 1927. Heisenberg's career spanned the Nazi period of German history. He sympathized with right-wing policies. While a youth he was involved in street fights with communists and, during the period of Hitler's leadership, he accepted gradually more prominent positions, including leadership of the German atomic bomb project. After the Second World War he became Director of the Max Planck Institute and professor at Berlin. During his time at Leipzig, he developed a version of quantum mechanics that was later shown to be equivalent to the version that Schrödinger developed at about the same time. Heisenberg was awarded the 1932 Nobel Prize for physics for his contributions to the development of quantum mechanics. He was keenly interested in the moral and philosophical implications of scientific discovery.

Figure 2.11 Werner Heisenberg.

As you have seen in this subsection, the uncertainty principle arises from the wave packet description of a particle — the attempt to reconcile the wave and particle properties of matter.

3.4 Some examples of the uncertainty principle

Physical measurements

By far the best way to see how the uncertainty principle affects measurements in physics is to look at some examples. First, try Question 2.2, which will give you some feel for the numerical values involved in using the uncertainty principle.

Question 2.2 (a) In a single-slit electron diffraction experiment (Figure 2.12) the electrons in the beam have momentum of magnitude p and are incident normally on a slit of width w.

The restriction imposed by the slit means that the electrons' position is known to an accuracy of w in the y-direction. Show that the resulting uncertainty in the y-component of the electrons' momentum produces an angular spread θ in the beam that is within about an order of magnitude of λ/w, i.e. the known width of the central maximum of the single-slit diffraction pattern for waves.

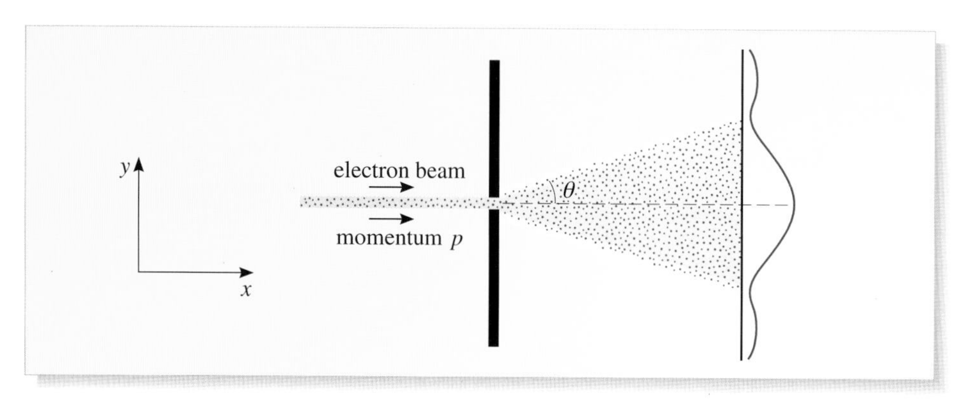

Figure 2.12 A beam of electrons of magnitude of momentum p is diffracted by a slit of width w.

(b) Assuming that quantum mechanics can be applied to macroscopic objects such as bullets, estimate the angular spread of a 'beam' of bullets each of mass 0.1 kg travelling with a speed of $1500\,\mathrm{m\,s^{-1}}$ which passes through a slit just wide enough to allow the bullets through — say, about 1 cm. ■

It should be clear to you from your answers to Question 2.2 that because Planck's constant, h, has such a small value, the indeterminacy imposed by the uncertainty principle is only significant in the realm of atomic and subatomic physics. This is why this limitation does not manifest itself in everyday measurements, where it is the practical limit of experimental error that restricts the precision for macroscopic measurements. In everyday situations, we can conveniently forget these quantum effects.

Atomic measurements

We can now turn our attention to the question of the size and the stability of atoms — problems that had perplexed Thomson, Rutherford and Bohr. By doing the following series of linked questions, you can use the uncertainty principle to give a plausible argument for the approximate radius r_H of the hydrogen atom.

Question 2.3 Write down the expressions for the electrostatic potential energy E_{pot} of the electron in a hydrogen atom if the electron–proton distance is represented by r, and for the kinetic energy E_{kin} of the electron in terms of the magnitude of its momentum, p, and its mass. Hence write down the equation for the total energy E_{tot} of the electron. ■

We don't yet know the actual position of the electron; all we know is that it is within a distance r of the nucleus. Thus the components of the electron's position from the nucleus have uncertainties

$$\Delta x = \Delta y = \Delta z \approx r.$$

Now the uncertainty principle tells us that this confinement in space requires a minimum spread in each component of the electron's momentum. For example

$$\Delta p_x \approx \hbar/2\Delta x \approx \hbar/2r.$$

We don't know the actual momentum components p_x, p_y or p_z; we only know that their magnitudes are somewhere in the range Δp_x, Δp_y and Δp_z. It is therefore a reasonable approximation to assume that, for example

$$|p_x| \approx \Delta p_x \approx \hbar/2r.$$

Question 2.4 Using the equation you obtained for E_{tot} in Question 2.3 and the above approximations for the magnitude of the momentum components, derive an expression for E_{tot} at r in terms of e, r, ε_0, $\hbar$ and m_e only. ■

The expression you obtained in Question 2.4 depends on the unknown quantity r. However, we do know how the total energy of the electron depends on its distance from the nucleus. Figure 2.13 shows the form of the two contributions (kinetic E_{kin} and potential E_{pot}) to the total energy of the electron, and the purple line shows how the total energy, E_{tot}, varies with r. The electron will tend to minimize its energy and hence the atomic radius, r_H, will be that value at which the graph of E_{tot} against r goes through its minimum. At this point, the gradient of the graph is equal to zero.

Question 2.5 By differentiating your equation for E_{tot} with respect to r and finding the value of r for which the gradient is zero, determine an approximate value for the radius r_H of the hydrogen atom. ■

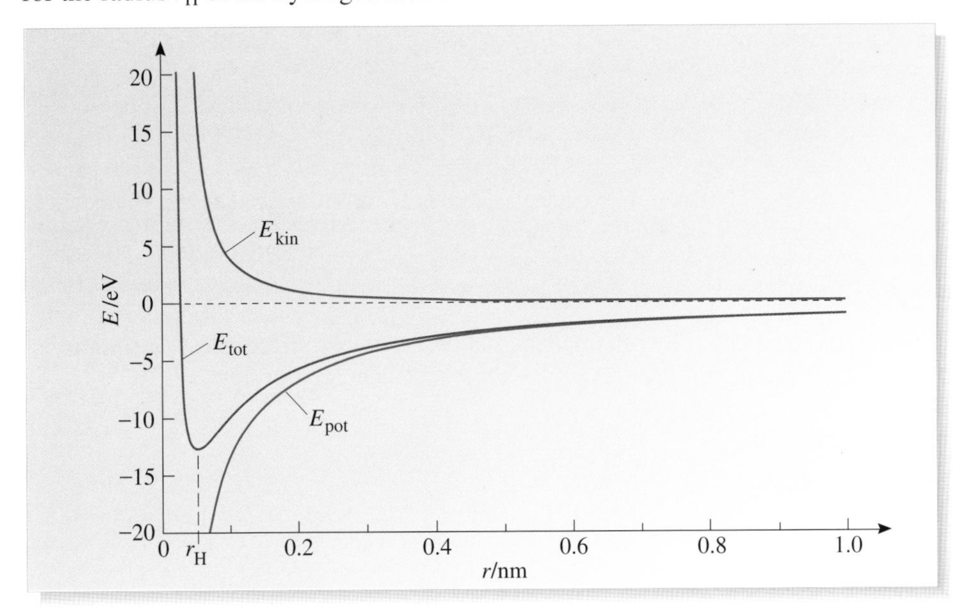

Figure 2.13 The contributions (potential energy and kinetic energy) to the total energy of the electron in a hydrogen atom. Here the kinetic energy is calculated using the uncertainty principle. (For use with Question 2.5).

The answer to Question 2.5 agrees very well with the size of the orbit of an electron in the ground state of the hydrogen atom. This estimate is very close to the Bohr radius for the hydrogen atom ($a_0 = 0.53 \times 10^{-10}$ m), which you met in Chapter 1.

This argument not only predicts the approximate size of the hydrogen atom, but also explains why atoms are stable. You may recall that, classically, there was a fundamental difficulty concerning the stability of the motion of an electron in a Bohr orbit. According to ideas based on Maxwell's equations, the negatively charged electron should continuously emit electromagnetic radiation since it is continuously accelerating in its circular orbit (Chapter 1). The electron should, therefore, spiral towards the nucleus. However, Heisenberg's uncertainty principle asserts that this cannot happen. If the electron *were* to spiral in towards the nucleus, its position would become more and more precisely known. If the electron were nearly at the nucleus, its position would be known to within about 10^{-15} m, a factor of 10^5 smaller than the radius of the hydrogen atom, and hence the uncertainty in the electron's momentum would have to increase by a factor of 10^5. This, in turn, implies that the electron's kinetic energy (which is proportional to p^2) would increase by a factor of 10^{10}. There is no way for the electron to gain this amount of energy: the loss in electrostatic potential energy is nowhere near enough.

3.5 Another form of the uncertainty principle

There is another important form of Heisenberg's uncertainty principle, which involves limitations to the precision with which the energy of a particle can be measured. It is again possible to see intuitively how these limitations arise as soon as matter is described in terms of probability waves.

Consider the question 'How much time is needed for us to specify the energy of a free particle?' To answer this question, notice that since $p = mv$, where v is the particle's speed, then

$$E = \frac{p^2}{2m}$$

where we have assumed that the potential energy of the free particle is zero everywhere. Thus, if we have an exact knowledge of the particle's energy, we must also have an exact knowledge of the magnitude p of its momentum. Furthermore, if we have an exact knowledge of the magnitude p of the particle's momentum, we must also have an exact knowledge of its de Broglie wavelength $\lambda_{dB} = h/p$ (Equation 1.15). You saw in Section 3.1 that we can know the exact wavelength of a particle's probability wave *only* if the wave is infinitely long (Figure 2.8a). However, if the wave is localized (Figure 2.8b), then the arguments used in Section 3.1 tell us that the wave packet is 'built up' from many waves of different wavelengths. Hence, in order to establish that a particle's energy is known exactly, it must be confirmed that its probability wave is infinitely long, and it would take an infinitely long time to make such a check!

The Heisenberg uncertainty relationship that describes this effect can also be derived from the mathematical properties of wave packets. The relationship is

$$\Delta E \, \Delta t \geq \hbar/2 \tag{2.8}$$

where ΔE is the uncertainty in the energy and Δt is the time interval during which the energy is known to be within the range ΔE. This idea can be stated informally as follows:

Heisenberg's uncertainty principle for energy and time:
There is a fundamental limit to our knowledge of a particle's energy E over a finite time interval.

Equation 2.8 tells us that a precise specification of energy (*small* ΔE) requires a long time (*large* Δt). Alternatively, over a very short time interval (*small* Δt), the energy of a particle cannot be accurately specified (*large* ΔE).

This indeterminacy in energy might make you wonder about the law of conservation of energy. If there is an inherent uncertainty in the specification of any energy, how can we check that energy is conserved? This apparent difficulty disappears in practice since all tests of the law of energy conservation extend over comparatively long time intervals. Thus Δt *is* sufficiently large to allow ΔE to be infinitesimal compared with other inherent errors of experimental measurement.

You can now check that you have understood the two forms of Heisenberg's uncertainty principle by trying Questions 2.6 to 2.8.

Question 2.6 An experimenter (who has not studied quantum mechanics) decides to make an extremely accurate study of the behaviour of electrons that are accelerated from an electron gun towards a positively charged plate. The position of each electron can be specified with respect to a hypothetical x-axis, along which the electron gun is pointing.

The experimenter tells you that he is going to measure *to arbitrarily high accuracy* the position, momentum *and* energy of an electron at a particular instant of time.

(a) Using the appropriate form of Heisenberg's uncertainty principle, state why the experimenter is being overambitious.

(b) The experimenter then puts forward a more modest proposal — to measure *to arbitrarily high accuracy* the position and the momentum of the particle *at different times*. Might he, in principle, succeed this time?

Question 2.7 An electron in an excited atom typically remains in an excited state for about 10^{-8} s before it loses the excess energy by emitting electromagnetic radiation. Use the energy–time uncertainty relationship to estimate the indeterminacy that this implies for the energy of the excited atomic states. In physical terms, what effect do you think this indeterminacy will have on measurements of spectral lines?

Question 2.8 In a series of accurate measurements of the mass of a new elementary particle, the results showed a spread in the measurements of about $20m_e$, where m_e is the electron's rest mass. The imprecision introduced into the measurements due to limitations of the apparatus used was about $2m_e$. What conclusion can you draw about the lifetime of the particle from this observation? (*Hint*: Use Einstein's equation, $E = mc^2$, and assume the particle had negligible relativistic momentum when the measurements of its mass were made.) ■

3.6 Summary of Section 3

In this section, you have seen some of the consequences of describing the behaviour of matter using probability waves. Section 3.1 showed that if the behaviour of an unconfined particle is described using a wave packet, we can understand how the particle can be localized despite the extended nature of the individual probability waves.

Further examination of this idea led to two different formulations of Heisenberg's uncertainty principle

$$\Delta x\, \Delta p_x \geq \hbar/2 \qquad \text{(Eqn 2.7)}$$

and

$$\Delta E\, \Delta t \geq \hbar/2. \qquad \text{(Eqn 2.8)}$$

This extremely important principle specifies fundamental limits to our simultaneous knowledge of x and p_x for a particular particle at a particular time, and to our knowledge of the energy of a particle if it is measured in a finite time interval.

4 The Schrödinger equation

4.1 Introduction

Much of the credit for laying the foundations of a proper theoretical understanding of atomic matter belongs to the Austrian physicist Erwin Schrödinger (Figure 2.14). Schrödinger used de Broglie's ideas as the starting point for the systematic development of **wave mechanics**, one of the earliest and simplest formulations of quantum mechanics. (There are several mathematically equivalent ways of expressing the principles of quantum mechanics; wave mechanics is one of them.) In developing wave mechanics, Schrödinger abandoned Bohr's primitive ideas of 'orbiting' and 'jumping' electrons, and went far beyond de Broglie's simple association of waves and particles. In particular, in 1926, Schrödinger published an equation that did for quantum mechanics what Newton's laws of motion had done for classical mechanics nearly 250 years earlier. This tremendous step forward in our understanding of Nature is the main subject of this section.

In Schrödinger's theory, probability waves arise as variations in a quantity Ψ (the upper case, Greek letter psi) that is referred to as a **wavefunction**. In Schrödinger's wave mechanical description of the behaviour of a particle, Ψ is a function of position and time, so it may be written as $\Psi(\boldsymbol{r}, t)$. At this stage it's best to think of Ψ as the analogue of the quantities that vary in other types of wave motion — the displacement y of a plucked string from its equilibrium position for example, or, in the case of an electromagnetic wave, the electric field $\mathcal{E}(\boldsymbol{r}, t)$ or the magnetic field $\boldsymbol{B}(\boldsymbol{r}, t)$, both of which oscillate as the wave propagates. It is not possible to ascribe any direct physical meaning to the quantity Ψ, but, as will be explained later, wavefunctions can be related to probabilities of measured results, just as the amplitudes of probability waves were in Section 2.

Figure 2.14

Erwin Schrödinger (1887–1961)

Born in Vienna, Erwin Schrödinger (Figure 2.14) succeeded Max Planck to the chair of theoretical physics in Berlin. With the rise of Nazism, he left Germany in 1933, and spent the period from 1939 to 1956 as Director of the School of Theoretical Physics at the Institute of Advanced Studies in Dublin. In developing quantum mechanics, he was strongly influenced by de Broglie's idea that a wave is associated with each free particle. Schrödinger's attempts to extend this idea to *confined* particles led him to formulate his famous equation (Figure 2.15). He shared the 1933 Nobel Prize for physics with Paul Dirac for their important contributions to 'new and productive forms of atomic theory'. He was a man of very broad scientific interests: in addition to his definitive work in atomic theory, he wrote papers in the fields of colour perception, X-ray diffraction, statistical mechanics and general relativity.

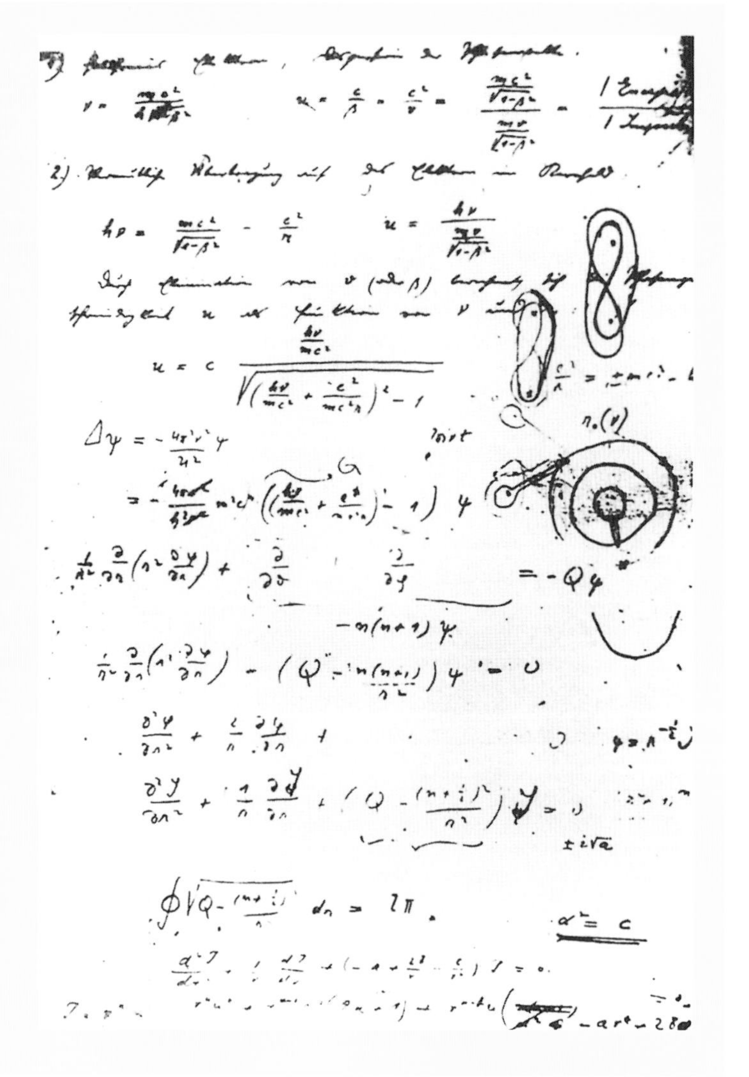

Figure 2.15 The point in Schrödinger's notes at which his celebrated equation first appears.

The wavefunction Ψ that describes the behaviour of any particular quantum system is found by solving the appropriate form of Schrödinger's equation. Now, there is an inherent difficulty in writing down Schrödinger's equation and pursuing the ideas of quantum mechanics at the mathematical level of this course. It is an unfortunate fact that the mathematical language required is generally quite complicated. Nevertheless, by restricting our study of quantum mechanics to sufficiently simple situations, it is possible to develop an insight into the subject, and this is what we shall endeavour to do in this section. In particular, we shall mostly restrict the discussion to just one dimension — usually taken to be the x-direction. Even so, you should still be able to see how Schrödinger's equation provides a description of particle behaviour in terms of a wavefunction $\Psi(x, t)$, and why this description leads naturally to the quantization of the energy of particles in certain circumstances.

In what follows we shall be concerned with essentially two types of waves, *travelling waves* and *standing waves*. A travelling wave is characterized by the fact that it is associated with some particular direction of propagation, and the profile of the wave moves in that direction as time progresses. In contrast, a standing wave has no direction of propagation, though its profile does change with time. In the case of a standing wave there are generally several fixed points called *nodes*, at which the displacement from equilibrium is zero at all times; between the nodes the displacement increases and decreases periodically as each part of the wave oscillates about its equilibrium position. A sequence of snapshots of a travelling wave and a standing wave are shown in Figure 2.16. Both waves have the same amplitude and period, and both have the same wavelength λ.

Now, as you know, waves can have any wavelength when they are free to propagate over an effectively infinite distance — an infinitely long string for example, or a very large expanse of water. On the other hand, when a wave is confined in a

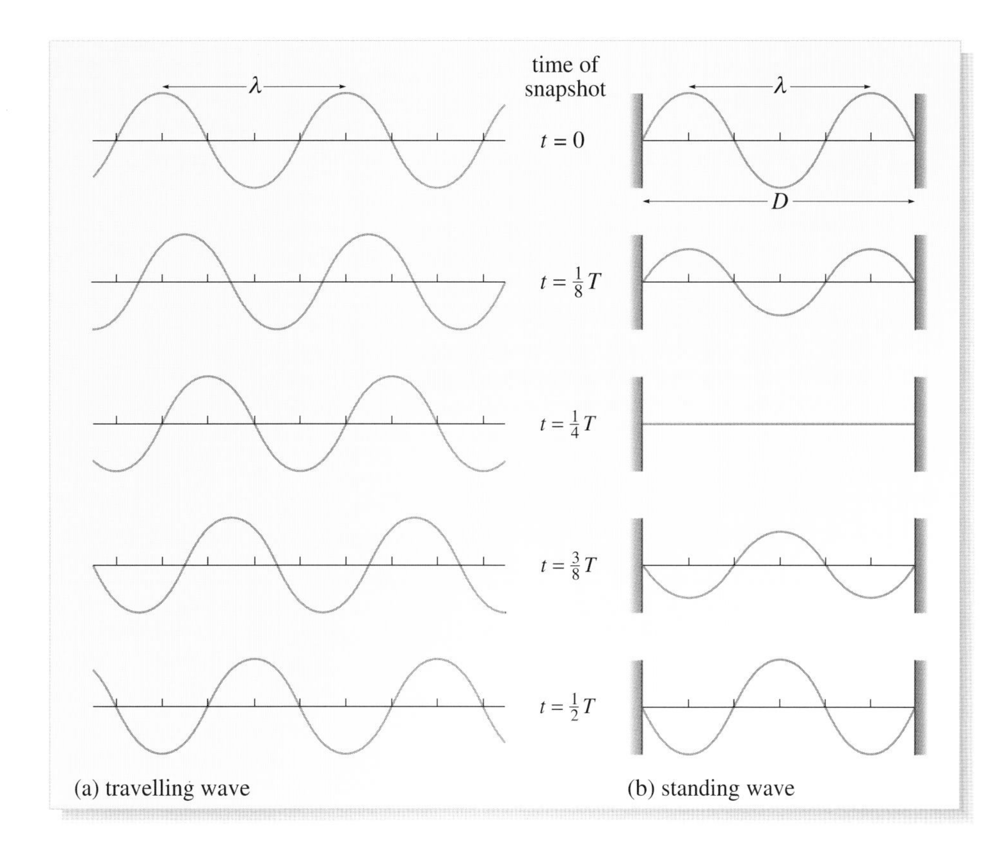

Figure 2.16 Travelling waves and standing waves. (a) A sequence of snapshots, taken over half a period, of a travelling wave on an infinitely long string. (b) A similar sequence of snapshots, taken over half a period, of a standing wave, of wavelength $\lambda = 2D/3$, on a finite string of length D with fixed endpoints.

restricted space — a finite length D of string fixed at both ends for example, or an organ pipe, the wave may take the form of a standing wave with a wavelength determined by the size of the system (see Figure 2.16). Similar behaviour occurs in situations described by quantum mechanics: when a particle is free to move over an unrestricted region of space, its wavefunction is characterized by the (de Broglie) wavelength h/p, which can take any value, since the magnitude of the momentum p is unrestricted. If, however, the particle is confined in some way, by being trapped in a 'box' of some sort, its wavefunction will involve standing waves of a particular wavelength and there will be corresponding restrictions on its movement and energy.

Whether or not a particle is 'free' or 'confined' is determined by the energy of the particle, and by the way its potential energy varies throughout the region in which the particle might be found. The *potential energy function* of a particle, $E_{\text{pot}}(\boldsymbol{r})$, is simply the expression that relates the particle's potential energy, E_{pot}, to its position in space, $\boldsymbol{r}$. If the particle's motion is restricted to the x-direction then it can simply be written $E_{\text{pot}}(x)$. Not surprisingly, the potential energy function turns out to be a very important quantity in Schrödinger's equation. By including it in the equation we can be assured that Schrödinger's equation takes a form that is relevant to the problem being considered, whatever that might be. Because of this, before we go on to look at Schrödinger's equation in detail, it is worth considering what we mean by 'free' or 'confined' in terms of the potential energy of a particle. This is the subject of the next subsection. In the subsequent subsections, we will consider Schrödinger's equation for a particle in a variety of circumstances and show how its solution in each of those circumstances describes the corresponding behaviour of the particle.

4.2 Free and confined particles

In everyday terms, a particle is confined, or bound, within a certain region if there is some sort of barrier preventing its escape from that region. Think of a squash ball bouncing about in a squash court. The ball has quite a lot of kinetic energy but it will never be able to break through the barrier constituted by the walls that confine it. The ball would be destroyed before it could acquire enough energy to penetrate any wall of the court. If the walls were more flimsy however, it is possible to imagine that the ball might acquire enough kinetic energy to break through and escape. If its energy is less than this threshold the ball will be confined, if it is greater then it will not be confined.

In more technical terms, a particle is confined, or **bound**, if its total energy is insufficient to allow it to overcome the barrier of potential energy that is surrounding it. If the particle's total energy is greater than the potential energy barrier, the particle can escape and is then said to be **unbound**. An unbound particle for which the potential energy may be taken to be constant everywhere is said to be a **free particle**: if E_{pot} is constant, the particle has no forces acting on it.

Let us think about a concrete example, using the form of potential energy that is probably most familiar to you — gravitational potential energy. At the surface of the Earth, we can confine an object in a gravitational 'potential well' simply by digging a hole (or well) and dropping the object in with insufficient energy for it to bounce out again. The potential energy function for such a particle is shown in Figure 2.17a, where the zero of potential energy has been taken to be at the bottom of the hole. In this case the hole has vertical sides, a depth h and a width d. As long as the particle's total energy (E_1 say) in the hole is less than mgh, it will never escape. If it is given more energy (E_2), then it can escape but of course its *kinetic* energy after its escape will be reduced to $E_2 - mgh$ since it must use mgh in getting out of the well.

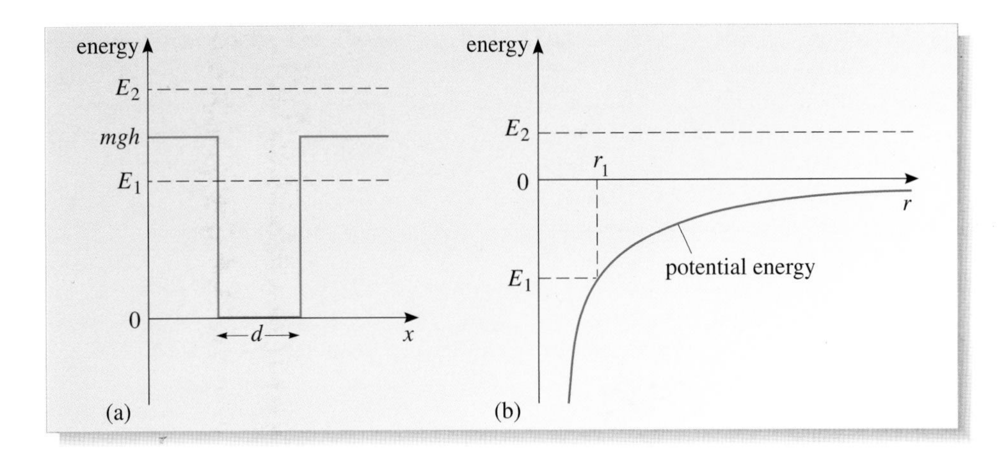

Figure 2.17 Two possible potential energy functions.

Thinking now in more general terms, you should be able to see that the Earth as a whole acts as a three-dimensional gravitational potential well for any other massive object in the vicinity. The gravitational potential energy of a particle of mass m at a distance r from another particle of mass M is given by

$$E_{\text{grav}} = -\frac{GMm}{r}$$

where G is Newton's gravitational constant. The shape of this potential energy curve is shown in Figure 2.17b, where, as usual, the zero of potential energy is taken to correspond to $r = \infty$. In the case of the Earth, if an object has a negative total energy, say E_1 (as shown on the diagram), then it will be unable to escape from the gravitational attraction of the Earth. It is bound to the Earth. The maximum distance the object will be able to reach from the centre of the Earth is r_1, at which distance its energy will be entirely potential energy. At any distance less than r_1, some of the energy of the object will be in the form of kinetic energy. If, on the other hand, an object has total energy E_2 which is greater than zero (also shown on the diagram), it will be able to escape from the Earth altogether. These considerations are what give rise to the idea of the 'escape speed' of an object from the surface of the Earth. The gravitational potential energy of an object of mass m at the surface of the Earth is $-GM_{\text{E}}m/R_{\text{E}}$, where M_{E} is the mass of the Earth and R_{E} its radius. If we wish to launch this object from the surface of the Earth, and if it is to escape the gravitational attraction of the Earth, then it must be given a kinetic energy of at least $GM_{\text{E}}m/R_{\text{E}}$ at the moment of launch.

Now, quantum mechanics is a theory that comes into its own in the description of the behaviour of very small systems such as electrons in atoms. On such scales, the dominant interactions are those between fundamental particles, notably the electromagnetic interaction. If we consider the electron in a hydrogen atom, we know that its potential energy at a distance r from the proton is given by $-e^2/4\pi\varepsilon_0 r$, and this potential energy curve has just the same shape, but on a smaller scale, as the gravitational curve in Figure 2.17b. Exactly the same energy arguments apply, however. If the electron has a total energy less than zero, it will be trapped within the atom. The formal way of making this statement is to say it is confined, or bound, by an electrostatic potential energy well (or, more concisely in this context, a **potential well**). If the particle's energy is greater than zero, it will be able to escape from the potential well and be unbound.

As was pointed out in the introduction to this section, whether a particle is free, bound or unbound has a profound effect on the wavefunction of the particle. If a particle's total energy is greater than its potential energy over an effectively unlimited region, then it is unbound and its wavefunction will involve travelling waves. If a particle is confined in some sort of potential well, then its wavefunction will involve standing waves. It will be helpful if you can bear this in mind as we come to consider the wavefunctions that satisfy Schrödinger's equation for a variety of potential energy functions in the following subsections.

4.3 The time-dependent Schrödinger equation

The full expression of Schrödinger's equation, usually referred to as the **time-dependent Schrödinger equation** or Schrödinger's wave equation, involves a level of mathematics that is beyond the scope of this course. Nonetheless, you should be able to appreciate the following points about the equation:

- Schrödinger's time-dependent equation is a differential equation, so its solutions, known as **time-dependent wavefunctions**, are functions of position and time. These solutions are the wavefunctions $\Psi(\boldsymbol{r}, t)$ that we have been discussing.
- In the context of a specific problem, Schrödinger's equation involves the potential energy function that is appropriate to that problem. Consequently, the detailed form of Schrödinger's equation (and of the wavefunctions that satisfy it) will reflect the problem in hand.
- In the particular case of a free particle moving in one dimension, where $E_{\text{pot}}(x)$ is constant, the wavefunction $\Psi(x, t)$ obtained by solving Schrödinger's equation involves travelling waves with an angular wavenumber k $(= 2\pi/\lambda)$ that may take *any* positive value.
- In the case of a confined particle, the wavefunction $\Psi(x, t)$ obtained by solving Schrödinger's equation will be a standing wave and the allowed values of the angular wavenumber k will be restricted to certain values.
- In the case of a free particle the value of the angular wavenumber k is determined, by the Schrödinger equation, to be

  $$k = \frac{\sqrt{2mE_{\text{kin}}}}{\hbar} \tag{2.9}$$

  where the kinetic energy of the particle is given by its total energy minus the potential energy, i.e.

  $$E_{\text{kin}} = E_{\text{tot}} - E_{\text{pot}}. \tag{2.10}$$

  Thus a free particle can have *any* positive value of kinetic energy. Equation 2.9 is equivalent to de Broglie's relation $k = p/\hbar$ (since $p = \sqrt{2mE_{\text{kin}}}$). As you will see, for a confined particle Equation 2.9 also applies but the allowed values of k, and therefore of E_{kin}, are restricted.

It should be noted that, just as simple de Broglie waves may be added together to form wave packets, so it is possible to form wave packets from the travelling wave solutions to Schrödinger's equation. These wave packet solutions provide a means of describing free particles when they are known to be *localized* to some extent.

You should pay particular attention to the fact that quantum mechanics predicts a continuous range of possible energies for the free particle since there is a widespread misconception that quantum mechanics is all about systems in which energy is 'quantized' into discrete energy levels. While it is certainly true that quantum physics originated with Planck's concept of the quantum of energy, and while it is also true that quantum mechanics provides a natural account of energy quantization, it is certainly *not* true that quantum mechanics is *restricted* to cases in which energy is discretely quantized. The free particle is clearly a case in which the Schrödinger equation can be written down and solved (even though we have not done so here), and in which the solutions predict a continuous range of possible total energies. The system can be treated using quantum mechanics, yet its energy is not 'quantized'. When trying to grasp the essence of quantum mechanics, the fact that we learn about the possible values of the energy of a system by solving the Schrödinger equation and considering the relevant wavefunctions is more significant than the issue of whether or not those possible energy values are quantized.

Here is a summary of the main result of this subsection:

The fundamental equation of quantum mechanics is Schrödinger's time-dependent equation. For a free particle, this differential equation has travelling wave solutions with angular wavenumber given by $k = \sqrt{2mE_{\text{kin}}}/\hbar$ where k can have any positive value. For bound particles the solutions are standing waves where k is also given by $\sqrt{2mE_{\text{kin}}}/\hbar$ but where only certain values of k are allowed.

4.4 Schrödinger's time-independent equation

As was noted in the introduction to this section, when quantum mechanics is used to study confined particles we can expect the relevant solutions to involve standing waves. One of the characteristics of a standing wave, such as that shown earlier in Figure 2.16b, is that it is quite simple to separate its spatial variation from its temporal variation. From a mathematical point of view this means that a wavefunction describing a standing wave can be written as a product of two independent functions that depend separately on position and time. Thus, in one dimension, for a system described by standing waves:

$$\Psi(x, t) = \psi(x)\phi(t) \tag{2.11}$$

where ψ, the lower case, Greek letter psi, is used to indicate a function of position that describes the 'shape' or profile of the standing wave as it varies with position, x, while ϕ, the lower case, Greek letter phi, indicates a function of time that causes the overall form of the wave to vary with time.

Note that the upper case psi (Ψ) has horizontal bars at the top and bottom of the character. There are no such bars on the lower case psi (ψ).

In the case of a particle of mass m and total energy E_{tot}, moving in one dimension, where the potential energy function is $E_{\text{pot}}(x)$, Schrödinger's time-dependent equation implies that the function $\psi(x)$ must satisfy a differential equation of the form

$$\frac{d^2\psi}{dx^2} + \frac{2m}{\hbar^2}(E_{\text{tot}} - E_{\text{pot}}(x))\psi = 0. \tag{2.12}$$

Now, nothing in this equation depends on time, so the equation is usually referred to as the **time-independent Schrödinger equation**, and its solutions $\psi(x)$ are often referred to as **time-independent wavefunctions**. Much of the everyday work for which quantum mechanics is used boils down to solving Equation 2.12, with some particular form of the potential energy function $E_{\text{pot}}(x)$, to find the relevant time-independent wavefunction $\psi(x)$ for the given problem. This is so much the case that $\psi(x)$ is often referred to simply as a 'wavefunction', even though waves are inherently time-dependent and are more properly described by time-dependent functions such as $\Psi(x, t)$. Similarly, Equation 2.12 is sometimes referred to as the Schrödinger equation, even though it only represents one particular aspect of the full Schrödinger equation.

The time-independent wavefunctions $\psi(x)$ that satisfy Equation 2.12 are said to describe **stationary states** of the confined particle, since all of the properties to which they relate are independent of time. For a particle confined by some particular potential well, $E_{pot}(x)$, there may be several of these stationary states, each corresponding to a different time-independent wavefunction, just as a string stretched between two fixed points may support any of a number of different standing waves. Each of the solutions to Equation 2.12 (for a given potential function $E_{pot}(x)$) will correspond to a particular value of E_{tot}. As you will see in the next subsection, when the particle is confined in a potential well of the kind described in Section 4.2, these possible values of E_{tot} will generally take the form of separate and distinct energy levels. In the case of confined particles, the Schrödinger equation is therefore able to account for the phenomenon of energy quantization.

It is worth noting that it is possible for a confined particle to be in a state that is not a stationary state and which, therefore, does not correspond to any particular solution of Equation 2.12, nor to any particular value of the total energy E_{tot}. We shall not be concerned with such non-stationary states in this chapter but we shall return to them in Chapter 4, where you will see that they can be represented by combinations of stationary states, much as a musical sound can be thought of as a combination of harmonics.

Here is a summary of the main points to emerge from this subsection:

Schrödinger's time-dependent equation for confined particles has standing wave solutions that correspond to particular values of the total energy. The spatial profiles of these solutions in one-dimension are described by *time-independent wavefunctions* $\psi(x)$ that satisfy the *time-independent Schrödinger equation*:

$$\frac{d^2\psi}{dx^2} + \frac{2m}{\hbar}(E_{tot} - E_{pot}(x))\psi = 0. \qquad \text{(Eqn 2.12)}$$

By using this equation to determine the values of E_{tot} for which wavefunction solutions exist, predictions can be made of the possible outcomes of measurements of the total energy of a particle confined by a potential well with potential energy function $E_{pot}(x)$.

4.5 The Schrödinger equation for a particle in a one-dimensional infinite square well

In this subsection, we are going to consider in detail the case of a *confined* particle, and we will see exactly how the solutions to Schrödinger's equation indicate that the particle has discrete energy levels determined by the allowed wavefunctions.

Let us consider the simplest possible case for a confined particle. A particle of mass m is restricted to one dimension and is confined between two infinitely high walls a distance D apart (Figure 2.18). We want to find the time-independent wavefunctions, ψ, that satisfy the time-independent Schrödinger equation for this particle. So, we start from Equation 2.12:

$$\frac{d^2\psi}{dx^2} + \frac{2m}{\hbar^2}(E_{tot} - E_{pot}(x))\psi = 0. \qquad \text{(Eqn 2.12)}$$

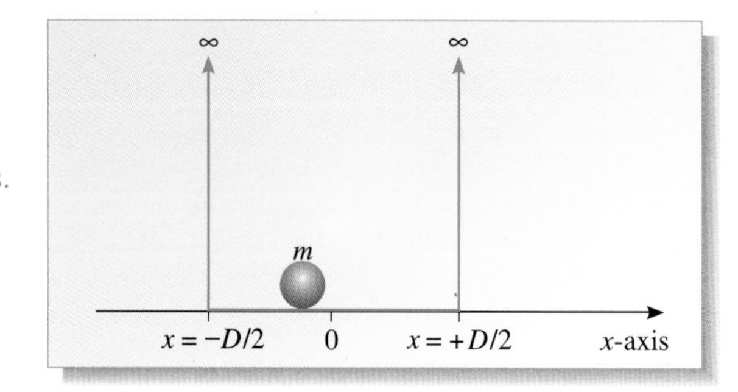

Figure 2.18 A particle of mass m that moves between two infinitely high, rigid, vertical walls. Since the particle's position can always be specified using only the x-coordinate, we say that the particle is restricted to one dimension.

● Now, what do we need to know about the particle before we can set about solving the Schrödinger equation?

❍ We need to know its potential energy function $E_{pot}(x)$. That is, we need to know the potential energy of the particle at every point in space. Once we have the potential energy function, the solutions of the Schrödinger equation will give the allowed stationary state wavefunctions ψ and the energy of the particle, E_{tot}, associated with each wavefunction. ■

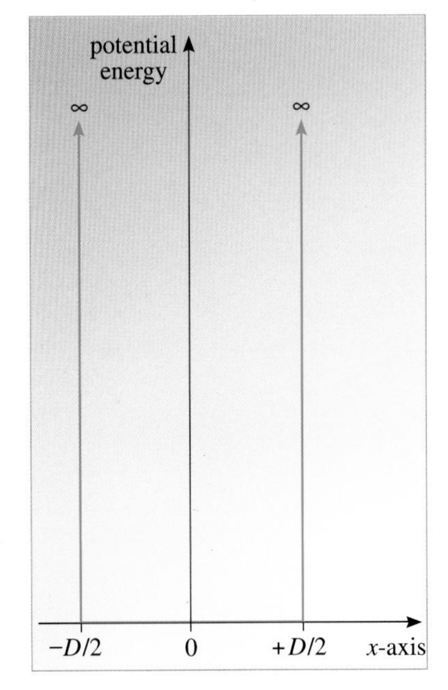

Figure 2.19 This potential energy function specifies at every point in space the potential energy of the particle shown in Figure 2.18. Because of the similarity of the shape of this function with that of a well, this potential energy function is said to be a *potential well*.

So, what is the particle's potential energy function in this case? The particle is free to move between the confining walls and its potential energy is constant in this region. Hence, we can define the particle's potential energy to be zero when it is *between* the walls. Since the confining walls are supposed to be infinitely high, it would take an infinite amount of energy to remove the particle from between the walls. Hence, the potential energy of the particle must be infinite outside the walls. Thus, the particle's potential energy function is that shown in Figure 2.19. The shape of this graph resembles an infinitely deep well, so for this reason the confined particle is said to be in an *infinite square well*. The easiest way to write down the potential energy function is to write a separate equation for each of the different regions. Thus, we have:

$$\left.\begin{array}{ll} E_{\text{pot}}(x) = 0 & \text{for } -D/2 < x < D/2 \\ E_{\text{pot}}(x) = \infty & \text{for } x \leq -D/2 \text{ and } x \geq D/2. \end{array}\right\} \tag{2.13}$$

Now that we have the particle's potential energy function, we can set about solving the Schrödinger equation. The particle's wavefunction must satisfy Schrödinger's equation at all points in space, and since we have specified the potential energy function separately for the different regions, it is sensible also to solve Schrödinger's equation piecemeal.

Remember that ψ in this context is a function of x, i.e. $\psi(x)$ but we shall often, for simplicity, just write ψ.

First, we will consider the regions where the potential energy is infinite. Now, it is fairly easy to see that ψ must take the value zero in these regions. From a purely physical point of view, we can see that, in order to move a particle into a region where its potential energy is infinite, we would have to supply the particle with an infinite amount of energy. Since this is not possible, we must assume that the particle can never be found in this region, and this in turn implies that $\psi = 0$ in this region.

Now we must find the solution *inside* the well, where $E_{pot} = 0$. When we substitute $E_{pot} = 0$ into Schrödinger's equation, we obtain

$$\frac{d^2\psi}{dx^2} + \frac{2m}{\hbar^2}E_{\text{tot}}\psi = 0$$

which we may write as

$$\frac{d^2\psi}{dx^2} + k^2\psi = 0 \tag{2.14}$$

where $k = \sqrt{2mE_{tot}}/\hbar$. Note that since the potential energy function is zero within the well, the particle's total energy (E_{tot}) is the same as its kinetic energy ($E_{kin} = E_{tot} - E_{pot}$). If the potential energy function within the well were to have some finite value other than zero, then we would have

$$k = \frac{\sqrt{2mE_{\text{kin}}}}{\hbar} = \frac{\sqrt{2m(E_{\text{tot}} - E_{\text{pot}})}}{\hbar}.$$

Now, Equation 2.14 is a differential equation that arises in many areas of physics. (It was first introduced in *Describing motion*, as the equation of simple harmonic motion.) We know from our previous encounters with this equation that the time-independent functions ψ that satisfy it are sinusoidal functions and that they include

$$\psi = \psi_0 \sin kx \quad \text{and} \quad \psi = \psi_0 \cos kx. \tag{2.15}$$

However, here we have an extra condition to fulfil. In general, for wavefunctions ψ to be acceptable solutions to Schrödinger's equation, both the wavefunction ψ and its first derivative $\mathrm{d}\psi/\mathrm{d}x$ must be continuous functions. By 'continuous' we mean that they must vary smoothly and have no sudden changes in value. In this case, this means that the value of the wavefunction inside the well must match its value outside the well when they meet at the boundary. This is called a *boundary condition*, and is required to ensure that the wavefunction has a unique value at every point. This means that the only solutions (Equation 2.15) that are allowed are those for which $\psi = 0$ at $x = -D/2$ and $x = D/2$. The requirement that $\mathrm{d}\psi/\mathrm{d}x$ be continuous need not be met in this case because we have stipulated an infinite potential energy function outside the well.

These solutions are very similar to the standing waves on a stretched string of finite length (say a violin or guitar string) that were introduced in Chapter 2 of *Dynamic fields and waves*. A wave-like motion can be set up on the string but it is constrained by the fact that at the fixed ends of the string the oscillation must go to zero. The result is that the allowed standing waves are those for which the length of the string is a whole number of half-wavelengths. The same is true for the stationary state wavefunctions of a particle in an infinite square well. From the possibilities expressed in Equation 2.15, only those that have a wavelength $\lambda = 2D/n$ are allowed, where n is an integer. Since $k = 2\pi/\lambda$, this means that $k = n\pi/D$. So finally, the allowed stationary state wavefunctions for a particle in an infinite square well are

$$\psi = \psi_0 \sin kx, \quad \text{where } k = \frac{n\pi}{D} \text{ and } n = 2, 4, 6, \text{ etc.} \tag{2.16a}$$

and

$$\psi = \psi_0 \cos kx, \quad \text{where } k = \frac{n\pi}{D} \text{ and } n = 1, 3, 5, \text{ etc.} \tag{2.16b}$$

all of which have $\psi = 0$ for $x \leq -D/2$ and $x \geq D/2$.

The wavefunctions for $n = 1, 2, 3$ and 4 are shown in Figure 2.20. Also shown are the values of the energy E_{tot} of the particle associated with each wavefunction. Since $k = \sqrt{2mE_{\text{tot}}}/\hbar$, the restriction on the allowed values of k implies a restriction of the allowed values of E_{tot} of the particle.

Rearranging the expression for k, we have

$$E_{\text{tot}} = \frac{\hbar^2 k^2}{2m}, \tag{2.17}$$

and substituting for k from Equation 2.16, we find

$$E_{\text{tot}} = \frac{n^2 h^2}{8mD^2}, \quad \text{where } n = 1, 2, 3, \text{ etc.} \tag{2.18}$$

Thus the confined particle has *energy levels* — its energy is *quantized* — and the number n that labels each wavefunction and its associated energy is called a **quantum number**. For a particle with a set of quantized energy levels like this, the state with the lowest possible value of the energy, in this case the $n = 1$ state, is usually referred to as the *ground state*. The states with higher energies ($n \geq 2$) are referred to as *excited states*.

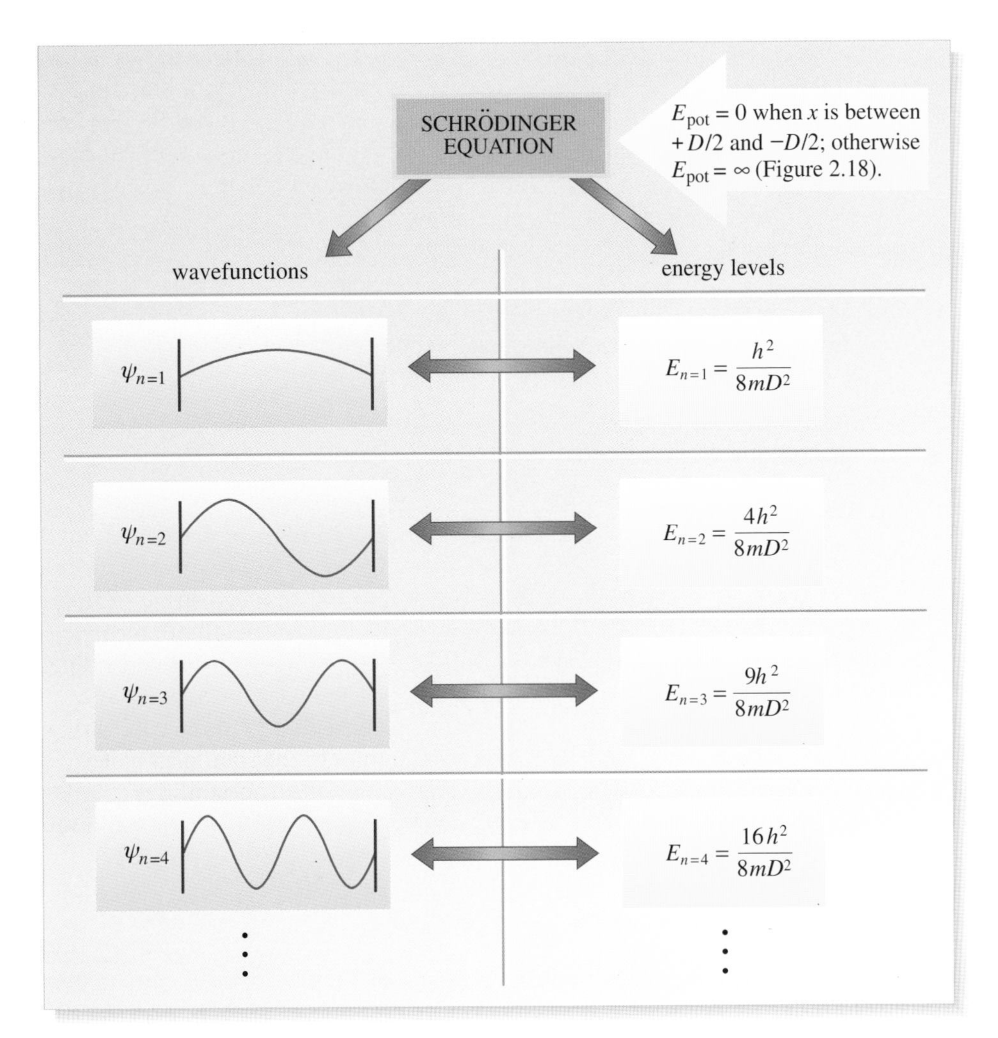

Figure 2.20 The way in which the time-independent Schrödinger equation is used to investigate the behaviour of the confined particle shown in Figure 2.18.

Let us recap on what we have done. In order to investigate the behaviour of a particle in a one-dimensional infinite square well, we have

1 Inserted the particle's potential energy function into the time-independent Schrödinger equation.

2 Found the time-independent wavefunctions, ψ, that are the solutions of Schrödinger's time-independent equation for this situation, subject to the constraint that $\psi = 0$ at the edges of the well.

3 Calculated the value of the particle's energy which is associated with each of these time-independent wavefunctions.

We have therefore found the spatial part of each of the possible stationary state wavefunctions of the particle (Equations 2.16a and 2.16b). Each of these wavefunctions corresponds to a *definite* value of the total energy given by Equation 2.18. So if the total energy of the particle is measured when it is in a stationary state with quantum number n, the value $E_{tot} = n^2h^2/8mD^2$ will certainly be obtained. The fact that only certain energies are allowed is completely at odds with classical physics in which a particle in an infinite square well can have any value of energy.

Another important feature of the quantum-mechanical description of a particle confined in an infinite square well is that it *cannot* have zero kinetic energy. The lowest allowed value of n for the particle in the infinite square well is 1. This, also, is in contrast to classical mechanics in which it is perfectly allowable for a particle to have zero kinetic energy, when it is stationary.

We can take two further points from this discussion.

- In solving the time-independent Schrödinger equation for a particle in an infinite square well, we have seen how the relationship, $k = \sqrt{2mE_{\text{kin}}}/\hbar$, between the angular wavenumber k and the particle's kinetic energy E_{kin} (or its total energy E_{tot} where $E_{\text{pot}} = 0$) arises. For a free particle we cannot separate the time and space variation of the travelling wave wavefunction as in Equation 2.11. Nevertheless, the solutions to Schrödinger's time-dependent equation for a free particle give rise to the same relationship in the same way.
- If the potential energy function $E_{\text{pot}}(x)$ is not constant, but varies with x, then the wavefunctions (for both free and confined particles) will reflect this in a varying value of k, which, at any point x, will be that appropriate to E_{pot} at that point according to

$$k(x) = \frac{\sqrt{2m(E_{\text{tot}} - E_{\text{pot}}(x))}}{\hbar}.$$

  So, if E_{pot} varies with x, then k and E_{kin} will vary accordingly, but the value of the particle's total energy, E_{tot}, remains fixed.

Question 2.9 Criticize and suggest an improved rephrasing of the following statement about quantum mechanics. 'Schrödinger's approach to quantum mechanics is all about systems with quantized energy, i.e. systems in which the energy can only have certain discrete values.' ■

4.6 Interpretation of the wavefunction

It was stated earlier that quantum-mechanical wavefunctions do not have any *direct* physical interpretation. Nonetheless, they play a central role in Schrödinger's approach to quantum mechanics and they have a very great deal of indirect significance. In fact, according to the conventional interpretation of quantum mechanics (which will be discussed further in Chapter 4), each wavefunction of a quantum system represents a particular state of the system and contains all the information about that state that it is possible to have.

By the end of this section you will see how the wavefunction may be related to the probability of detecting the particle at any position, but first, here's a reminder of Section 2.3.

In Section 2.3 you saw that, when a particle is associated with a simple de Broglie wave, the probability of detecting the particle at some position is proportional to the square of the amplitude of the wave at that position. Now, for a simple travelling wave — such as a sinusoidal wave shown in Figure 2.21 — the amplitude has the same value at all points along the wave. The wave is travelling from left to right along the x-axis, so at any point the wave is oscillating sinusoidally in time with an amplitude A. The probability of detecting the particle is therefore proportional to A^2, and is the same at every point along the x-axis. Of course, if this de Broglie wave described a particle with precisely known momentum, and therefore precisely known wavelength, the wavetrain would have to be infinitely long; then the probability of finding it would be spread so thin that it would be effectively zero everywhere!

In Section 3, we argued that localized particles are generally better represented by a wave packet of the type shown in Figure 2.8b which is drawn again here in

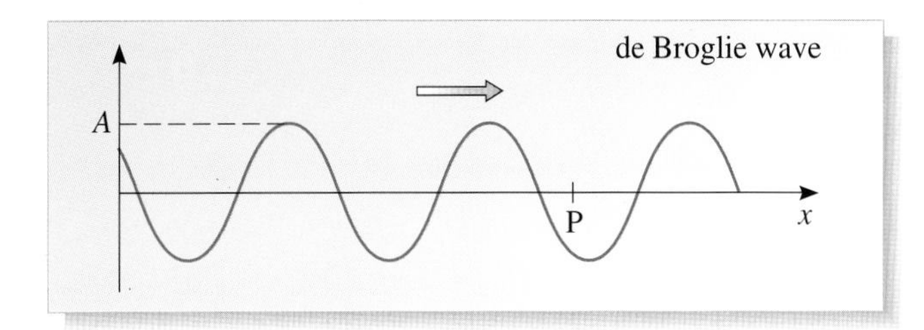

Figure 2.21 An infinite sinusoidal travelling de Broglie wave, travelling from left to right.

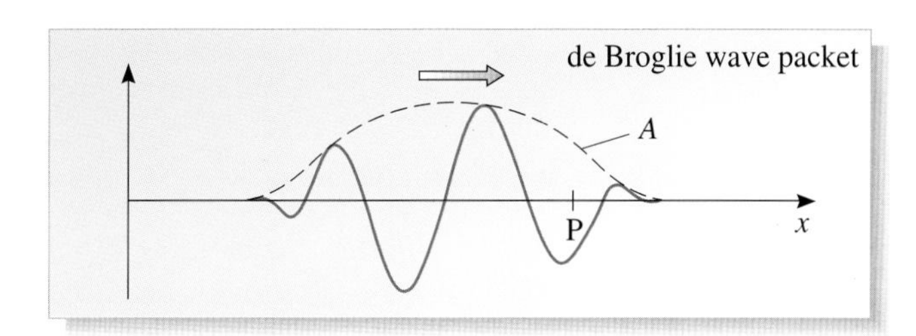

Figure 2.22 A travelling de Broglie wave packet. The dashed line shows how the amplitude A of oscillation varies across the wave packet.

Figure 2.22. This wave packet also propagates along the x-axis, but at any point P on the axis the amplitude A of the oscillation increases from zero to a maximum as the wave packet approaches then falls to zero again after the wave packet has passed. Consequently, the probability of detecting the particle at P will rise from zero, go through a peak for a short time then fall to zero again. It will do this for all points on the x-axis, but at different times according to the value of x. Nonetheless, it is still the case that at any point at any time, the probability of detecting the particle is still related to the square of the amplitude of the oscillation.

Extending these ideas to quantum mechanics, where the primitive notion of a de Broglie wave, or a packet of de Broglie waves, is replaced by that of a wavefunction $\Psi(x, t)$ is fairly straightforward. This is especially true in the case of stationary states of confined particles, where the wavefunction takes the form of a standing wave and its spatial dependence is described by a time-independent wavefunction $\psi(x)$.

Consider, for example, the time-independent wavefunctions depicted in Figure 2.23a. These are the allowed solutions to the time-independent Schrödinger equation, for $n = 1$ and $n = 5$, for a particle in a one-dimensional infinite square well. What can be said about the position of a particle represented by the standing probability waves that correspond to these particular time-independent wavefunctions?

As in the case of the travelling wave packet discussed above, the important feature of the standing probability wave will be the amplitude of the oscillation at each value of x. Now, you can think of the time-independent wavefunctions in Figure 2.23a as snapshots of those standing waves at a moment when each part of the wave has its

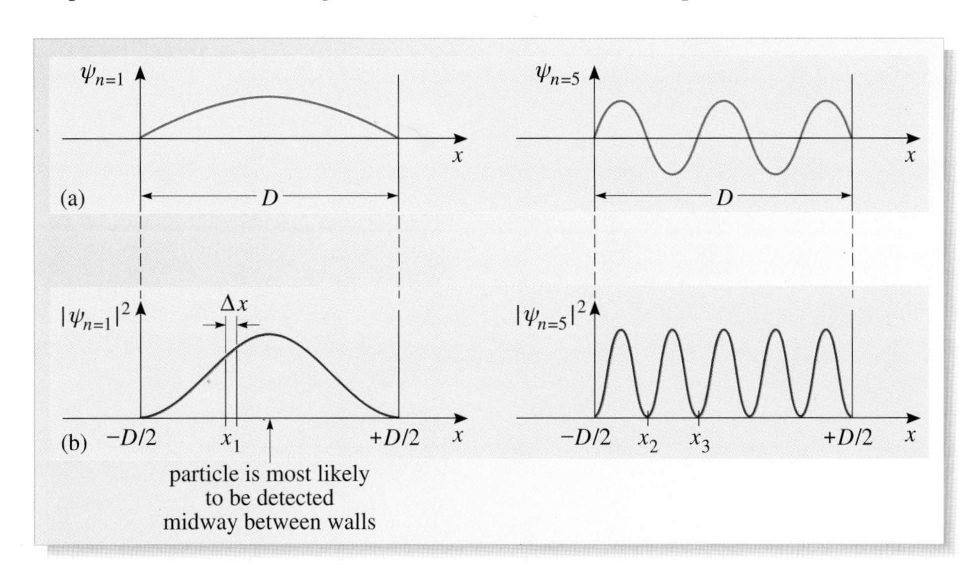

Figure 2.23 (a) Two of the standing probability waves, $\psi_{n=1}$ and $\psi_{n=5}$, that can describe the particle in a one-dimensional infinite square well. (b) The relative probability of detecting the particle in different regions, when the particle is described by the standing probability waves $\psi_{n=1}$ and $\psi_{n=5}$.

greatest positive or negative value. Half a period later, a snapshot of either wave would be an inverted version of that shown in Figure 2.23a, and after another half-period, the wave would again have the form shown in the figure. (The information for this time dependence is, of course, contained in the function $\phi(t)$. Remember, for a standing wave, we can write $\Psi(x, t) = \psi(x)\phi(t)$.)

Bearing this in mind, we can see that the value of the amplitude of oscillation at any point along the x-axis in Figure 2.23a is simply proportional to the value of $|\psi|$ at that point. Thus if we square the value of $|\psi|$ at each point x in Figure 2.23a, we will obtain the relative probabilities of detecting the particle at different positions, and this is shown in Figure 2.23b.

According to this interpretation, $|\psi|^2$ is proportional to the probability of detecting the particle. For example, when the particle is described by the $n = 1$ wavefunction, the particle is *most likely* to be detected midway between the walls since $|\psi_{n=1}|^2$ has its maximum value at this value of x (Figure 2.23b).

What is the probability of detecting the particle in the small region of width Δx around the position $x = x_1$ in Figure 2.23b? The probability P is proportional to $|\psi(x_1)|^2$ where $|\psi(x_1)|$ is the value of the amplitude of the wavefunction at the position $x = x_1$. (We assume that Δx is so small that we can ignore the variation of ψ over the region.) Thus

$$P \propto |\psi(x_1)|^2. \qquad (2.19)$$

The probability of detecting the particle is also proportional to the width Δx of the region. Hence,

$$P \propto \Delta x. \qquad (2.20)$$

These two proportionalities may be combined to give:

$$P \propto |\psi(x_1)|^2 \Delta x. \qquad (2.21a)$$

The most convenient choice is to make P *equal to* $|\psi(x_1)|^2 \Delta x$ and we therefore usually write

$$P = |\psi(x_1)|^2 \Delta x. \qquad (2.21b)$$

For this to be true, the *sum* of all the $|\psi(x)|^2 \Delta x$ segments over the width of the box, i.e. the total area under the curve, must be equated to the total probability of finding the particle *somewhere* between the walls. This probability is, by convention, equal to one. The values of ψ must therefore be scaled to make the area under the $|\psi|^2$ curve equal to one. In mathematical terms this means that we must have

$$\int_{-\infty}^{\infty} |\psi(x)|^2 \, dx = 1.$$

This scaling process is known as **normalization**, and is the mathematical way of ensuring that the total probability of finding the particle somewhere between the confining walls is one. The relative chance of finding it in the region described by a particular slice is then exactly equal to the area of that slice of the curve. So, to summarize:

For a particle in a stationary state described by the normalized time-independent wavefunction $\psi(x)$, the probability of finding the particle in a narrow range Δx, centred on x, is $P(x) = |\psi(x)|^2 \Delta x$.

Now look at the part of Figure 2.23b that shows the probability of detecting the particle in different regions between the walls when the particle is described by the $n = 5$ wavefunction. In this case, the total probability of detecting the particle between x_2 and x_3 is exactly 1/5 because the area under the curve between these limits is 1/5 of the total area between $x = -D/2$ and $x = +D/2$. Use this observation to help you answer the following question.

Question 2.10 An electron is confined between the infinitely high, rigid walls shown in Figure 2.19.

(a) Sketch the time-independent wavefunction that describes the electron when it has a total energy of $E_{\text{tot}} = 9h^2/(8m_eD^2)$.

(b) Indicate on your sketch the positions where the electron is most likely to be detected when it has a total energy of $E_{\text{tot}} = 9h^2/(8m_eD^2)$.

(c) What would happen if the electron made a transition from the energy level $E_{\text{tot}} = 9h^2/(8m_eD^2)$ to the energy level $E_{\text{tot}} = h^2/(8m_eD^2)$? (*Hint*: Think back to Chapter 1.) ■

The relationship between wavefunctions and probabilities may be generalized. It applies to all sorts of stationary states, not just those of a particle in an infinite square well, and it may even be extended to situations in which the behaviour of a particle is described by a time-dependent wavefunction $\Psi(x, t)$. In fact, we can say quite generally that:

> For a particle described by a normalized wavefunction $\Psi(x, t)$, the probability of finding the particle in some narrow range Δx, centred on x, at time t is
>
> $$P(x, t) = |\Psi(x, t)|^2 \Delta x.$$

This does *not* mean that the probability $P(x, t)$ passes through zero every half-cycle. You must still think of $|\Psi(x, t)|$ as the amplitude of the wave. You will see how this arises when you read Box 2.1 on the next page.

Because of this general relationship, the wavefunction of a particle is sometimes said to be the *probability amplitude* for finding the particle in the neighbourhood of a point.

This physical interpretation of a wavefunction was introduced by the German physicist Max Born (Figure 2.24) in 1926 and has thus come to be known as the **Born interpretation**. Prior to that, the significance of the wavefunction had been unclear, even though it had been used to solve a number of problems.

Before leaving the interpretation of wavefunctions there is a point that needs to be cleared up concerning our use of notations such as $|\psi|^2$ or $|\Psi|^2$ when you might have expected ψ^2 or Ψ^2 to be used. This involves a mathematical subtlety concerning wavefunctions that is explained in Box 2.1. You can regard the box as a sort of extended aside.

Figure 2.24 Max Born (1882–1970) was born in Breslau (now Wroclaw, Poland). After teaching in Berlin and Frankfurt, he became head of the Physics Department at Göttingen University, which, under his administration, became a centre of theoretical physics rivalled only by the Niels Bohr Institute in Copenhagen. Hitler's politics forced Born to leave Germany in 1933, and until 1953 he spent his time at Cambridge, Bangalore and Edinburgh. In 1954 he was awarded the Nobel Prize for physics (shared with Walther Bothe) for the work he had done much earlier in laying the foundations for a statistical interpretation of electron wavefunctions.

Box 2.1 The complex nature of the wavefunction

Generally, the wavefunctions that satisfy the Schrödinger equation are what mathematicians describe as *complex* quantities. That is to say, for given values of x and t, the quantity $\Psi(x, t)$ is not generally a simple number but is more like a two-component vector. This is indicated by writing its value as

$$\Psi = a + \mathrm{i}b$$

where a and b are ordinary decimal numbers (called *real* numbers in this context) while i is an algebraic symbol representing a quantity with the property

$$\mathrm{i}^2 = -1.$$

Now, no real number can satisfy the equation $\mathrm{i}^2 = -1$, so i is said to be an *imaginary number* and is usually referred to as the 'square root of minus one'. Any number that involves i, such as $\Psi = a + \mathrm{i}b$, is said to be a *complex number.* Schrödinger's time-dependent equation explicitly involves i, so it is not surprising that its solutions involve complex numbers.

In the context of quantum mechanics, although we can generally write $\Psi = a + \mathrm{i}b$, it is often more useful to think of Ψ as a two-dimensional vector with components a and b as indicated in Figure 2.25. This suggests that yet another way of representing a complex quantity such a Ψ is in terms of a length and an angle as is also indicated in Figure 2.25. In the language of complex numbers, the length $|\Psi|$ is referred to as the *modulus* of Ψ.

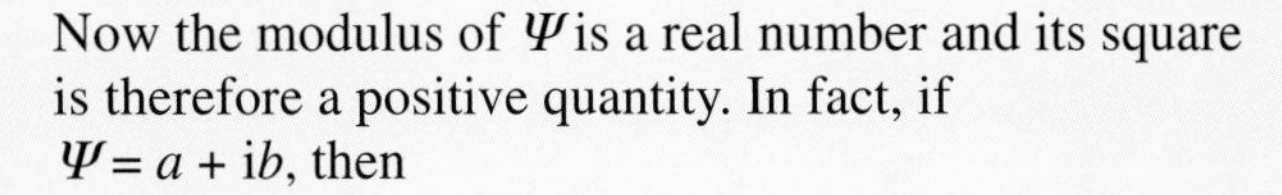

Now the modulus of Ψ is a real number and its square is therefore a positive quantity. In fact, if $\Psi = a + \mathrm{i}b$, then

$$|\Psi|^2 = a^2 + b^2.$$

Since $|\Psi|^2$ is always positive it can be interpreted as a probability (which must always be positive). Hence the use of $|\Psi|^2$ in Born's interpretation.

We shall not be much concerned with complex numbers in this book, even though they are crucial to quantum physics as a whole. However, we shall continue to use the modulus notation whenever it is appropriate to do so.

Incidentally, now that you have been introduced to complex numbers you might be interested in Figure 2.26 which shows a snapshot of the kind of complex travelling wave that might represent a free particle in quantum mechanics. The real part of the wave looks like a simple sinusoidal wave (as does the imaginary part), but the full wave is more like the thread of a bolt. In this case the time-dependent wavefunction would be

$$\Psi(x, t) = A\cos(\omega t - kx) + \mathrm{i}A\sin(\omega t - kx)$$

where A is a constant. Then $|\Psi(x, t)|^2 = A^2$, showing that there is an equal probability of finding the particle at any point.

In the case of a standing wave, you can simply think of the whole wave profile rotating about the x-axis.

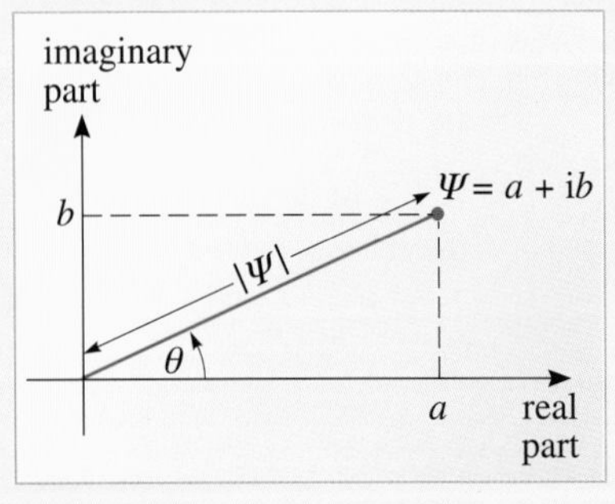

Figure 2.25 A complex number $\Psi = a + \mathrm{i}b$ may also be represented in terms of its modulus $|\Psi|$ and an angle θ.

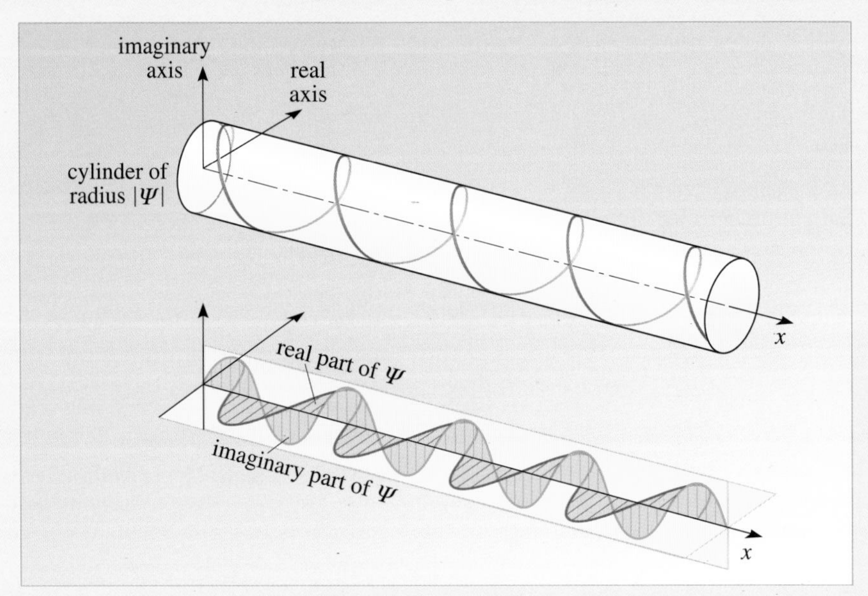

Figure 2.26 A snapshot of the kind of complex travelling wave that might represent a free particle in quantum mechanics.

4.7 The Schrödinger equation for a particle in a one-dimensional finite square well

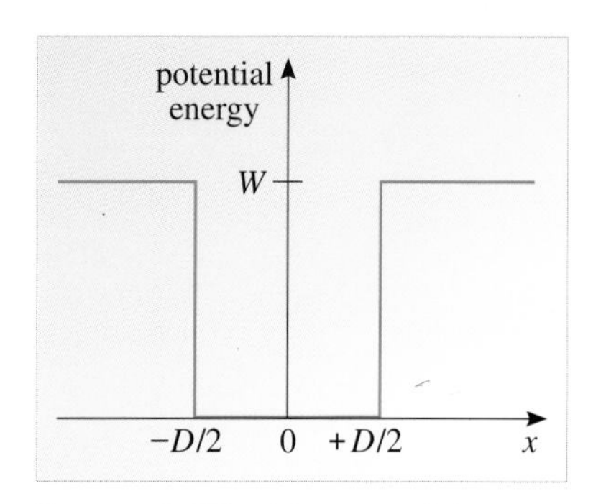

Figure 2.27 The potential energy function of a finite square well.

One of the most interesting results of quantum mechanics is its prediction for the behaviour of a particle in a region where the total energy of the particle is *less* than its potential energy. Classically, of course, a particle can never enter a region where its energy is less than the value of its potential energy function. Quantum mechanics, however, says otherwise. We can illustrate this by considering a potential well similar to the one described in the previous subsection, but this time having a finite depth W. This well is illustrated in Figure 2.27; for obvious reasons it is called a *finite square well*. The potential energy function for this well is

$$\left.\begin{array}{ll} E_{\text{pot}}(x) = 0 & \text{for } -D/2 < x < D/2 \\ E_{\text{pot}}(x) = W & \text{for } x \leq -D/2 \text{ and } x \geq D/2 \end{array}\right\} \qquad (2.22)$$

Within the well, we have the same situation as for the infinite well, and the time-independent wavefunctions ψ that satisfy Schrödinger's equation here are oscillatory, standing wave type functions of the form $\psi = \psi_0 \sin kx$ or $\psi = \psi_0 \cos kx$ where $k = \sqrt{2mE_{\text{tot}}}/\hbar$ as before. But the allowed values of k and therefore of E_{tot} are *not* the same as for the infinite well because this time the wavefunction does *not* have to go to zero at the walls of the well. In fact, the wavefunction decays away exponentially in the region outside the well, and to satisfy the requirement that ψ and $\mathrm{d}\psi/\mathrm{d}x$ should be continuous, the oscillatory part of the wavefunction, and its slope, within the well have to match at the boundary with those of the exponentially decaying part outside.

This is rather complicated to do mathematically, and at the level of this course we cannot get the exact mathematical solutions. However, *qualitatively*, it is easy to see what the time-independent wavefunctions must look like. The first three solutions of lowest energy are shown in Figure 2.28. In each case you can see the oscillatory region inside the well for $-D/2 < x < D/2$ and the exponentially decaying region outside the well. The requirements on the form of the wavefunction in the different regions and the matching at the boundaries means that in this case, as in the case of the infinite well, only certain values of k and therefore certain values of E_{tot} are allowed: the particle has energy levels. This is true for *all* confined particles, however shallow the potential well.

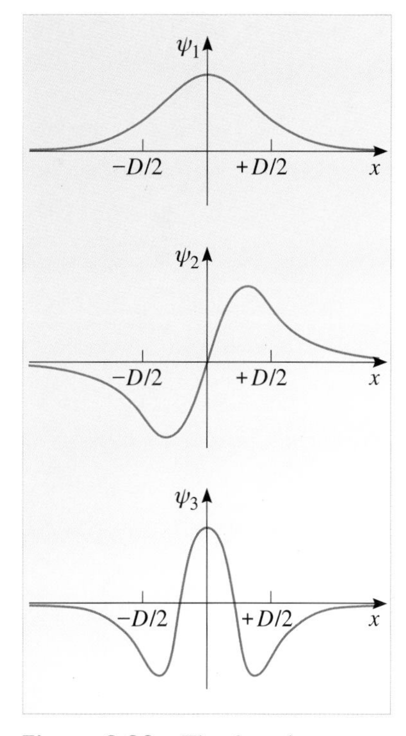

Figure 2.28 The three lowest energy wavefunctions for a particle in a finite square well. Within the well ($-D/2 < x < D/2$) the wavefunction is oscillatory. Outside the well, it decays exponentially.

The distance that the wavefunction penetrates into the region outside the well depends on the height of the walls above the energy of the particle (i.e. $W - E_{\text{tot}}$). This region is often referred to as the 'classically forbidden region' since, according to classical mechanics, a particle with total energy E_{tot} can never penetrate into a region where its potential energy function is greater than E_{tot} since this implies a negative kinetic energy.

Figure 2.29 shows qualitatively how the wavefunction and the energy of the lowest allowed energy level vary as the depth of the well is increased. The need for the wavefunction to match at the boundary means that for shallower wells the wavelength of the oscillatory part of the wavefunction (inside the well) is longer and thus the energy level lower than for deeper wells.

Before leaving the finite square well, we will pause to consider the solutions to Schrödinger's equation for the case in which the particle's energy E_{tot} is *greater* than W, the height of the walls of the well. In this case the particle is *unbound* and the solutions are travelling waves. In each region separately, the wavefunction has an angular wavenumber given by $\sqrt{2m(E_{\text{tot}} - E_{\text{pot}})}/\hbar = \sqrt{2mE_{\text{kin}}}/\hbar$ and, as for the case of the confined particle, the different parts of the wavefunction must match at

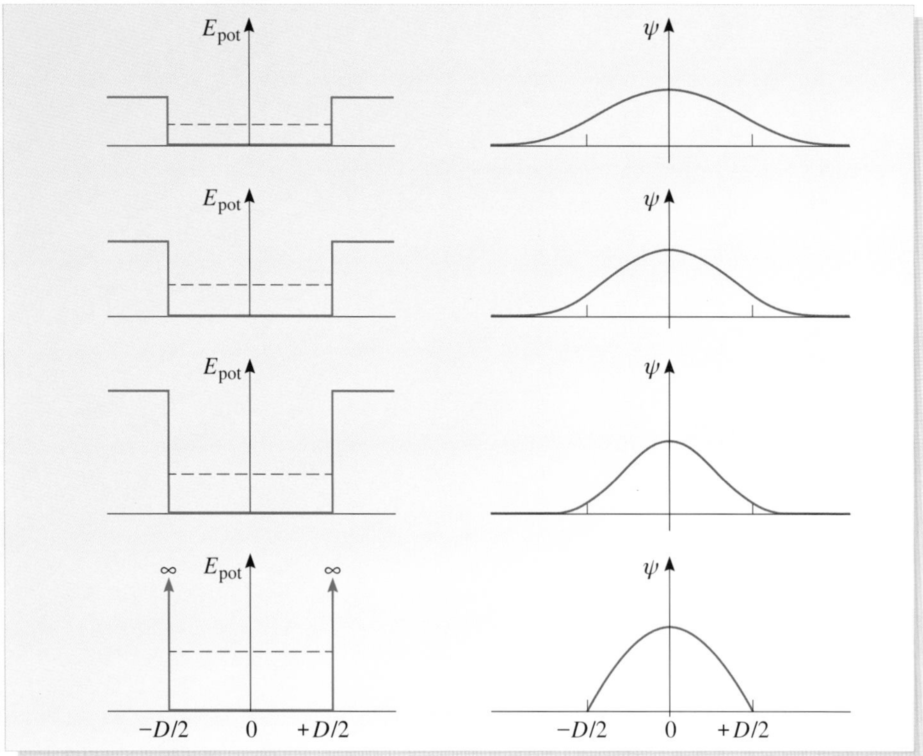

Figure 2.29 Sketches of the lowest energy time-independent wavefunction for a particle confined in a potential well of various heights. The wavefunction penetrates further into the 'classically forbidden' region in a potential well of lower height. The dashed line indicates the energy of the ground state (state of lowest energy) in each case.

the boundaries. Even so, the particle can have any value of E_{tot} and a snapshot of the wavefunction will look similar to that shown in Figure 2.30.

Question 2.11 What is the total energy of the particle in Figure 2.30 when it is in a region outside the well? ■

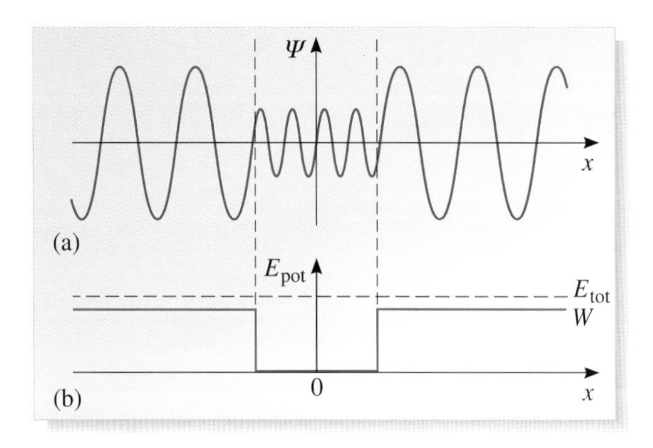

Figure 2.30 (a) A possible wavefunction of an unbound particle in the vicinity of (b) a finite square potential well. The wavelength in each region is inversely proportional to the magnitude of the momentum of the particle, where $p = \hbar k = \sqrt{2m(E_{\text{tot}} - E_{\text{pot}})}$.

4.8 Barrier penetration

In the last subsection, you saw that the wavefunction of a particle can actually have a finite value in a region that is classically forbidden. This means, of course, that there is a finite probability of finding the particle in this region. One extraordinary consequence of this is that if the potential energy function were not just a well but a barrier of finite width *and* height, as in Figure 2.31b, then the wavefunction can still have a finite value on the *outside* of the barrier. We can interpret this as meaning that there is a finite probability of the particle escaping from its confinement even though

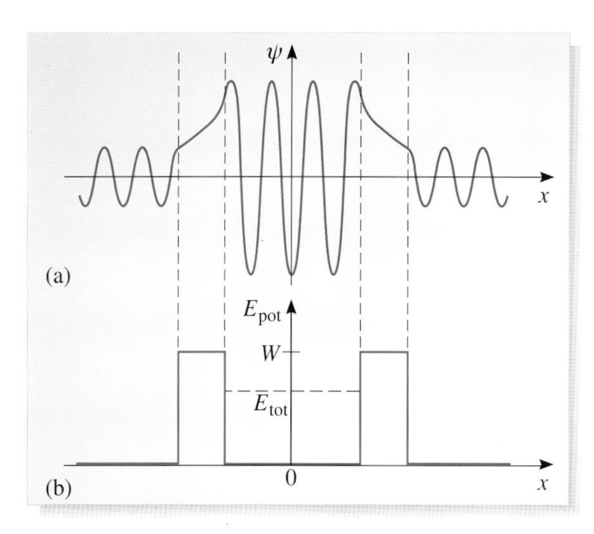

Figure 2.31 (a) A possible wavefunction for a particle confined by (b) potential energy barriers of finite width and height. The wavefunction penetrates the classically forbidden region and has a finite value on the outside of the barrier.

its total energy is nowhere near enough (in classical terms) for it to surmount the potential energy barrier. This phenomenon is known as *barrier penetration* or **tunnelling** and has many important applications in physics.

The wavefunction illustrating barrier penetration is shown in Figure 2.31a. You should be able to see how the wavefunction satisfies Schrödinger's equation in each of the three regions (inside the well, within the barrier and outside the well). In the region where E_{pot} is zero the wavefunction is oscillatory and the angular wavenumber k will be given, as usual by $\sqrt{2mE_{tot}}/\hbar$. Thus, the wavefunction has the same wavelengths inside and outside the well. In the 'classically forbidden region' within the barrier, the wavefunction decays exponentially at a rate which will depend on $W - E_{tot}$. So for walls much higher than E_{tot} or walls which are wide, the wavefunction will have decayed to a smaller value than for lower or narrower walls. Notice also that the continuity requirements are also met at all the boundaries. The wavefunction and its first derivative (slope) match (are continuous) at all points.

Table 2.1 summarizes the main results of this chapter for a particle moving in one dimension.

Question 2.12 An electron is confined in one dimension within a potential well with infinitely high walls a distance 0.5 nm apart. The potential energy of the electron within the well is zero. Calculate the lowest possible energy of this electron in electronvolts and sketch its wavefunction. Now suppose that the walls of the well are lowered (while still trapping the electron) until they have a finite potential energy of 2 eV. Sketch the new wavefunction of the confined electron in its lowest energy level. Use your sketches and an argument based on the value of the angular wavenumber k to decide whether the energy of the lowest energy level for the electron is smaller in the finite or in the infinite well. ■

This is quite a long question, so you may like to tackle it using the three-stage problem-solving technique as outlined in the revision chapter of *Predicting motion*.

Table 2.1

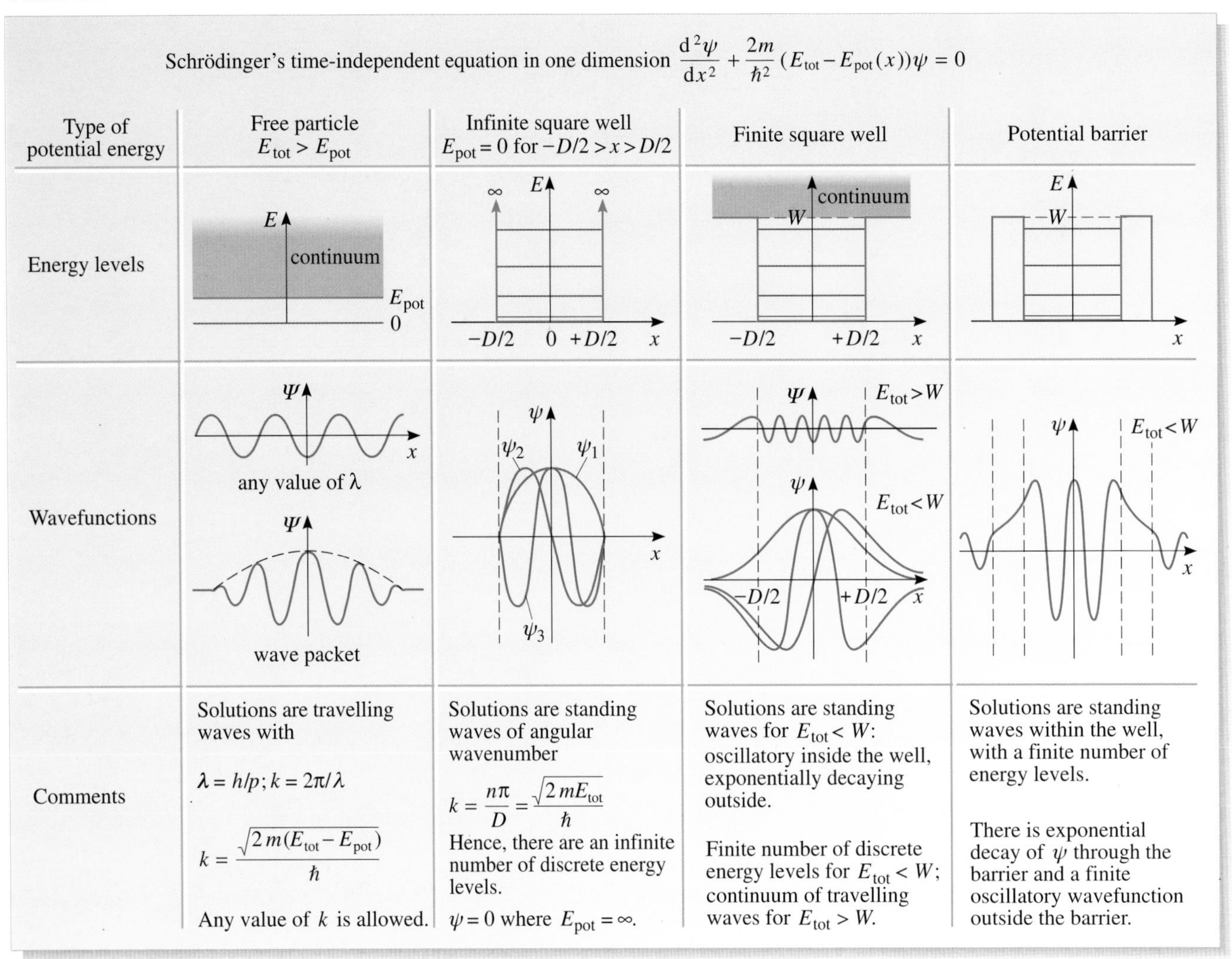

Schrödinger's time-independent equation in one dimension $\frac{d^2\psi}{dx^2} + \frac{2m}{\hbar^2}(E_{tot} - E_{pot}(x))\psi = 0$

Type of potential energy	Free particle $E_{tot} > E_{pot}$	Infinite square well $E_{pot} = 0$ for $-D/2 > x > D/2$	Finite square well	Potential barrier
Energy levels	E; continuum; E_{pot}; 0	∞; E; ∞; $-D/2$; 0; $+D/2$; x	continuum; W; $-D/2$; $+D/2$; x	E; W; x
Wavefunctions	ψ; x; any value of λ; ψ; wave packet	ψ; ψ_2; ψ_1; x; ψ_3	ψ; $E_{tot} > W$; ψ; $E_{tot} < W$; $-D/2$; $+D/2$; x	ψ; $E_{tot} < W$; x
Comments	Solutions are travelling waves with $\lambda = h/p; k = 2\pi/\lambda$ $k = \frac{\sqrt{2m(E_{tot} - E_{pot})}}{\hbar}$ Any value of k is allowed.	Solutions are standing waves of angular wavenumber $k = \frac{n\pi}{D} = \frac{\sqrt{2mE_{tot}}}{\hbar}$ Hence, there are an infinite number of discrete energy levels. $\psi = 0$ where $E_{pot} = \infty$.	Solutions are standing waves for $E_{tot} < W$: oscillatory inside the well, exponentially decaying outside. Finite number of discrete energy levels for $E_{tot} < W$; continuum of travelling waves for $E_{tot} > W$.	Solutions are standing waves within the well, with a finite number of energy levels. There is exponential decay of ψ through the barrier and a finite oscillatory wavefunction outside the barrier.

4.9 A particle confined in three dimensions — degeneracy

When a particle is confined in *one* dimension between two walls, the solutions of Schrödinger's equation in Sections 4.5 and 4.7 showed that the particle should have *energy levels*, characterized by a single integer n (Equation 2.18). These solutions were very similar to the solutions of the wave equation for the allowed standing waves on a stretched string that you met in *Dynamic fields and waves* Chapter 2. It is fairly easy to extend this example to two or three dimensions. Suppose we have a two-dimensional surface, such as a drum consisting of a skin stretched over a square frame. In this case, it is reasonable to expect that the kind of standing waves that can be set up must satisfy *independently* the condition that a whole number of *half*-wavelengths can fit into each of the *two* coordinate directions, with nodes at the edges. This means that *two* independent integers are needed to specify the various standing waves that can exist. Figure 2.32a shows some of the possible standing waves that can be set up on such a surface.

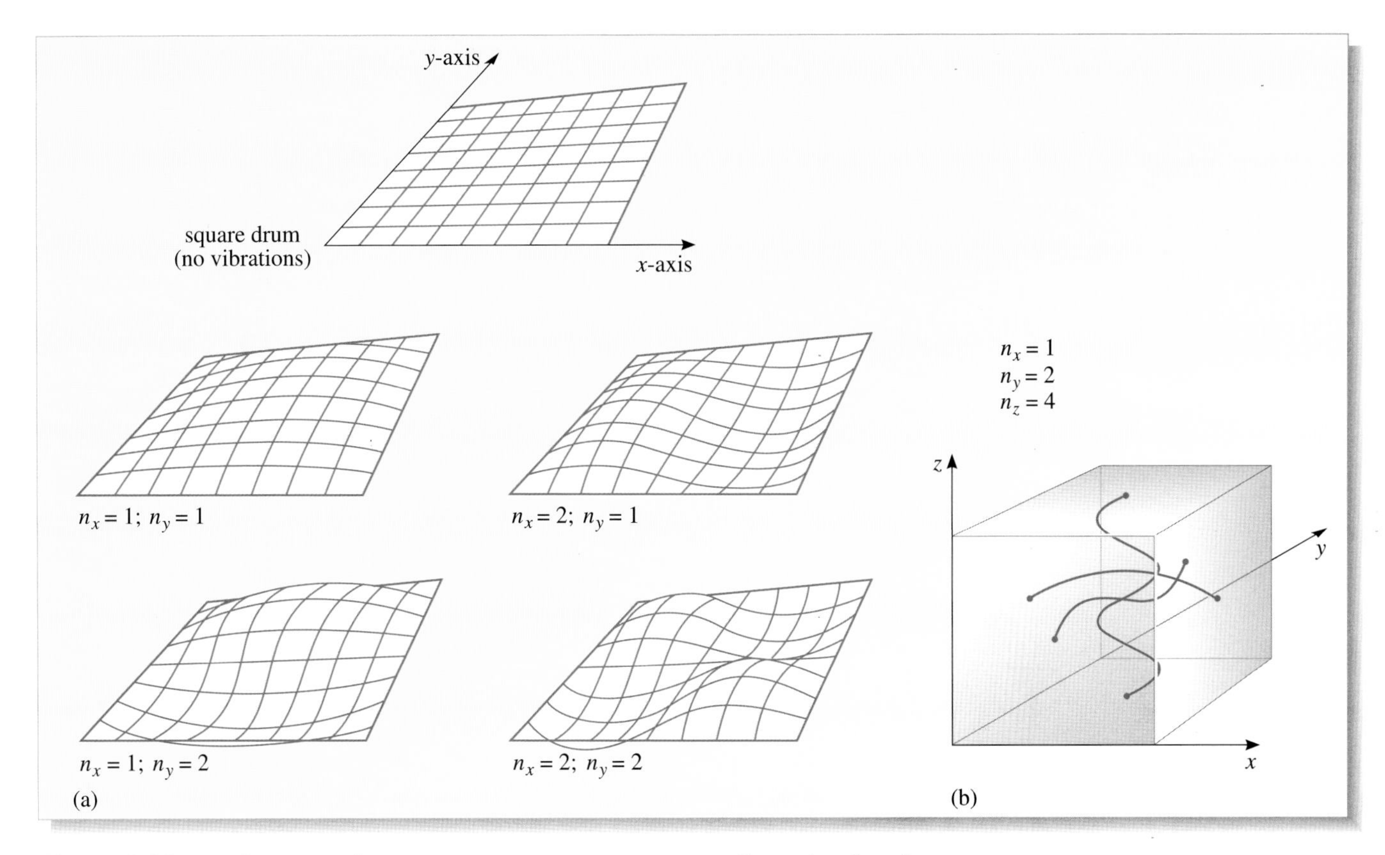

Figure 2.32 (a) Some standing wave patterns on a square, two-dimensional surface. The numbers n_x and n_y indicate the number of half-wavelengths in the x- and y-directions, respectively. (b) Standing waves can exist when waves are confined to the shape of a cube in three dimensions.

This idea can be extended to *three* dimensions (Figure 2.32b), in which case *three* independent integers now specify the allowed standing waves. When we come to solve the Schrödinger equation for a particle confined in an infinite square well in two dimensions or three dimensions, we find the equation must be solved for each direction separately, satisfying the condition that $\psi = 0$ on all of the boundaries. So, in the same way as we found for the one-dimensional case, we find that the allowed wavefunctions have wavelengths of only certain, definite values (by analogy with the waves shown in Figure 2.32b). Thus a particle confined in a hollow cubical box will have only certain, definite values of energy, i.e. energy levels. What are these values of energy?

You saw in Section 4.5 that the energy levels of a particle confined in one dimension are given by

$$E_{\text{tot}} = \frac{n^2h^2}{8mD^2}, \quad \text{where } n = 1, 2, 3, \text{etc.} \qquad \text{(Eqn 2.18)}$$

The corresponding formula for the energy levels of a particle confined in *three* dimensions (in a hollow cubical box) has a similar form:

$$E_{\text{tot}} = \frac{h^2}{8mD^2}(n_1^2 + n_2^2 + n_3^2); \quad \begin{cases} n_1 = 1 \text{ or } 2 \text{ or } 3, \text{etc.} \\ n_2 = 1 \text{ or } 2 \text{ or } 3, \text{etc.} \\ n_3 = 1 \text{ or } 2 \text{ or } 3, \text{etc.} \end{cases} \qquad (2.23)$$

The integers n_1, n_2 and n_3 (equivalent to n_x, n_y and n_z) specify the number of half-wavelengths that are fitted into each of the three directions parallel to the edges of the box.

An intriguing new effect now comes into play. The particle confined in three dimensions can have the same total energy even if it is described by *different* sets of integers n_1, n_2 and n_3. For example, the total energy is the same when $n_1 = 3$, $n_2 = 4$, $n_3 = 10$ as it is for $n_1 = 4$, $n_2 = 3$, $n_3 = 10$, and indeed for $n_1 = 8$, $n_2 = 6$ and $n_3 = 5$, as you can check for yourself using Equation 2.23.

This phenomenon is known as **degeneracy**. When the same energy value arises from different sets of integers n_1, n_2, n_3, the different combinations are said to be *degenerate* in energy. Each different set of quantum numbers n_1, n_2, n_3, specifies a particular **quantum state** of the system. Thus a particular energy level will be degenerate if it corresponds to more than one quantum state. It is important to be aware of this distinction between *quantum states* and *energy levels*. The concept of degeneracy is very important, and although we shall not pursue it further at present, you will see in Chapter 3 that it plays a crucial role in interpreting the allowed energy levels of the electron in the hydrogen atom.

While discussing the three-dimensional case, it is useful to extend another idea that you have met already in Section 4.6. In the one-dimensional example of the particle confined between two walls, the probability of finding the particle in a region of width Δx was given by

$$P = |\psi(x)|^2 \Delta x \qquad \text{(Eqn 2.21b)}$$

assuming that the normalization process has been carried out. This idea can be extended naturally to three dimensions, so that the probability of finding a particle represented by a three-dimensional standing wave described by the time-independent wavefunction $\psi(x, y, z)$ in a small volume element ΔV is similarly given by

$$P = |\psi(x, y, z)|^2 \Delta V. \qquad (2.24)$$

Again, it is necessary to normalize ψ so that the total probability of detecting the particle somewhere in the three-dimensional volume in which it is confined is equal to one.

Example 2.1

A particle of mass $m = 1.6 \times 10^{-29}$ kg and total energy $E_{tot} = 2.3$ eV has the potential energy function shown in Figure 2.33; this consists of a square potential energy barrier of height $W = 4.4$ eV and width $L = 3.0 \times 10^{-10}$ m. Schrödinger's equation for this particle takes the form

$$\frac{d^2\psi}{dx^2} + \frac{2m}{\hbar^2}(E_{tot} - E_{pot}(x))\psi = 0,$$

where $E_{pot}(x)$ is a suitable potential energy function. In the region from $x = 0$ to $x = L$, the wavefunction describing the particle can be written in the form

$$\psi(x) = Ae^{-\kappa x},$$

where A and κ are constants.

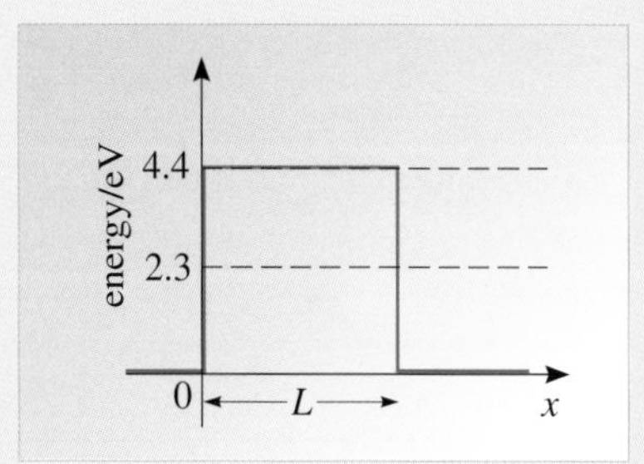

Figure 2.33 For Example 2.1.

(a) By differentiation, show that this wavefunction does satisfy Schrödinger's equation in the region from $x = 0$ to $x = L$. What is the value of the constant κ?

(b) Suppose that the probability of finding the particle in a small region near $x = 0$ is 0.2. What is the probability of finding the particle in a similar region near $x = L$?

Solution

In this kind of question, the first step is to see whether the time-independent wavefunction satisfies the time-independent Schrödinger equation. This will require us to differentiate ψ twice.

Preparation (a) We know that within the range $0 < x < L$, the potential energy has a constant value $W = 4.4\,\text{eV}$. So Schrödinger's equation becomes

$$\frac{d^2\psi}{dx^2} - \frac{2m}{\hbar^2}(W - E_{tot})\psi = 0. \qquad (2.25)$$

In this region, we are given

$$\psi(x) = Ae^{-\kappa x}.$$

To differentiate this, we will need the standard result

$$\frac{d(e^{-\kappa x})}{dx} = -\kappa e^{-\kappa x}.$$

(b) For this part of the problem, we need to find the probability P.

$$P \propto |\psi|^2 \Delta x.$$

We do not know whether the wavefunction is normalized or not, so we should stick to a proportionality, rather than an equality, here.

Working (a) Differentiating the suggested wavefunction twice:

$$\frac{d\psi}{dx} = -A\kappa e^{-\kappa x} \quad \text{and} \quad \frac{d^2\psi}{dx^2} = A\kappa^2 e^{-\kappa x}.$$

Substitution into Equation 2.25 gives

$$A\kappa^2 e^{-\kappa x} - \frac{2m}{\hbar^2}(W - E_{tot})Ae^{-\kappa x} = 0.$$

So the given wavefunction can satisfy Schrödinger's equation with a value of κ such that

$$\kappa = \frac{\sqrt{2m(W - E_{tot})}}{\hbar}$$

$$= \frac{\sqrt{[2 \times 1.6 \times 10^{-29} \times (4.4 - 2.3) \times 1.6 \times 10^{-19}]\,\text{kg J}}}{1.06 \times 10^{-34}\,\text{J s}}$$

$$= 3.1 \times 10^{10}\,\text{m}^{-1}.$$

W and E_{tot} are quoted in eV, so have to be converted into joules. Convince yourself that $\sqrt{(1\,\text{kg} \times 1\,\text{J})}/(1\,\text{J s})$ does indeed equal $1\,\text{m}^{-1}$.

(b) The ratio of the probability P_2 of finding the particle at $x = L$ to the probability P_1 of finding it at $x = 0$ is

$$\frac{P_2}{P_1} = \left| \frac{Ae^{-\kappa L}\,\Delta x}{Ae^{-\kappa 0}\,\Delta x} \right|^2.$$

Notice that the factor Δx cancels, because we are dealing with equally sized regions in the two cases. The factor A also cancels, so it doesn't matter whether the wavefunction is normalized or not.

But $e^0 = 1$, so at $L = 3.0 \times 10^{-10}$ m,

$$\frac{P_2}{P_1} = \left| \frac{\exp(-3.1 \times 10^{10} \times 3.0 \times 10^{-10})}{1} \right|^2$$
$$= |e^{-9.3}|^2 \approx 8.4 \times 10^{-9}.$$

Since $P_1 = 0.2$,

$$P_2 \approx 0.2 \times 8.4 \times 10^{-9} \approx 1.7 \times 10^{-9}.$$

Checking The units provide a good check on part (a): κ has come out in reciprocal metres, as expected.

If κL were smaller, there would be a greater chance of tunnelling through the barrier. According to our equations, this could be achieved if m were smaller or if $(W - E_{tot})$ were smaller. (It seems reasonable that the tunnelling should increase for lighter particles that are closer to the top of the barrier. Such particles would have longer de Broglie wavelengths and display more obvious quantum-mechanical effects.) Alternatively, tunnelling is increased if L is made smaller; it is easier to tunnel through a narrow barrier.

4.10 The correspondence principle

In the previous subsections, you saw that the effect of confining a particle is to limit its allowed energy to certain discrete values. Why, then, don't we see evidence for such quantization in the everyday world?

The classical and quantum-mechanical viewpoints can be reconciled by remembering that, for everyday objects, the energies involved are huge in comparison with $h^2/(8mD^2)$ (Equation 2.18). This implies that the quantum numbers, n, that specify these energies are correspondingly huge. In such cases, there is a very large number of half-wavelengths in the standing probability wave that describes a particle confined in one dimension between two reflecting walls (Figure 2.34a). The probability of detecting the particle in a small region between the walls is then *independent* of the region's location (Figure 2.34b). This agrees with expectations based on classical physics — if an experiment to find the position of the confined particle is performed at random, it is found that the particle is equally likely to be detected at any position between the walls. The conclusion is that when the number n characterizing the particle's probability wave (and energy level) is very *large*, the quantum-mechanical description of the particle becomes almost identical to its classical description.

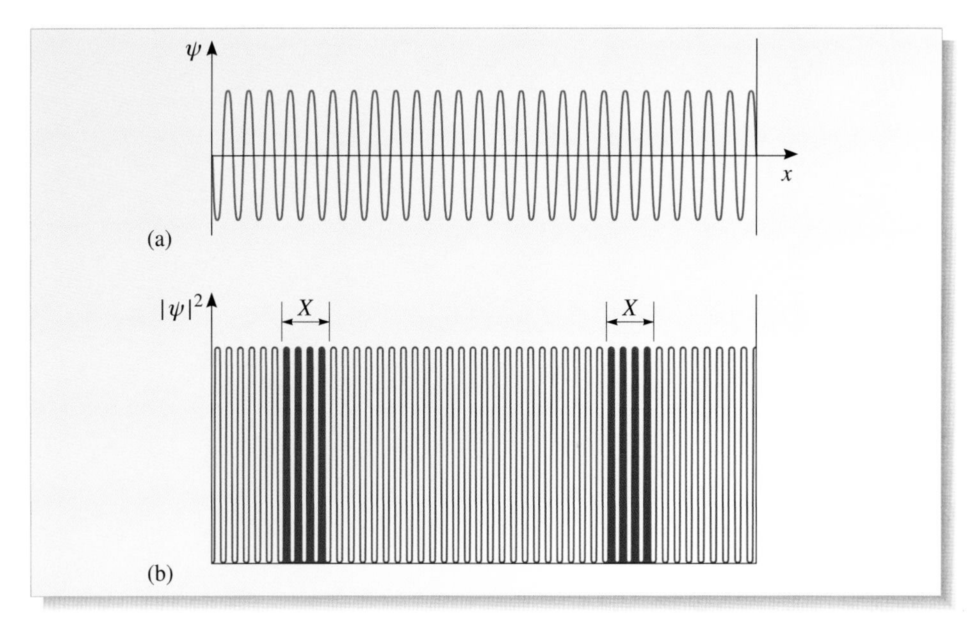

Figure 2.34 When the behaviour of a particle confined in one dimension (Figure 2.18) is described by a probability wave characterized by a *high value* of n (Equation 2.18), the likelihood of detecting the particle in a region of a given size X is approximately independent of the location of the region between the walls. As n tends to infinity, the classical limit is approached, i.e. the probability of detecting the particle in a region of a given size is independent of the region's location. This illustrates the correspondence principle.

Now try Question 2.13, in order to examine some more predictions that quantum mechanics makes about macroscopic objects.

Question 2.13 Calculate the *minimum* kinetic energy that a small glass bead of mass 1×10^{-3} kg can have when it is confined between two rigid walls that are 0.1 m apart (Figure 2.18). Assuming that we can interpret this energy as the kinetic energy of the bead as it moves with a speed v, calculate this speed and hence the time the bead would take to cross the box. ■

The speed you have calculated in your answer to Question 2.13 is infinitesimally small, and this means that a macroscopic object, such as a glass bead, is effectively at rest when it occupies its lowest energy level.

Question 2.14 Suppose that the glass bead in Question 2.13 is travelling at a speed such that it travels between the walls in 0.5 s. What is its total energy? Assuming it to be in a stationary state with this energy, to what value of n does this energy correspond? Comment on your result. (Assume that $E_{\text{pot}} = 0$ between the walls.) ■

Note *You should read the answer to Question 2.14 carefully before you move on.*

Your answers to Questions 2.13 and 2.14 should have convinced you that the discrete effects predicted by quantum mechanics (e.g. the energy levels of a *confined* particle) are unobservable when they are applied to everyday problems — quantum effects can safely be neglected in such situations. To put it more broadly, *the predictions of quantum mechanics become practically identical with those of classical mechanics when they are applied to problems about the macroscopic world.* This is analogous to the way in which the predictions of special relativity become practically identical to the corresponding predictions of non-relativistic mechanics at speeds much less than the speed of light (*Dynamic fields and waves* Chapter 4).

Quantum mechanics is believed to provide a more general and correct description of matter than classical mechanics, but it is nevertheless true that quantum mechanics contains the more familiar ideas of Newtonian (classical) mechanics within its framework. In Questions 2.13 and 2.14, you have encountered two one-dimensional examples (the motion of a particle and of a glass bead between the reflecting walls) in which the quantum-mechanical description of the system is indistinguishable from the corresponding classical predictions. The idea that quantum mechanics must have a classical limit is embodied in the so-called **correspondence principle**. In fact, as shown in Question 2.14, the classical limit is equivalent to the limit in which the number n in Equation 2.18 becomes very large.

The correspondence principle

In the limit of large quantum numbers, known as the classical limit, the predictions of quantum mechanics are in agreement with those of classical mechanics.

This is a profoundly important principle. It says, in effect, that in the limit of large quantum numbers, the predictions of quantum mechanics must be indistinguishable from those of classical (Newtonian) mechanics. We should expect this to be the case since classical mechanics is an extremely successful theory when applied to matter on the large scale.

Open University students should leave the text at this point and study the multimedia package *Stepping through Schrödinger's equation* that accompanies this book. When you have completed the package (which should take about 2 hours), you should return to this text.

4.11 Summary of Section 4

- In this section, you have had your first acquaintance with one of the most important equations in physics — the Schrödinger equation. The solutions to Schrödinger's equation are called wavefunctions and are usually denoted by Ψ. They are of central importance in Schrödinger's wave mechanical approach to quantum mechanics.
- The full version of Schrödinger's equation was not written down, but you were introduced to the time-independent Schrödinger equation in one dimension. The solutions of this equation are called time-independent wavefunctions (denoted $\psi(x)$) and they correspond to standing wave solutions of Schrödinger's equation that describe the so-called stationary states.
- In order to find the solutions of the time-independent Schrödinger equation, it is necessary to know the potential energy function that describes the problem in hand.
- You have seen the solutions to the time-independent Schrödinger equation for the case of particles confined in an infinitely deep square potential well. You have also seen, qualitatively, the form of the solutions in a finite square potential well. Each of these solutions describes a particular stationary state of the system, and each one has an associated energy.
- For free and unbound particles, any value of total energy is allowed.
- For confined particles, the energy can only take certain discrete values, that is, a confined particle has energy levels.
- The wavefunctions and energy levels for a particle in one dimension are summarized in Table 2.1 (Section 4.8).

- For particles confined in two or three dimensions most of the energy levels are degenerate, that is, there is more than one quantum state (wavefunction) for a given energy level.

Question 2.15 In a crude model of the hydrogen atom, it is assumed that the electron in the atom is confined in a hollow, cubical box whose sides each have a length of 3×10^{-10} m.

(a) Using Equation 2.23, write down a formula that gives the energy levels of the electron in the hydrogen atom according to this model.

(b) What is the spacing of the two lowest energy levels of the electron according to this model? How well does this prediction of the model compare with the experimentally measured spacing of 1.6×10^{-18} J?

(c) Give at least one reason why this model of the hydrogen atom is inadequate.

Question 2.16 A particle of mass m is confined by infinitely high walls so that it moves in two dimensions within a square whose sides have length D. By referring to the form of the equations for the allowed energy levels of standing waves in one dimension and three dimensions (Equations 2.18 and 2.23), write down the expression for the energy levels of this confined particle. How many degenerate states are there for each of the three lowest energy levels? ■

5 Closing items

5.1 Chapter summary

1. In order to interpret the behaviour of particles such as electrons, it is necessary to apply both wave and particle models. The two different models are required to explain different aspects of their behaviour.
2. In experiments involving just a single electron, it is impossible to predict the outcome of that experiment, only the probability of the various possible outcomes can be predicted.
3. The probability P of detecting a particle at a particular place is proportional to the square of the amplitude A of the particle's de Broglie wave at that place, $P \propto A^2$.
4. Quantum mechanics asserts that the behaviour of matter can be modelled using probability waves. A localized particle can be modelled as a wavepacket which can be produced by summing many infinitely long waves with a range of wavelengths and amplitudes.
5. The range of wavelengths required to build up a localized particle implies an uncertainty in our knowledge of the particle's momentum. The more tightly localized a particle is, the greater is this uncertainty. This idea leads to Heisenberg's uncertainty principle, which has two formulations:
$$\Delta x \, \Delta p_x \geq \hbar/2 \quad \text{and} \quad \Delta E \, \Delta t \geq \hbar/2. \qquad (2.7, 2.8)$$
6. According to Schrödinger's wave-mechanical approach to quantum mechanics, the information describing the behaviour of a particle is contained in its wavefunction Ψ, which is the solution to the time-dependent Schrödinger equation for that particle.

7 The solution to Schrödinger's equation for a free particle involves travelling waves characterized by an angular wavenumber k that may have any positive value. The relationship between k and the particle's kinetic energy is

$$k = \frac{\sqrt{2mE_{\text{kin}}}}{\hbar} \tag{2.9}$$

or, rearranging

$$E_{\text{kin}} = \frac{\hbar^2 k^2}{2m}.$$

So, for a free particle, any positive value of the kinetic energy (and hence total energy) is allowed.

8 For a particle of mass m in a stationary state, described by a wavefunction of the form $\Psi(x, t) = \psi(x)\phi(t)$, the time-independent wavefunction $\psi(x)$ will satisfy the time-independent Schrödinger equation

$$\frac{\mathrm{d}^2\psi}{\mathrm{d}x^2} + \frac{2m}{\hbar^2}(E_{\text{tot}} - E_{\text{pot}}(x))\psi = 0 \tag{2.12}$$

where E_{pot} is the potential energy function of the particle.

9 For a particle in a one-dimensional infinite square well, the time-independent wavefunctions take the form $\psi(x) = \psi_0 \sin kx$ and $\psi(x) = \psi_0 \cos kx$, but only certain values of k are allowed, and therefore only certain discrete values of the energy are allowed. That is, the particle's energy is quantized:

$$E_{\text{tot}} = \frac{n^2h^2}{8mD^2}, \quad \text{where } n = 1, 2, 3, \text{etc.} \tag{2.18}$$

10 In a finite one-dimensional square well the particle's energy is still quantized but the wavefunction penetrates to some extent into the classically forbidden region (outside the well where $E_{\text{pot}} > E_{\text{tot}}$). This leads to the phenomenon of barrier penetration.

11 The probability P of finding a confined particle in a small region Δx at position x in a one-dimensional well is $P = |\psi(x)|^2 \Delta x$, if the time-independent wavefunction ψ is normalized. More generally, for particles described by a normalized time-dependent wavefunction $\Psi(x, t)$, the probability is

$$P(x, t) = |\Psi(x, t)|^2 \Delta x.$$

12 For a particle confined in three dimensions the energy levels are given by

$$E_{\text{tot}} = \frac{h^2}{8mD^2}(n_1^2 + n_2^2 + n_3^2), \tag{2.23}$$

where n_1, n_2 and n_3 are positive integers. This leads to the phenomenon of degeneracy, where different combinations of the three quantum numbers n_1, n_2 and n_3 can lead to the same value for the energy, E_{tot}.

13 The correspondence principle states that, for large quantum numbers, the predictions of quantum mechanics must be in agreement with those of classical mechanics.

5.2 Achievements

After studying this chapter, you should be able to:

A1 Explain the meaning of all the newly defined (emboldened) terms introduced in this chapter.

A2 Describe and interpret experiments that provide evidence for the wave model of electrons.

A3 Recall and apply the de Broglie formula, $\lambda_{dB} = h/p$.

A4 State two forms of Heisenberg's uncertainty principle, $\Delta x \, \Delta p_x \geq \hbar/2$ and $\Delta E \, \Delta t \geq \hbar/2$, and use them to discuss simple problems involving measurements of the variables, x, p_x, E and t.

A5 Discuss briefly what are meant by free or unbound and bound or confined particles, and describe qualitatively the difference between the wavefunctions of particles in these conditions.

A6 Describe how the Schrödinger equation is used to study the behaviour of particles. Sketch the time-independent wavefunctions that are solutions to Schrödinger's time-independent equation for some simple cases.

A7 Relate the probability P of detecting a particle in a region to the value of its wavefunction in the region, e.g. $P \propto |\psi|^2 \Delta x$.

A8 Apply the formulae for the energy levels of a particle confined in one dimension between two infinitely high reflecting walls and confined in three dimensions in a hollow cubical box.

A9 Understand the implications of the correspondence principle and confirm it for simple cases.

5.3 End-of-chapter questions

Question 2.17 A certain free particle of mass 1.00×10^{-30} kg has a de Broglie wavelength of 7.00×10^{-11} m. Calculate the kinetic energy of the particle.

Question 2.18 After the radioactive decay of a radium nucleus, the ejected α-particle is observed to have a kinetic energy of 6.00 MeV. What is the de Broglie wavelength of the α-particle?

Question 2.19 A particle confined in a potential well is excited up to the $n = 2$ energy level. The particle remains in this energy level for an average of 1×10^{-9} s before making the transition back down to the ground state ($n = 1$). Calculate the uncertainty in the energy of the $n = 2$ level. (Take $\hbar = 1.1 \times 10^{-34}$ J s.)

Question 2.20 A particle of mass 1×10^{-10} kg is confined in one dimension between two infinitely high, rigid walls that are separated by 1×10^{-10} m. The total energy of the particle is 2.69×10^{-36} J.

(a) What is the quantum number that characterizes the wavefunction of the particle when it occupies this energy level?

(b) Sketch a graph that shows the relative probabilities of detecting the particle in different regions between the walls.

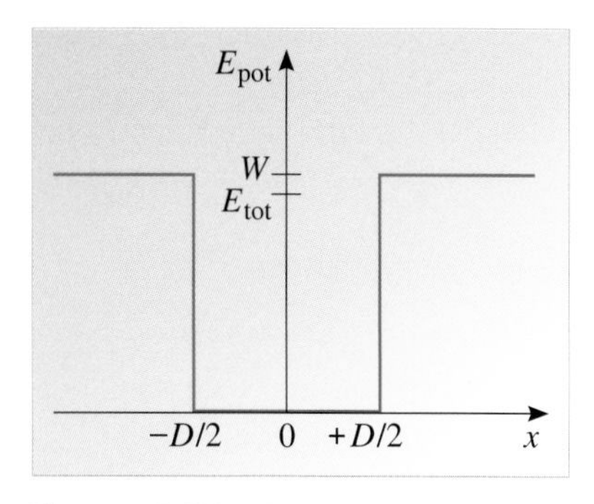

Figure 2.35 For use with Question 2.21.

Question 2.21 A particle is confined in a one-dimensional finite square well such as that shown in Figure 2.35. The depth of the well is W and the total energy of the particle is E_{tot}. Show that $\psi = Ae^{-\kappa x}$ is a solution to the time-independent Schrödinger equation for the particle in the region $x > D/2$.

Question 2.22 A particle of mass m is confined in a hollow cubical three-dimensional box, each edge of which has a length D, and for which the potential energy function is zero inside, and infinite outside, the box. How many quantum states are there with an energy of $19h^2/4mD^2$? ■

Chapter 3 Quantum mechanics in atoms

1 Lasers — a modern tool

Lasers are a relatively recent invention. Microwave lasers were in existence in the 1950s and the first visible-light laser appeared in 1960. In the early 1960s some people were of the opinion that lasers were just an interesting curiosity but with no significant use. Nowadays we find lasers in all sorts of everyday situations: CD players, rock concert light shows, bar code scanners, etc. They are also extensively used in technological applications such as eye surgery and welding: modern car plants use robot-controlled lasers to weld sheets of metal together (Figure 3.1) and they are used in experiments at the forefront of research in spectroscopy (Figure 3.2).

Lasers have found these many applications because of their particular properties, which are useful in a variety of situations. A laser beam is intense, monochromatic (has a very small range of wavelengths), narrow and well directed.

How do these features arise? Laser light originates in the transitions made by electrons in atoms from one energy level to a lower one. Its properties depend on many atoms being stimulated to make the same transition at the same time. So, laser operation depends crucially on the electronic structure of atoms as well as their interaction with electromagnetic radiation. In this chapter, we hope to show you how quantum mechanics can help us to understand atomic structure.

Figure 3.1 Cars built by robot-controlled lasers.

Figure 3.2 An experiment at the Open University, which uses infrared diode lasers to cool rubidium atoms to temperatures of 50 μK.

Quantum mechanics is a challenging, wonderful theory. After Chapter 2 you may nevertheless feel ill at ease with it! Probably this unease arises because quantum mechanics seems to be alien — far removed from intuitive notions of 'reasonable behaviour'. Discomfort of this kind is common in modern physics. However, you will see that with the power of reason and the evidence of experiments, science can transcend the limitations of intuition. For example, you may have begun this course with opinions about motion that were more Aristotelian than Newtonian. With effort and practice at using Newton's ideas, you may now see things more comfortably from a Newtonian perspective, but there will always be the risk that, at an unguarded moment, you may lapse into an Aristotelian mode of thought.

Despite the unease it engenders in the novice and expert alike, quantum mechanics *is* founded in experiment and it is *extraordinarily* successful. An understanding of many exciting technologies, such as the laser, relies upon quantum mechanics.

In Chapter 2 we laid the groundwork for a detailed study of the quantum world, that is, the microscopic world of things on the atomic scale where classical mechanics is no longer adequate. In this chapter, we apply these basic, quantum-mechanical ideas to atoms, starting with the hydrogen atom. Hydrogen is, of course, a very simple atom with just one electron and you would be forgiven for thinking that an appreciation of its electron energy levels cannot assist in an understanding of many-electron atoms. We hope to convince you otherwise.

We introduce some profound new ideas in this chapter, for example, *electron spin* (Section 4), and the *Pauli exclusion principle* (Section 5), which have no classical counterpart. Alongside this, in Section 5 we shall be able to harness these remarkable ideas, together with the discussions of Section 2, to provide an account of the electronic structure of heavy atoms (their energy levels) and hence the Periodic Table of the chemical elements.

The atomic electron transitions that give rise to emission spectra are considered in Section 6. There we will see that the quantum mechanics of atomic spectra constitutes the essential physics of the operation of lasers.

2 Hydrogen — the simplest atom

2.1 Introduction

The hydrogen atom is a relatively simple quantum system consisting of a single electron bound by the electrostatic force to a nucleus (a single proton in this case) and thus provides an excellent test case to which Schrödinger's equation can be applied. We will not, in this chapter, solve Schrödinger's equation for the hydrogen atom analytically — that is beyond the scope of this course, but the quantum mechanics you have studied in Chapter 2 provides you with all the information you need to understand the qualitative aspects of the hydrogen wavefunctions.

In Chapter 1, you saw that Bohr's model gave a reasonable description of many of the observed properties of the hydrogen atom, even though the underlying assumptions of the theory were rather contrived and empirical. But Bohr's model had many shortcomings. It did not provide a satisfactory explanation for the stability of hydrogen atoms. Also, it said nothing about *which* transitions take place from one energy level to another, or about the probability with which such transitions take place. Clearly, a quantum-mechanical description of the hydrogen atom, if it is to supersede the Bohr model, must not only reproduce the successful predictions of that model, but it must also explain effects that were *not* accounted for by Bohr's ideas.

This section begins with a short discussion of how the interaction between the electron in the hydrogen atom and the nucleus (a proton) is incorporated in the Schrödinger equation. The electron's wavefunctions that are predicted by the equation are then described in outline, and their associated quantum numbers are described in some detail. Finally, you will see how quantum mechanics accounts for the observed line spectra of hydrogen, including the effect of an external magnetic field on the pattern of the lines.

2.2 Using the Schrödinger equation to study the electron in the hydrogen atom

The dominant force that affects an electron in a hydrogen atom is the electrostatic force attracting it to the proton which constitutes the nucleus of the atom. The hydrogen atom has no net electrical charge, since the charge of the proton is $+e$, and that of the electron is $-e$.

● What piece of information must be known about the electron in the hydrogen atom before its behaviour can be studied using the Schrödinger equation?

❍ It is the electron's *potential energy function* that must be known. ■

We know that the electrostatic force between the electron and the proton is described by Coulomb's law, and that the electron's electrostatic potential energy is given by

$$E_{\text{el}} = \frac{-e^2}{4\pi\varepsilon_0 r} \tag{3.1}$$

where r is the instantaneous distance of the electron from the proton.

Remember that the presence of the minus sign in Equation 3.1 implies that energy must be *transferred to* the confined electron in order to move it from the vicinity of the proton (i.e. the nucleus) to infinity where the potential energy is zero.

Figure 3.3 shows how the electron's potential energy E_{el} varies with its radial distance r from the nucleus. You can see that the shape of the potential energy curve could be said to resemble the shape of one wall of a deep well. For this reason, the electron is sometimes said to be in a Coulomb potential well.

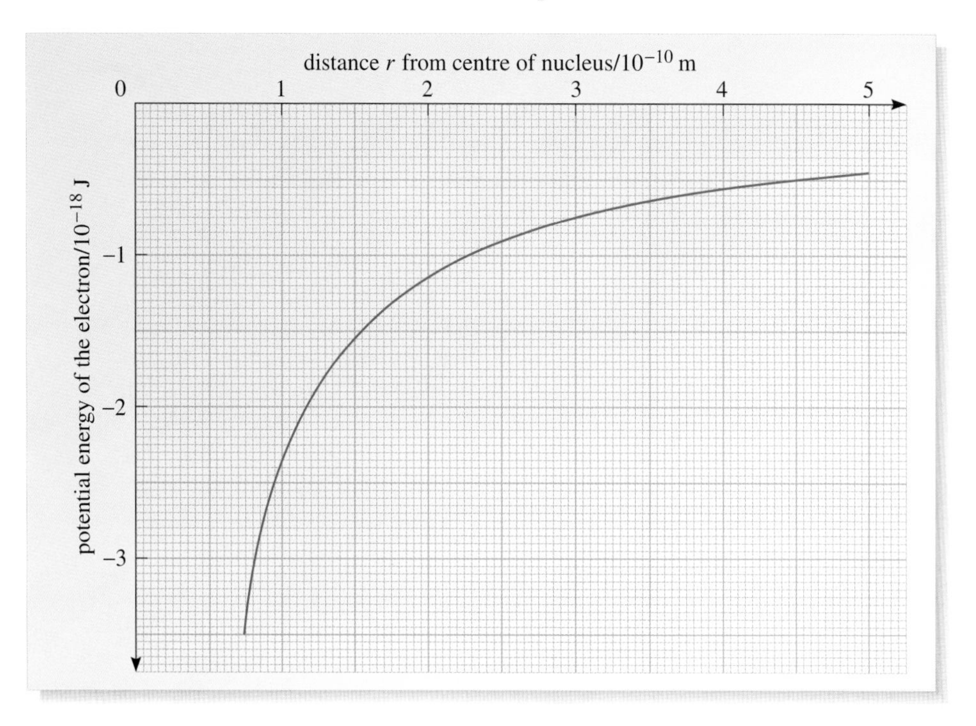

Figure 3.3 The potential energy function $E_{el} = -e^2/4\pi\varepsilon_0 r$ of the electron in a hydrogen atom.

The electron in the hydrogen atom is in a similar situation to the particle confined in a hollow, cubical box (Chapter 2, Section 4.9), although it is more difficult to envisage how the Coulomb potential well of Figure 3.3 has the same confining influence as the three-dimensional box. Nevertheless, it is reasonable to expect that the effect of the confinement will be the same in both cases — the wavefunctions that describe the confined particle should be localized standing waves and the confined particle should have discrete energy levels.

Equation 3.1 shows that the potential energy function of the electron in the hydrogen atom depends only on the variable r. If we want only to find the stationary state wavefunctions and the energy levels of the electron, we need to use the time-independent Schrödinger equation which was introduced in Chapter 2 (Equation 2.12). You may remember from Chapter 2, that when we came to solve the Schrödinger equation for a particle in a three-dimensional 'box', we solved it for each of the x-, y- and z-components separately, which gave us the separate quantum numbers n_1, n_2 and n_3 appropriate to each of the three independent directions. This was an appropriate approach in the case of a square well in three dimensions since the symmetry of the system makes x, y and z suitable coordinates to use. The symmetry of the hydrogen atom is different, however. Equation 3.1 shows that the potential energy function depends only on r, the distance from the nucleus. That is, the potential energy function has **spherical symmetry** and this means it is much more efficient to use a different coordinate system to describe the position of the electron in the hydrogen atom. We shall now pause to discuss this coordinate system.

Spherical polar coordinates

In Chapter 2 of *Describing motion*, we described how, in a simple (Cartesian) coordinate system, a point in space can be exactly specified by means of the x-, y-, z-coordinates that refer to three mutually perpendicular axes in space (Figure 3.4).

However, there is nothing special about this particular coordinate system. Many other systems can be devised, and some coordinate systems are particularly suitable for describing particular physical situations because of the symmetries that these situations possess. As you have already seen, the potential energy function of the electron in the hydrogen atom is *spherically* symmetric, and because of this symmetry, the Schrödinger equation is far easier to solve, and the solutions far simpler, if we use a **spherical polar coordinate system**.

Figure 3.5 shows that in a spherical polar coordinate system, the position of a point in three-dimensional space is specified by means of a length r and two angles θ and ϕ (see also Box 3.1). The coordinate r is simply the radial distance of the point in question from the origin. The angle θ is the angle between the position vector of the point and the z-axis. The angle ϕ is the angle between the direction of the positive x-axis and the direction of the projection of the position vector on to the xy-plane. This system has a built-in advantage for describing the electron in the hydrogen atom — the distance r, which determines the electron's potential energy function (Equation 3.1), is one of the three coordinates (r, θ, ϕ) that are used directly to specify the position of the electron.

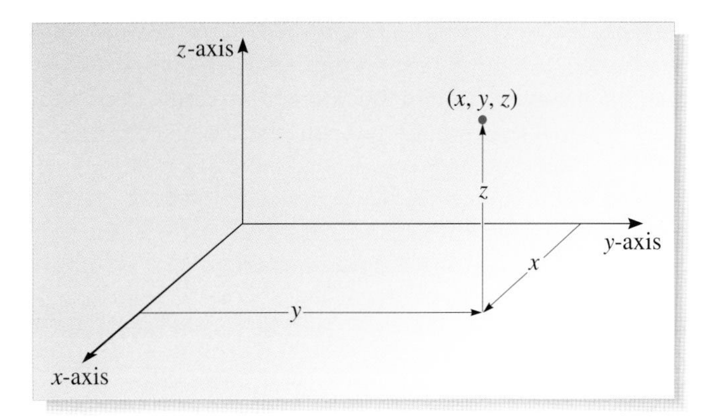

Figure 3.4 The position of a point in space can be specified relative to the origin by using a Cartesian coordinate system.

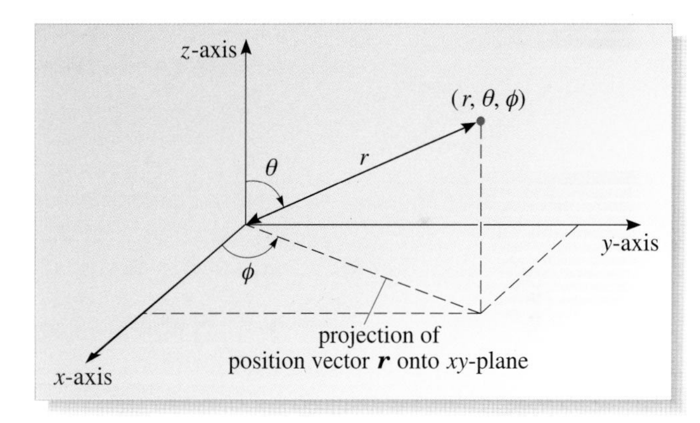

Figure 3.5 The position of a point in space can also be specified relative to the origin by using a spherical polar coordinate system.

Box 3.1 Latitude and longitude

You are probably more familiar with the use of spherical polar coordinates than you think. How do we specify the position of any location on the surface of the Earth? We give the latitude and longitude of that location. The latitude of a place is related to the angle θ in spherical polar coordinates, though in this case the zero of latitude is taken to be at the Equator rather than at the pole, so, latitude = $90° - \theta$. Longitude is just the angle ϕ, and in this case, the zero of longitude is taken to be the Greenwich meridian.

Now check that you understand how spherical polar coordinates are used by trying Question 3.1.

Question 3.1 By referring to Figure 3.5, write down the spherical polar coordinates for the four points whose Cartesian coordinates (x, y, z) are:

(a) (0, 0, 0) m; (b) (2, 0, 0) m; (c) (1, 1, 0) m; (d) (1, 1, 1) m. ■

Remember that *stationary states* are described by the time-independent, standing-wave wavefunctions that satisfy Schrödinger's equation for a confined particle.

When a spherical polar coordinate system is used to specify the instantaneous position of the electron in the hydrogen atom, there is a bonus. It turns out that each allowed stationary state wavefunction ψ of the electron, which, in principle, can depend on r, θ and ϕ, can be expressed as a *product* of three time-independent wavefunctions that each depend on only *one* of the coordinates r, θ or ϕ. In mathematical language, we can express this by writing

$$\psi(r, \theta, \phi) = \psi_1(r) \times \psi_2(\theta) \times \psi_3(\phi), \quad (3.2)$$

where $\psi_1(r)$ is a function only of r, and not of θ or ϕ; $\psi_2(\theta)$ is a function only of θ and not of r or ϕ; and $\psi_3(\phi)$ is a function only of ϕ and not of r or θ.

This means that, when solutions have been found for the wavefunctions corresponding to each variable, the total wavefunction ψ is simply the product of the three separate wavefunctions for r, θ and ϕ. We shall be using this property when we describe the distribution patterns for atomic electrons at the beginning of Section 3.

Although we are now using a different coordinate system, the basic procedure for solving Schrödinger's equation is exactly the same as before. We insert the relevant potential energy function (Equation 3.1) into the general time-independent Schrödinger equation, which we then solve to obtain the allowed wavefunctions and their associated energy levels. This is mathematically too complicated an operation to be carried out in this course, but in the next subsection we shall discuss the resulting predictions for the electron's energy levels. We can then go on to define the quantum numbers that specify each of the electron's possible wavefunctions. In Section 3, you will study the shapes of some of these wavefunctions.

2.3 The energy of the electron in the hydrogen atom

As you learnt in Chapter 2, the solutions of Schrödinger's equation give the stationary state wavefunctions for a confined particle and the allowed energies, E_{tot}, that are associated with those wavefunctions. Now, the electron in a hydrogen atom is in a potential well which, though a rather different shape, has some qualitative similarities with the finite square potential well discussed in Chapter 2. Bearing this in mind, consider first the behaviour of the electron when it has an energy, E_{tot}, greater than the value of the potential energy at the top of the well. In the case of the electron in the hydrogen atom, this means $E_{tot} > 0$, the electron is unbound (the atom is ionized) and can have any positive value of energy. In other words, a *continuum* of positive energy values is available to the electron. Secondly, for values of $E_{tot} < 0$, the electron is bound to the proton (thus forming the hydrogen atom) but only certain discrete values of energy are allowed, i.e. the energy is quantized. This occurs for essentially the same reason in the hydrogen atom as for the particle in the finite square well: the wavefunction is oscillatory within the well (where $E_{tot} > E_{pot}$) and decays away exponentially at larger values of r where $E_{tot} < E_{pot}$. The constraints on the wavefunction and the matching at the boundaries mean that only certain wavefunctions will 'fit' with the potential energy function and therefore only certain energies are allowed.

After solving the Schrödinger equation for the electron in the hydrogen atom, it is found that the energy levels of the confined electron are given by

$$E_{\text{tot}} = -\frac{1}{n^2}\left\{\frac{m_e e^4}{8h^2\varepsilon_0^2}\right\}$$

which gives

$$E_{\text{tot}} = -\frac{13.6}{n^2}\,\text{eV}, \tag{3.3}$$

where the number n can be equal to 1 or 2 or 3, etc. These energy levels are shown in Figure 3.6. You may well recognize Equation 3.3 — it is exactly the same as the Bohr model's prediction for the electron's energy levels, which you met in Chapter 1! Bohr obtained the result in a more straightforward manner, but, as you will soon see, a much more detailed description of the electron can be given using the quantum-mechanical approach.

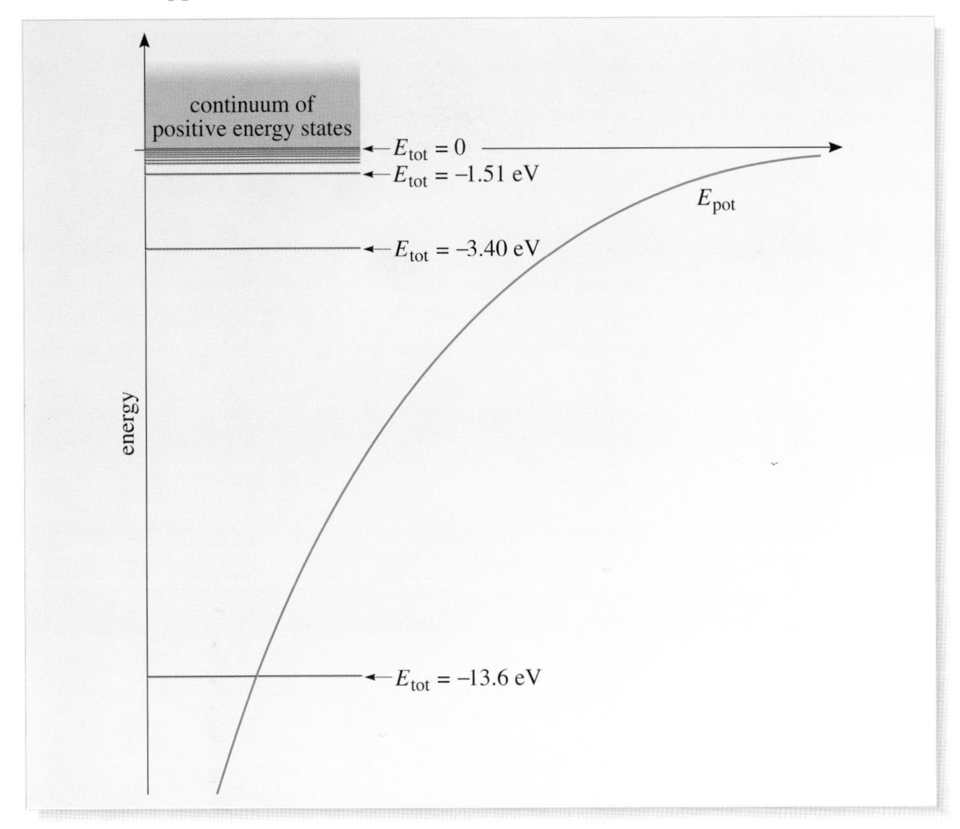

Figure 3.6 The possible values of energy of the electron in a hydrogen atom. When the electron is *confined* to the atom, it has quantized energy levels, but when it is *unbound*, it can have any positive value of total energy, i.e. it has a continuum of positive energy states.

It is not possible, within the scope of this course, to solve Schrödinger's equation for the electron in the hydrogen atom, or, even, to give the full algebraic expressions for the wavefunctions that are the stationary state solutions. We will, however, in Section 3, describe the shapes of the wavefunctions in qualitative terms and, in the next subsection we discuss the quantum numbers that emerge on solving Schrödinger's equation for the hydrogen atom.

2.4 The quantum numbers that specify the wavefunctions of the electron in the hydrogen atom

In Chapter 2, you saw that three *quantum numbers* (n_1, n_2 and n_3) are necessary to specify the wavefunctions of a particle confined in a hollow cubical box. It is not surprising, then, that three integers are also required to specify each wavefunction of the electron in the hydrogen atom. However, it turns out that these quantum numbers are not completely independent of each other, as was the case in the simple three-dimensional box.

The three quantum numbers of the electron in the hydrogen atom are designated by the letters n, l and m_l. (The reasons for this nomenclature will become clear shortly.) The various quantum numbers and their allowed values are, of course, fixed by the allowed solutions of Schrödinger's equation. It is worth emphasizing that in this regard there is an important difference between quantum mechanics and Bohr's theory. In the latter, the integer values for Bohr's quantum number n came from an arbitrary assumption about the quantization of the electron's angular momentum. *In quantum mechanics, on the other hand, the quantum numbers emerge naturally from the solutions to Schrödinger's equation.*

Let us now discuss the three quantum numbers for the electron in hydrogen, their possible values and how they are related to physical quantities such as the energy or angular momentum of the electron.

The principal quantum number

You have already met this quantum number, since it occurs in Equation 3.3, which is the Schrödinger equation's prediction for the energy levels of the electron confined in the hydrogen atom. This quantum number is therefore the one required to specify each of the energy levels of the confined electron. It is because the quantum number n is chiefly responsible for determining the electron's energy levels that it is called the **principal quantum number**.

Although the results obtained by Bohr and Schrödinger appear to be identical at first sight, you will see shortly that there are some important differences between them. In particular, *Bohr's assumption for the role of n turns out to be quite different from that predicted by quantum mechanics.* Whereas Bohr's quantum number n specified the magnitude of the electron's angular momentum, the quantum number *n cannot* be interpreted in this way according to the Schrödinger equation. You should think of the principal quantum number simply as the quantum number that determines the *energy* of the electron. (As you will see later, when the hydrogen atom is subject to a magnetic field, the energy of the electron also depends on the value of the m_l quantum number. However, the dependence of the energy on n is very much greater than its dependence on m_l.)

The orbital angular momentum quantum number

So far, it has emerged that the total energy of an atomic system is quantized. But what about its *angular momentum?* The confined electron is subject to an electrostatic force directed towards the nucleus of the atom, and, since forces in a radial direction (i.e. through the centre of mass) do not entail torques, it follows that the angular momentum of the electron should be constant. You should, therefore, not be too surprised by the fact that at least one of the quantum numbers of the electron in the hydrogen atom is associated with the electron's angular momentum.

The Schrödinger equation predicts that each of the electron's wavefunctions is characterized by a value of the so-called **orbital angular momentum quantum number**, l.

In terms of this number, the magnitude L of the electron's orbital angular momentum is given by

$$L = \sqrt{l(l+1)}\,\hbar \quad \text{where } l = 0 \text{ or } 1 \text{ or } 2 \text{ or } 3\text{, etc.} \tag{3.4}$$

Remember $\hbar = h/2\pi$.

The allowed values of the orbital angular momentum quantum number l are determined by the allowed solutions of the Schrödinger equation, and it turns out that l must be restricted to the following values:

$$l = 0, 1, 2, \ldots, n-1 \tag{3.5}$$

when the electron is described by a wavefunction with the principal quantum number n. In words, Equation 3.5 states that if the electron's wavefunction is characterized by the principal quantum number n, the possible values of the electron's orbital angular momentum quantum numbers are 0, 1, 2, up to a maximum of $n - 1$.

Equation 3.4 implies that the magnitude L of the electron's orbital angular momentum is *quantized.* In Question 3.2, you can determine the allowed values of angular momentum for a particular principal quantum number n.

Question 3.2 The electron in a hydrogen atom is described by a wavefunction that is characterized by the principal quantum number $n = 3$. What are the possible values of the magnitude L of the electron's angular momentum? ■

There is an intriguing result implicit in the allowed values for l given by Equation 3.5, which is that the magnitude L of the electron's angular momentum can be zero, when $l = 0$. *This is a new result, which contradicts the corresponding prediction of the Bohr model.* (You saw in Chapter 1 that the Bohr model predicts that the *minimum* value for the magnitude L of the electron's angular momentum is $h/2\pi$.)

Indeed, the result is all the more intriguing because intuitively one expects an *orbiting* electron to have some finite angular momentum. Once again, we can see the danger of adopting a simple *classical* picture of orbiting electrons.

Having defined l, we are now in a position to discuss a fundamental difference between Bohr's quantum number n, which was introduced in Chapter 1, and the quantum numbers n and l that characterize the *wavefunctions* of the electron in Schrödinger's theory. Can you see a contradiction between the roles of the principal quantum number n in Schrödinger's theory and in Bohr's theory? Well, Bohr assumed that the angular momentum of the electron in the hydrogen atom was quantized, $L = n\hbar$, and that the *same* quantum number n determined the electron's energy levels ($E_{tot} = -(13.6/n^2)$ eV). But the analysis based on the Schrödinger equation shows that the principal quantum number n that specifies the energy levels does *not*, in fact, determine the value of the angular momentum. It is remarkable that Bohr was able to make a false assumption about the electron's angular momentum and yet still derive correctly an equation for its energy levels!

The orbital magnetic quantum number

Schrödinger's equation not only predicts that the *magnitude* L of the electron's angular momentum is quantized, it also predicts that the *orientation* of the electron's angular momentum vector $\boldsymbol{L}$ is subject to certain restrictions: in other words, the direction of $\boldsymbol{L}$ is *quantized in space*. The electron's **orbital magnetic quantum number** m_l determines this directional quantization.

In order to describe the way that m_l governs the allowed orientations of $\boldsymbol{L}$, we must first choose a reference direction in space. Of course, so far as the hydrogen atom is concerned, there is no preferred direction in space, so the orientation of this chosen direction is completely arbitrary. However, once a direction has been chosen we can make it the z-axis of a system of coordinates. We will then find that no matter which direction was chosen as the z-direction, the z-component of the angular momentum vector $\boldsymbol{L}$ of the electron in the hydrogen atom is given by

$$L_z = m_l \hbar \qquad (3.6)$$

where m_l is the orbital magnetic quantum number. You will see *why* the quantum number has this name shortly.

The values that m_l can take are restricted: if the wavefunction of an electron in a hydrogen atom is characterized by the orbital angular momentum quantum number l, the possible values of the electron's orbital *magnetic* quantum number m_l are

$$m_l = -l, -l+1, -l+2, \dots, 0, \dots, l-2, l-1, l \qquad (3.7a)$$

or, more concisely,

$$m_l = 0, \pm 1, \pm 2, \dots, \pm l. \qquad (3.7b)$$

For example, if the orbital angular momentum quantum number of an electron in a hydrogen atom is $l = 3$, Equation 3.7a tells us that the possible values of m_l are -3, -2, -1, 0, $+1$, $+2$ and $+3$. Equation 3.6 then tells us that a measurement of the z-component, L_z, of the electron's angular momentum vector will give one of the results $-3\hbar$, $-2\hbar$, $-\hbar$, 0, $+\hbar$, $+2\hbar$ or $+3\hbar$. This example is illustrated diagrammatically in Figure 3.7. You can see from this picture that, in general, for a given value of l, the number of possible values of L_z (or equivalently the number of possible values of m_l) is given by $2l + 1$.

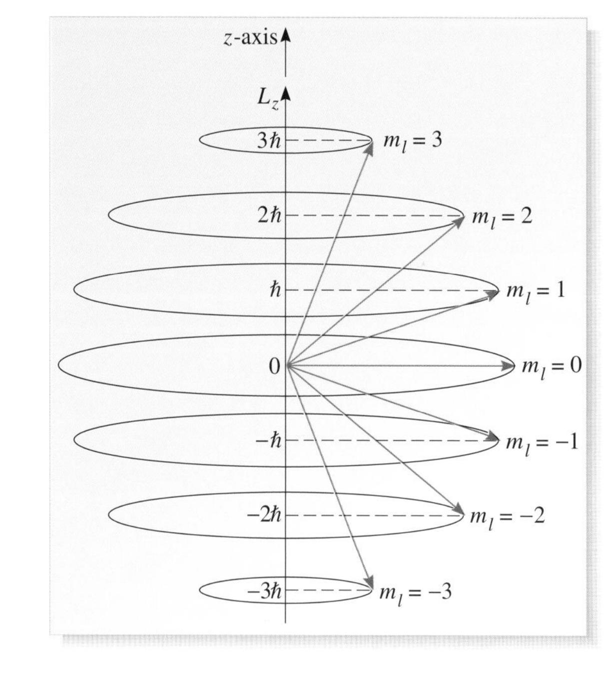

Figure 3.7 When the electron in a hydrogen atom has the orbital angular momentum quantum number $l = 3$, its z-component of angular momentum, L_z, can have the values $3\hbar$, $2\hbar$, $\hbar$, 0, $-\hbar$, $-2\hbar$ or $-3\hbar$. For clarity, all of the vectors have been shown in the plane of the page. However, for each value of m_l, the angular momentum vector can lie in any direction along the surface of a cone whose central axis is aligned along the z-axis.

Once again, the quantum nature of matter has emerged. The quantization of the z-components of angular momentum of the electron in the hydrogen atom is something quite strange — it has no counterpart in our observations of everyday objects.

Box 3.2 summarizes the most important points of the discussion so far, showing the allowed values of the quantum numbers n, l, and m_l that, according to the Schrödinger equation, are required to distinguish each wavefunction of the electron in the hydrogen atom, and indicating briefly the role that each quantum number plays in specifying the electron's motion in the hydrogen atom. When you have carefully examined Box 3.2, answer Questions 3.3 and 3.4.

Box 3.2 The allowed stationary state wavefunctions of the electron in the hydrogen atom and the quantum numbers that specify them (according to the Schrödinger equation)

1 Each time-independent wavefunction ψ of the confined electron may be written as a function of the spherical polar coordinates r, θ, ϕ:

$$\psi(r, \theta, \phi).$$

2 The Schrödinger equation leads to three separate equations for the dependence of ψ on each of the spherical polar coordinates r, θ and ϕ. The total wavefunction ψ can be obtained by multiplying together the solutions to each of these three equations:

$$\psi(r, \theta, \phi) = \psi_1(r) \times \psi_2(\theta) \times \psi_3(\phi). \qquad \text{(Eqn 3.2)}$$

3 Each time-independent wavefunction is characterized by a set of three integer quantum numbers n, l and m_l which obey the following restrictions:

name and symbol	allowed values	role of quantum number
principal quantum number, n	$n = 1, 2, 3, \ldots$	determines the energy levels of the confined electron (Eqn 3.3)
orbital angular momentum quantum number, l	$l = 0, 1, 2, \ldots, n - 1$ (Eqn 3.5)	determines the magnitude of the electron's orbital angular momentum (Eqn 3.4)
orbital magnetic quantum number, m_l	$m_l = 0, \pm 1, \pm 2, \ldots, \pm l$ (Eqn 3.7a)	determines the component of the electron's angular momentum vector with respect to an arbitrarily chosen z-axis (Eqn 3.6)

Question 3.3 (a) Write down general expressions for the magnitude L, and the z-component, L_z, of the orbital angular momentum of the electron in the hydrogen atom.

(b) If the electron in a hydrogen atom has an orbital angular momentum quantum number $l = 2$, what is the magnitude L of its angular momentum vector and what are the possible values of L_z?

(c) If the electron's orbital angular momentum quantum number is $l = 2$, draw a sketch showing the possible inclinations of the electron's angular momentum vector to the z-axis.

Question 3.4 Consider the magnitude L and the z-component, L_z, of the angular momentum of the electron in a hydrogen atom.

(a) By looking at the limitations imposed by the quantum numbers on L and L_z, show that the orbital angular momentum vector can never be aligned exactly parallel or antiparallel to the z-axis.

(b) By considering what happens to the allowed values for L and L_z for large values of the quantum numbers l and m_l, show that effectively all orientations of the angular momentum vector with respect to the z-axis become possible when l and m_l become large. ■

2.5 The effect of a magnetic field on the energy levels of the electron in a hydrogen atom

One point may have been worrying you during the discussion of the possible orientations of the angular momentum vector $\mathbf{L}$ of an electron in a hydrogen atom. Since the allowed directions of $\mathbf{L}$ are specified with respect to a z-axis that points in an *arbitrary* direction, why should the orientation of the electron's angular momentum vector matter at all? After all, the Coulomb electrostatic force on the electron is spherically symmetric.

Moreover, you have seen that the energy levels of the electron in the hydrogen atom are not affected by the orientation of the electron's orbital angular momentum. The equation

$$E_{\text{tot}} = -\frac{13.6}{n^2}\,\text{eV} \qquad \text{(Eqn 3.3)}$$

tells us that the energy levels depend only on the principal quantum number that characterizes the electron's wavefunction, *not* on the quantum numbers l and m_l. This implies that two or more wavefunctions characterized by *different* sets of quantum numbers may describe the electron when it has a particular value of energy. As you saw in Chapter 2, this phenomenon is called *degeneracy.* The degeneracy of the energy levels of the electron in the hydrogen atom reflects the fact that the electron's energy does not depend on the magnitude or direction of the electron's orbital angular momentum. This is a direct consequence of the form of the Coulomb electrostatic potential energy function to which the electron is subject. *Why, then, is it useful to specify the quantized directions of the angular momentum vector* $\mathbf{L}$ *of the electron in the hydrogen atom, using the orbital magnetic quantum number* m_l*?*

The relevance of the orbital magnetic quantum number m_l becomes clear when the electron in the hydrogen atom is in a situation where it *does* matter in which direction its angular momentum vector is pointing. The situation, for example, in which the hydrogen atom is subject to an external magnetic field. Let us investigate the effect of such a field on the electron in the atom.

First, we'll return to the subject of magnetic fields in general. The quantum theory of magnetic effects is very complex so we shall take a more classical (i.e. Bohr) point of view. In Chapter 4 of *Static fields and potentials* we described how a current flowing in a loop of wire produces a magnetic field. Furthermore, we stated (in the context of magnetic materials) that an orbiting electron in an atom can be thought of as a tiny current loop which similarly produces a magnetic field. So, it is reasonable to assume that

- the orbiting electron in the hydrogen atom will generate a magnetic field, and
- the direction of the magnetic field will be related to the direction of the electron's angular momentum.

It is possible to think of the internal magnetic field due to the electron's motion as an equivalent *atomic* 'bar magnet' of a certain size and field strength. It is the effect of an externally applied magnetic field on this tiny atomic magnet that we shall discuss here. We will start from a classical picture of a magnet, as described in Chapter 4 of *Static fields and potentials*.

A small bar magnet immersed in an external magnetic field (Figure 3.8) tends to align itself with the direction of the external magnetic field $\boldsymbol{B}_{\text{ext}}$. Energy must be transferred to the magnet in order to rotate it from the aligned position, so a magnet that is pointing at an angle different from the $\alpha = 0$ position is a source of stored magnetic potential energy.

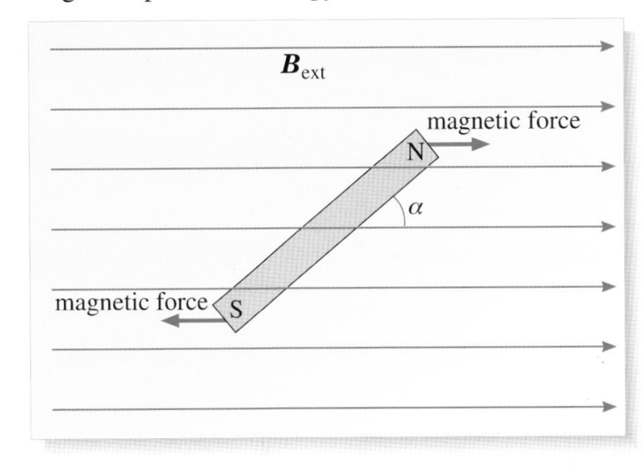

Figure 3.8 When a small bar magnet is in a uniform external magnetic field $\boldsymbol{B}_{\text{ext}}$, it tends to align itself with the direction of the magnetic field. This is why compass needles align themselves to point in the direction of the Earth's magnetic field.

So if a hydrogen atom is subject to a net external magnetic field $\boldsymbol{B}_{\text{ext}}$, its magnetic potential energy will depend on the orientation of the atomic bar magnet with respect to the field. However, the magnet's orientation is in turn determined by the direction of the electron's orbital angular momentum vector $\boldsymbol{L}$. *Hence the orientation of the electron's angular momentum vector will affect the electron's energy when it is subject to an external magnetic field.*

Using classical arguments, it can be shown that when an electron with orbital angular momentum $\boldsymbol{L}$ is subject to a net external magnetic field $\boldsymbol{B}_{\text{ext}}$, its magnetic potential energy is given by

$$E_{\text{mag}} = \left(\frac{e}{2m_{\text{e}}}\right)\boldsymbol{L}\cdot\boldsymbol{B}_{\text{ext}}$$

where $\boldsymbol{L}\cdot\boldsymbol{B}_{\text{ext}}$ represents the scalar product of the vectors $\boldsymbol{L}$ and $\boldsymbol{B}_{\text{ext}}$.

You have not met this equation before (and you are not required to remember it), but you should be able to manipulate it, using the standard rule for a scalar product, to get

$$E_{\text{mag}} = \left(\frac{e}{2m_{\text{e}}}\right)LB_{\text{ext}}\cos\theta \tag{3.8a}$$

where θ is the angle between the vectors $\boldsymbol{L}$ and $\boldsymbol{B}_{\text{ext}}$. This equation says that the magnetic potential energy of the electron will have its minimum (i.e. most negative) value when $\theta = 180°$, and its maximum value when $\theta = 0°$.

If we consider the magnetic field $\boldsymbol{B}_{\text{ext}}$ to be aligned along the z-direction, then $\boldsymbol{L}\cdot\boldsymbol{B}_{\text{ext}} = LB_{\text{ext}}\cos\theta = L_zB_{\text{ext}}$. Thus, we have

$$E_{\text{mag}} = \left(\frac{e}{2m_{\text{e}}}\right)L_zB_{\text{ext}}. \tag{3.8b}$$

Now, this result has been derived from an entirely classical description of the atom, but it has got us to an equation that includes the quantity L_z, the z-component of the electron's orbital angular momentum, the allowed values of which are determined by Schrödinger's equation. In other words, it is essential to have information about the allowed values of L_z in order to find the magnetic potential energy stored in a hydrogen atom that is subject to a magnetic field. And this is just the information supplied by the electron's orbital magnetic quantum number m_l. That is, $L_z = m_l\hbar$ (Equation 3.6). Substituting for L_z into the equation for E_{mag} therefore gives

$$E_{\text{mag}} = m_l\left(\frac{e\hbar}{2m_\text{e}}\right)B_{\text{ext}}. \tag{3.8c}$$

Equation 3.8c tells us that the *magnetic* potential energy of the electron in the hydrogen atom depends on the electron's orbital *magnetic* quantum number m_l. And of course that is exactly why m_l *is* called a *magnetic* quantum number — it determines the change in the electron's energy when the hydrogen atom is subjected to a magnetic field.

Figure 3.6 showed the energy levels of the electron in a hydrogen atom that was *not* subject to a magnetic field. As you know, the values of these energy levels depend only on the principal quantum numbers that characterize the electron's wavefunctions, *not* on the quantum numbers l and m_l. However, when the hydrogen atom *is* subject to a magnetic field, the energy levels will be different from those shown in Figure 3.6 — the new energy levels will depend on the orbital magnetic quantum numbers of the electron, as well as on the principal quantum number n.

This phenomenon was first observed experimentally by the Dutch physicist Pieter Zeeman in 1896. He found that when atoms were subject to a magnetic field, their spectral lines appeared to broaden very slightly. Later measurements with better resolution showed that the lines were not in fact broadened — the effect was due to there being several very closely spaced spectral lines around the spectral lines that were observed when the atoms were *not* subject to a magnetic field. This effect has come to be known as the *Zeeman effect*.

Question 3.5 Figure 3.9 shows two energy levels of an electron in a hydrogen atom that is *not* subject to a magnetic field. The wavefunctions that describe the electrons in these energy levels are characterized by the principal quantum numbers $n = 3$ and $n = 2$, and by the orbital angular momentum quantum numbers $l = 2$ and $l = 1$, respectively. Alongside the energy levels shown in Figure 3.9, sketch the separate, split energy levels that the electron would have if the hydrogen atom were subjected to an external magnetic field. Assume for the purposes of this question that it is only the electron's orbital angular momentum that gives rise to a magnetic interaction. (You will discover in Section 4 that this is not, in fact, the case, but you need not concern yourself with that here.) ■

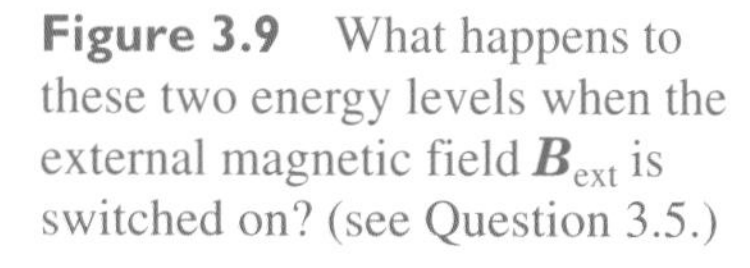

Figure 3.9 What happens to these two energy levels when the external magnetic field $\boldsymbol{B}_{\text{ext}}$ is switched on? (see Question 3.5.)

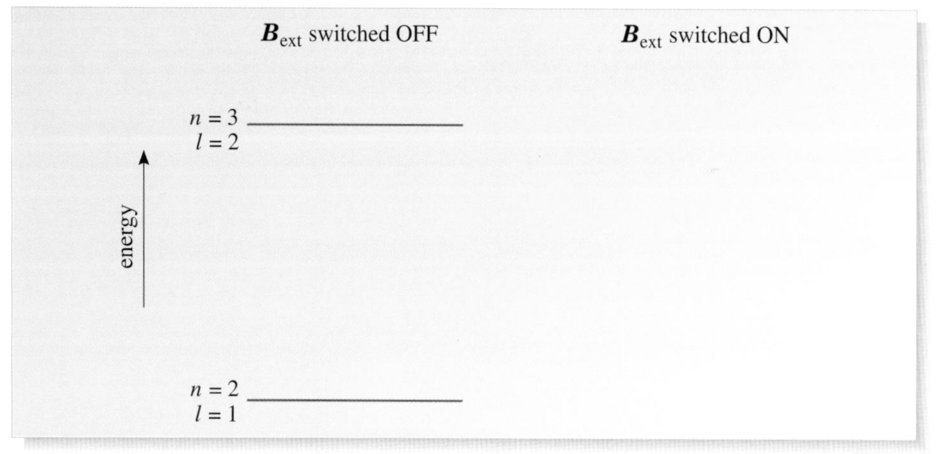

2.6 Spectroscopic notation

There is a standard **spectroscopic notation** that is universally used to characterize the wavefunctions (and energy levels) of electrons in atoms. You will find this notation very useful, not only for the hydrogen atom, but also for heavy atoms, which will be discussed later in the chapter. In essence, the notation is just a shorthand way of specifying the three quantum numbers that characterize the wavefunction of an electron in an atom.

We'll start with the single electron in the hydrogen atom. According to the Schrödinger equation, each of the electron's allowed stationary state wavefunctions is characterized by the three quantum numbers n, l and m_l. As you saw in Section 2.3, when the hydrogen atom is not subject to a net magnetic field, the energy levels of the electron are determined by the principal quantum number n (Equation 3.3), and the magnitude of the electron's angular momentum is determined by its orbital angular momentum quantum number l (Equation 3.4). In order to discuss the hydrogen atom when it is *not* in a magnetic field, it would therefore be useful to have a notation that succinctly expresses the values of n and l that characterize the electron's wavefunction.

The spectroscopic notation that is used today has links with the early days of spectroscopy. When spectral lines were first identified, they were classified empirically into series called *sharp*, *principal*, *diffuse* and *fundamental.* This nomenclature existed long before quantum mechanics. However, the initial letters of the names of the series are still used to indicate the orbital angular momentum quantum number, as you can see in Table 3.1.

Table 3.1 Spectroscopic notation.

orbital angular momentum quantum number, l	0	1	2	3
spectroscopic notation	s (sharp)	p (principal)	d (diffuse)	f (fundamental)

For $l = 4$ and beyond, the letters beyond f in the Roman alphabet are used in the usual order. Hence, g corresponds to $l = 4$, h corresponds to $l = 5$, and so on. The principal quantum number is written explicitly as a number placed before the letter describing the orbital angular momentum quantum number. Thus

2p is shorthand for $n = 2$, $l = 1$

4d is shorthand for $n = 4$, $l = 2$.

An electron that is described by a 2p wavefunction is said to be in one of the 2p *quantum states.* Each set of quantum numbers that characterizes a possible wavefunction of an electron in a hydrogen atom is said to characterize a possible quantum state of the electron. According to the Schrödinger equation, therefore, *three* quantum numbers (n, l and m_l) are required to characterize each of the electron's quantum states. Later, in Section 4, you will see that one more quantum number is required to specify *completely* each electron's quantum state, in order to take into account the electron's spin.

Now try Question 3.6, which is an important exercise in using this spectroscopic notation.

Question 3.6 In Table 3.2, the rows show five values of the principal quantum number of the electron in the hydrogen atom, and the columns show five values of its orbital angular momentum quantum number. Two of the electron's quantum states have already been filled in using spectroscopic notation.

(a) Use spectroscopic notation to complete Table 3.2, leaving blank any cells that correspond to an impossible combination of quantum numbers.

(b) Calculate the number of possible quantum states that correspond to each value of n. (Remember that if the electron has a value of the orbital angular momentum quantum number l greater than zero, it can have several different values of the orbital magnetic quantum number m_l.) ■

Table 3.2 Quantum states of the electron in the hydrogen atom. (For use with Question 3.6.)

		orbital angular momentum quantum number, l				
		0	1	2	3	4
principal quantum number, n	1	1s				
	2		2p			
	3					
	4					
	5					

2.7 Summary of Section 2

This section has introduced a number of new ideas and concepts, so it is a good idea to pause and take stock of the most important points that you have met.

The method used to apply the Schrödinger equation to the hydrogen atom problem was the one you met in Section 4 of Chapter 2. In order to use the Schrödinger equation to study the behaviour of the electron in the hydrogen atom, we must insert into it the electron's *potential energy function* (Equation 3.1). The Schrödinger equation predicts the energy levels and the corresponding allowed stationary state wavefunctions of the electron when it is confined in the hydrogen atom. Each wavefunction is characterized by a set of quantum numbers n, l and m_l, but the energy levels depend only on the principal quantum number n (Equation 3.3) in the absence of a magnetic field.

The quantum numbers n, l and m_l can have only certain values (see Box 3.2) and l and m_l have clear physical meanings. The orbital angular momentum quantum number l specifies the *magnitude* of the electron's orbital angular momentum (Equation 3.4) and the orbital magnetic quantum number m_l specifies the component of the electron's angular momentum in an arbitrarily chosen z-direction.

So far, we have discussed the *quantum numbers* that characterize each time-independent wavefunction ψ of the electron in the hydrogen atom, but we have not yet discussed the *shapes* of these wavefunctions. That is the subject of the next section.

Question 3.7 Figure 3.10 illustrates how the Schrödinger equation is used to study the behaviour of an electron confined in a hydrogen atom in zero magnetic field: note that ψ_{1s} is a shorthand notation for 'the time-independent wavefunction describing the electron in the 1s quantum state'. Fill in the empty boxes to indicate (a) the piece of information that must be inserted into the time-independent Schrödinger equation before it can be solved, (b) the energy of the second lowest level, and (c) the possible stationary state wavefunctions describing the confined electron in the three energy levels shown. ■

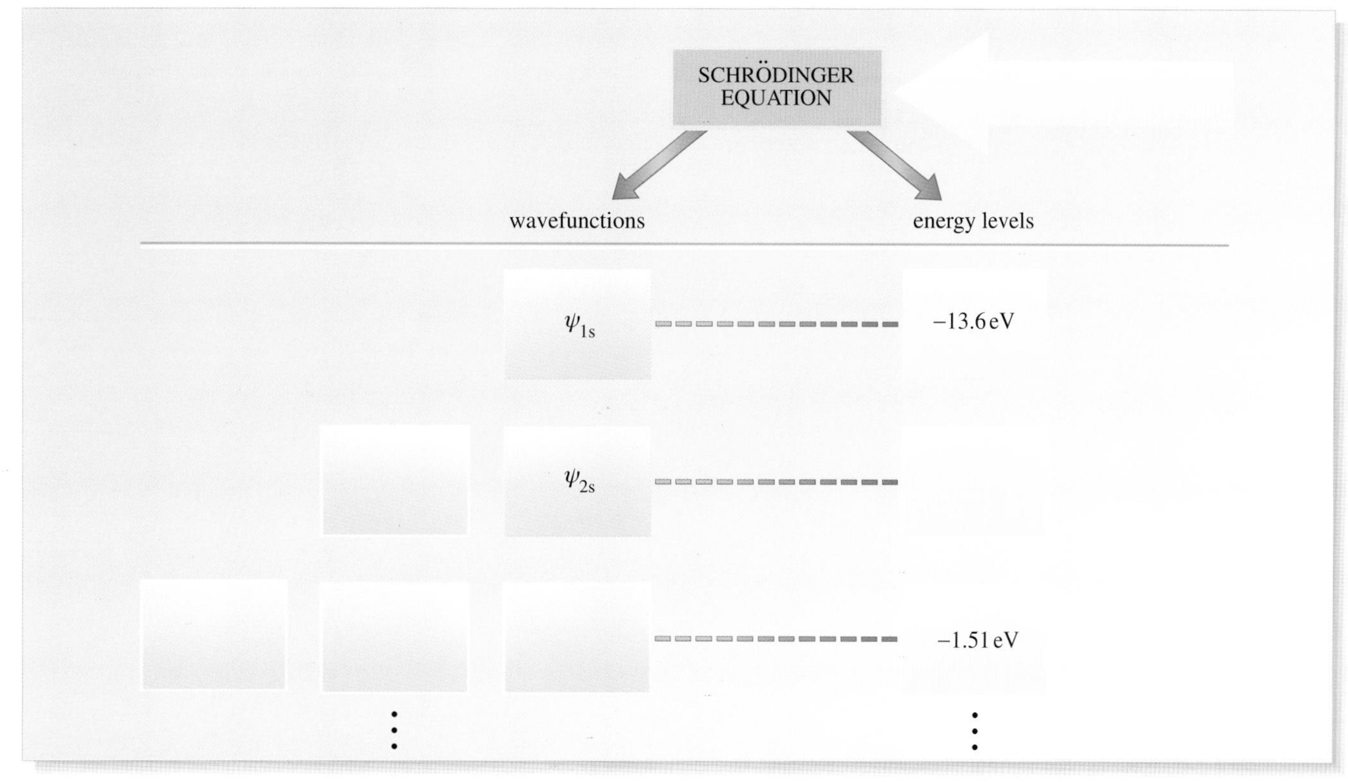

Figure 3.10 The way in which Schrödinger's time-independent equation is used to investigate the behaviour of an electron confined in a hydrogen atom. You should assume that the atom is *not* subject to a net external magnetic field. (See Question 3.7.)

3 Wavefunctions and transitions in hydrogen

3.1 Electron distribution patterns in hydrogen — the shape of the hydrogen atom

The spatial distribution of the electron in a hydrogen atom defines the shape of the hydrogen atom, and the Born interpretation (Chapter 2, Section 4.6) taught us that all the information obtainable about the location of a particle in a given state is contained in the wavefunction corresponding to that state.

The hydrogen atom is a three-dimensional object and it is not obvious that an electron distribution associated with the spherically symmetric Coulomb potential energy function could be anything other than spherically symmetric itself. But as we shall see, the probability distribution *can* vary with *direction* (which is specified by the polar angles θ, ϕ) as well as the distance r from the nucleus.

In Chapter 2, Section 4.9 we considered a particle confined in three dimensions and we learned that the probability P of finding the particle in a small volume element ΔV at a point defined by the Cartesian coordinates (x, y, z) is given by $|\psi(x, y, z)|^2\Delta V$, where $\psi(x, y, z)$ is the normalized time-independent wavefunction of the particle. When we are dealing with objects such as atoms, the wavefunction is usually expressed as a function of the polar coordinates (r, θ, ϕ) (Figure 3.11) but the principle is exactly the same. The probability of finding the electron in a small volume ΔV at a point defined by the coordinates (r, θ, ϕ) is $|\psi(r, \theta, \phi)|^2\Delta V$. The quantity $|\psi(r, \theta, \phi)|^2$ is the **probability density**, that is, the probability *per unit volume* of finding the particle at position (r, θ, ϕ). We saw in Section 2.2 that the wavefunction for the electron in hydrogen can be written as the product of three separate functions, one for each of the three polar coordinates. That is, $\psi(r, \theta, \phi) = \psi_1(r) \times \psi_2(\theta) \times \psi_3(\phi)$ (Equation 3.2). So the probability P of finding the particle in volume ΔV at position (r, θ, ϕ) is

$$P = |\psi(r, \theta, \phi)|^2 \Delta V = |\psi_1(r)|^2 \times |\psi_2(\theta)|^2 \times |\psi_3(\phi)|^2 \Delta V.$$

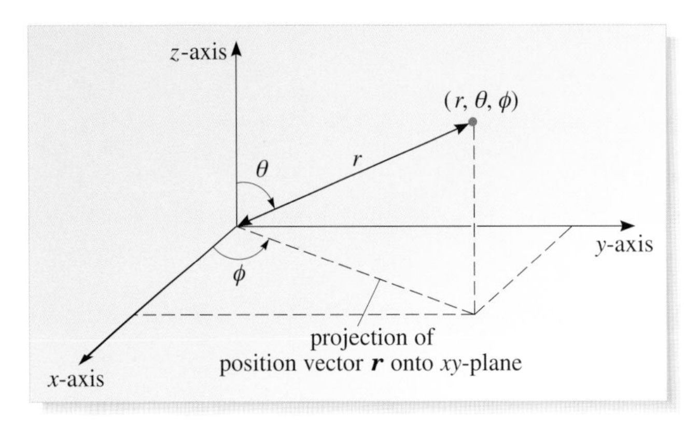

Figure 3.11 Spherical polar coordinates. (Repeat of Figure 3.5.)

Now, this separation helps us quite a lot when it comes to visualizing the hydrogen electron's probability distribution, because there are two important features that simplify matters.

- The first is that $|\psi_3(\phi)|^2$ is constant for *all* wavefunctions. This means that the probability distribution does not vary with the angle ϕ, (the angle in the xy-plane). As a result, we can show the atom in a two-dimensional cross-section containing the z-axis (e.g. in the xz- or the yz-plane), and then the full three-dimensional picture is obtained simply by rotating that cross-section about the z-axis.
- The second point to note is that $|\psi_2(\theta)|^2$ is constant for all s states, that is, all states for which $l = 0$. This means that all s states are spherically symmetric. Let's start by looking at the shape of some of these s states: states that have zero angular momentum.

Figure 3.12b is a computer-generated impression of the probability distribution for the electron in the 1s state, shown in cross-section in the xz- or yz-plane. This sort of picture is often called an electron cloud picture. The density of dots in any small region of the picture is proportional to the probability of detecting the electron in that region. As you can see, the probability distribution is, as stated earlier, spherically symmetric, that is, it depends only on r, not on θ or ϕ. This means that all the relevant information is contained in the radial part of the wavefunction, $\psi_1(r)$, which, for the 1s state, is shown in Figure 3.12a. You can see

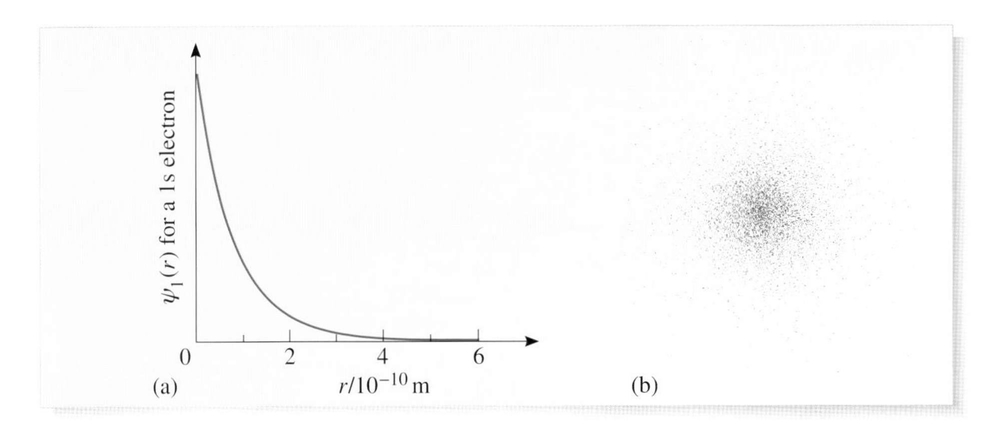

Figure 3.12 (a) The radial wavefunction for an electron in the 1s state in hydrogen. (b) The probability distribution (electron cloud picture) for a 1s electron in hydrogen.

that this *radial wavefunction* has its maximum value at $r = 0$ and then gradually diminishes with increasing r. This behaviour is reflected in the probability density ($\propto |\psi_1(r)|^2$) which is shown as an electron cloud picture for the 1s state (Figure 3.12b).

When the probability distribution is spherically symmetric, there is a further useful piece of information that can be extracted from the wavefunction, namely the *radial probability density* for the electron. This quantity indicates the relative probability of finding the electron at different radial distances from the nucleus. We can determine this function quite easily by judicious choice of our volume element ΔV. Imagine dividing the atom up into a series of thin concentric spherical shells of radius r and thickness Δr centred on the nucleus (Figure 3.13).

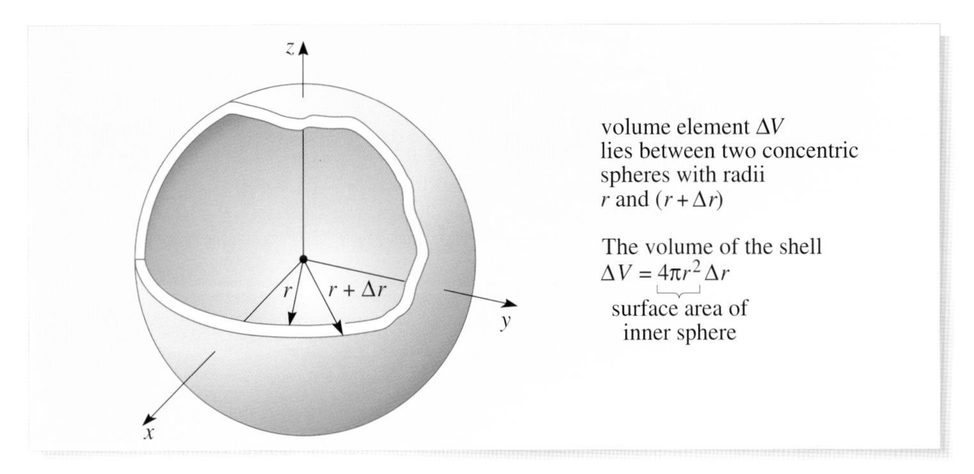

Figure 3.13 A thin spherical shell of thickness Δr, centred on the origin of coordinates, has a total volume ΔV given by (surface area of inner sphere) × (thickness of shell) = $4\pi r^2 \Delta r$.

As long as Δr is small enough, we can assume that the probability density has the same value at all points within the shell. The volume of a shell between two spheres of radius r and $r + \Delta r$ is

$$\Delta V = 4\pi r^2 \Delta r$$

where $4\pi r^2$ is the surface area of the inner sphere. The probability of detecting the electron in this volume is thus

$$P = |\psi|^2 \Delta V = |\psi|^2 4\pi r^2 \Delta r. \qquad (3.9)$$

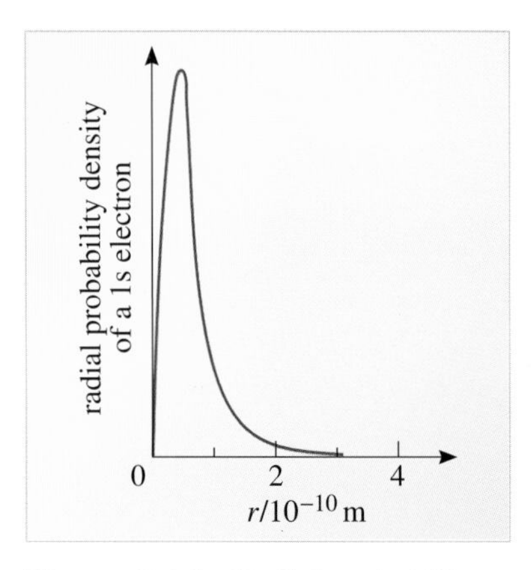

Figure 3.14 Radial probability density function for a 1s electron in hydrogen.

This implies that the probability per unit radial distance of finding the electron at radius r is given by the function $|\psi|^2 4\pi r^2$. This is the **radial probability density**, and since, for spherically symmetric s states $|\psi_2|^2$ and $|\psi_3|^2$ are constant, the wavefunction depends only on r.

Thus

$$\text{radial probability density} = 4\pi r^2 \times (\text{probability density}).$$

The radial probability density for a 1s electron is shown in Figure 3.14. Notice that this function has a maximum at a radial distance of about 0.5×10^{-10} m, which is very close to the radius a_0 of the lowest Bohr orbit. This is the distance from the nucleus at which the electron is most likely to be found. Notice also, however, that the maximum in the *probability density* itself ($\propto |\psi_1(r)|^2$) is at $r = 0$ as illustrated in Figure 3.12b.

The 2s and 3s radial wavefunctions, their corresponding radial probability densities and electron cloud pictures are shown in Figure 3.15a and 3.15b, respectively. There are two things to note in each case. First, that the number of maxima in the radial probability density function is equal to the value of the principal quantum number, n, and, second, that the value of r at which the maximum value of the *radial* probability density occurs is remarkably close to the radius of the Bohr orbit for that value of n.

When we come to look at states for which l is not zero the electron cloud picture becomes more complicated because here $|\psi_2(\theta)|^2$ is *not* constant. In fact, ψ_2 depends on both l and m_l. The electron cloud pictures of a 2p electron for $m_l = 0$ and $m_l = \pm 1$ are shown in Figure 3.16. As we go to higher values of angular momentum quantum

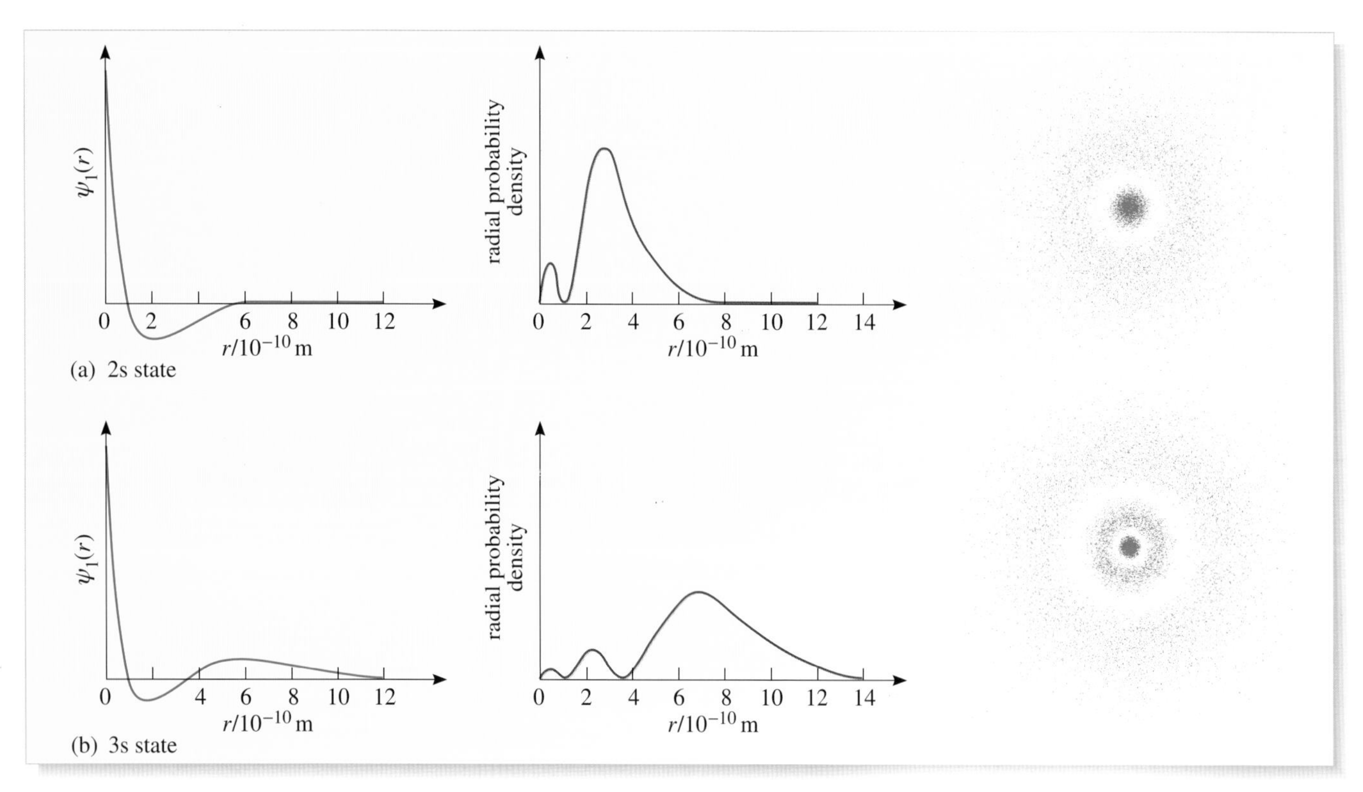

Figure 3.15 The radial wavefunctions, radial probability densities and electron cloud pictures for (a) the 2s and (b) the 3s states in hydrogen. Note that the scales of the electron cloud pictures are close to but not exactly the same as the scales for the graphs.

numbers, the electron probability distributions become even more complicated and beautiful. Some examples are shown in Figure 3.17.

Question 3.8 (a) What is the total energy of an electron in a 2p quantum state in the hydrogen atom?

(b) If an electron in a hydrogen atom has a total energy of –3.40 eV, is that electron necessarily in a 2p quantum state?

(c) When an electron in a hydrogen atom occupies its ground state, at what distance from the nucleus is it most likely to be detected? (*Hint*: Refer to Figure 3.14.)

(d) Contrast Bohr's incorrect description of the ground-state electron in a hydrogen atom with the description given by the Schrödinger equation. ■

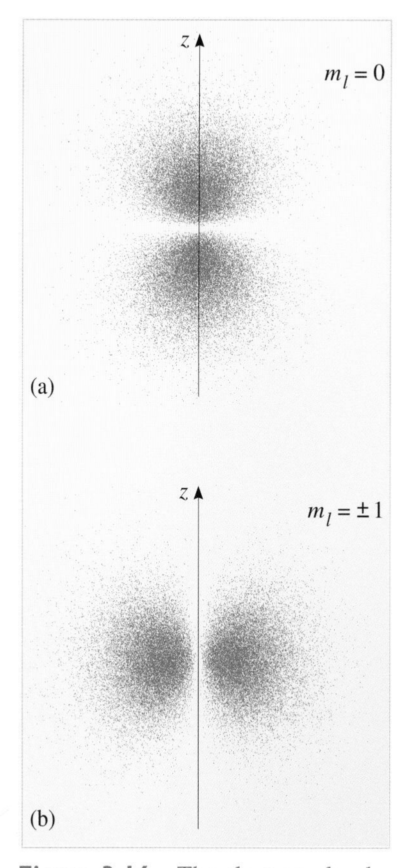

Figure 3.16 The electron cloud picture for a 2p electron in hydrogen showing (a) $m_l = 0$ and (b) $m_l = \pm 1$. (Note that the electron cloud pictures in this section are not all drawn to the same scale.)

3.2 Atomic electrons in transition — atomic spectra

So far in Sections 2 and 3, you have seen the *description* that quantum mechanics gives of the stationary states of the electron in the hydrogen atom. In fact, the theory goes much further, and provides detailed *predictions* about the transitions of the electron. In Bohr's theory, it was postulated that the electron radiates when it moves from one allowed orbit to another. But there was no explanation of why the spectral lines of atomic hydrogen had different intensities. This subsection looks at the account of atomic transitions given by quantum mechanics.

First, let us briefly review the way in which hydrogen atoms can interact with the outside world. Because the electron in a hydrogen atom has discrete energy levels (Figure 3.6), any external disturbance can change its energy (or angular momentum) only to values that correspond to allowed solutions of the Schrödinger equation. In other words, it is not possible to transfer arbitrary amounts of energy to the electron so long as it remains bound in the hydrogen atom.

If an interaction (such as a collision with another atom or particle) cannot supply sufficient energy to excite the electron at all (i.e. to change its quantum state), then the *total* kinetic energy of the hydrogen atom and its collision partner is unchanged. The collision is described as *elastic*. In that case, the electron stays in the same energy level, although it is possible that the kinetic energy of the whole hydrogen atom may change (with a compensating change in the kinetic energy of its collision partner). It is interesting to note that, at ordinary temperatures, the kinetic energy of atoms is so small that most collisions between atoms are elastic. Thus quantum-mechanical effects are not important in the collision process, and, so far as many properties are concerned, the interactions approximate very well to classical billiard-ball collisions. This is why it was possible to develop the kinetic theory of gases (*Classical physics of matter*) in the nineteenth century, before quantum mechanics was invented.

Let us now consider the interaction of *light* (or photons) with the electron in the hydrogen atom. There is an important difference between this case and the interactions of matter with atomic electrons that we have just described. A photon can be *completely absorbed* in a collision with an electron in a hydrogen atom, transferring *all* of its energy and momentum to the atom, and changing the electron's quantum state. The original photon no longer exists after this absorption process. If the photon's energy is equal to the difference in energy between two energy levels of the atom, then there can be a high probability that the photon will be absorbed, and the electron will make a transition to an excited state. However, the electron does not usually remain in an excited state for very long. It soon returns to a state of lower energy, with the *emission* of a photon. Transitions involving the absorption or emission of photons are often

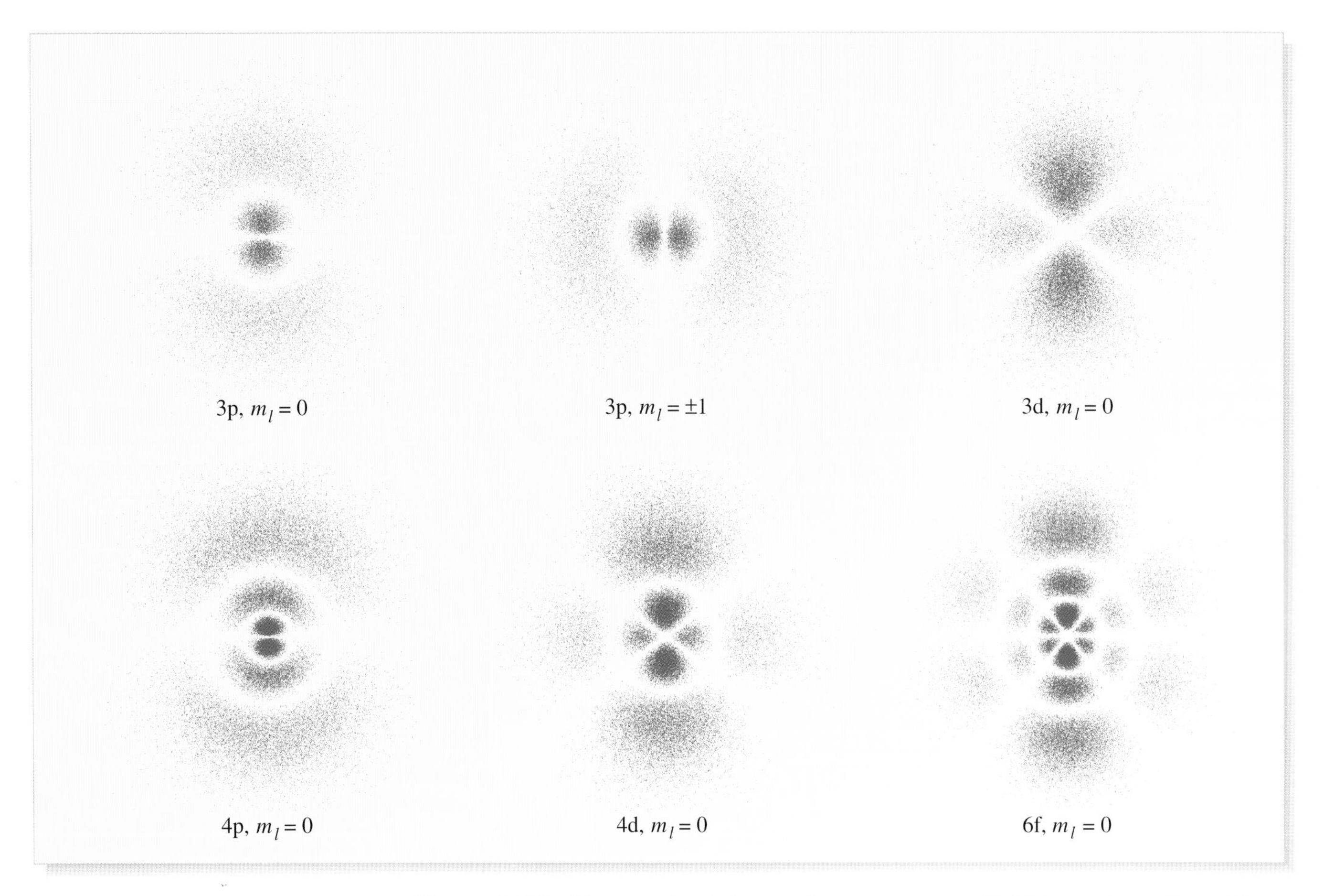

Figure 3.17 Electron cloud pictures for some states with higher values of n and l. (Note that the electron cloud pictures are not all drawn to the same scale.)

referred to as **radiative transitions**. Absorption and emission are illustrated schematically in Figure 3.18.

According to quantum mechanics, the frequency of the emitted photon is related, by Planck's law, to the energy difference of the quantum states between which the transition is made. The frequency is $f = (E_i - E_f)/h$, where E_i is the energy of the initial state and E_f is the energy of the final state. This formula is identical with the formula that was postulated in the simple Bohr theory in Chapter 1.

However, the account of the transition process given by Schrödinger's quantum mechanics is very much more detailed than the account given by the Bohr theory.

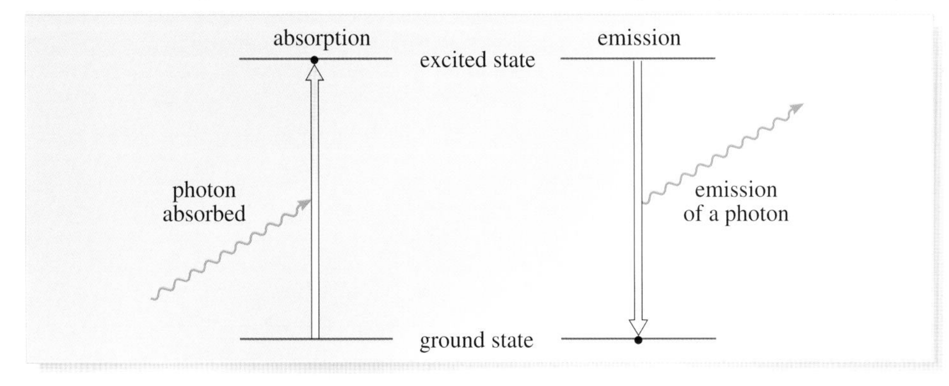

Figure 3.18 Schematic representation of the absorption of a photon by an electron in a hydrogen atom, and the subsequent emission when the electron returns to its ground state. These are radiative transitions.

First, Schrödinger's equation enables the *intensities* of the spectral lines to be calculated successfully, whereas Bohr's theory does not. Second, quantum mechanics predicts that only certain radiative transitions should occur, and the restrictions on the transitions that can take place depend on the initial and final quantum numbers of the electron undergoing the transition. These restrictions are formulated in terms of **selection rules**, which you should try to remember.

Some selection rules

If an electron is to make a transition between two quantum states, the change Δl in its orbital angular momentum quantum number must be

$$\Delta l = \pm 1. \tag{3.10}$$

Also, the change, Δm_l, in its orbital magnetic quantum number must be

$$\Delta m_l = 0 \text{ or } \pm 1. \tag{3.11}$$

There is no restriction on the change of the electron's principal quantum number, n.

Transitions that correspond to the selection rules are called **allowed transitions** and those that do not obey the selection rules are called **forbidden transitions**. In these latter cases, there is actually a very small probability that a radiative transition will occur — but it is so small that the corresponding spectral lines are not usually observable.

Question 3.9 Figure 3.19 shows the n = 1, 2, 3 and 4 energy levels for the hydrogen atom in the absence of a magnetic field. Using the selection rule given in Equation 3.10, draw in each of the allowed radiative transitions, and then list them, using spectroscopic notation. Two of the possible transitions have already been drawn in and labelled as examples.

Question 3.10 Refer back to the split energy levels that you drew on Figure 3.9 for Question 3.5 (or to Figure 3.46 in the answers). By applying the selection rules given in Equations 3.10 and 3.11, show on Figure 3.46 the transitions that are allowed to take place

(a) when the hydrogen atom *is* subject to a magnetic field,

(b) when the hydrogen atom is *not* subject to a magnetic field. ■

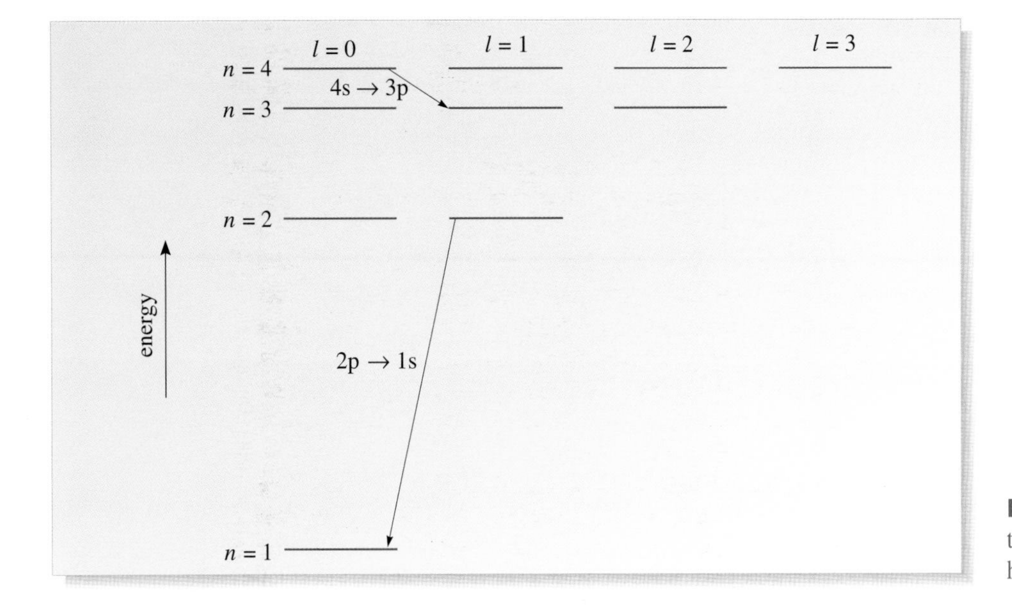

Figure 3.19 Two of the allowed transitions of the electron in the hydrogen atom. (Question 3.9.)

3.3 Summary of Section 3

In Section 3.1, you have seen how the wavefunctions that describe the electron in the hydrogen atom give extremely detailed information about the 'shape' of the atom. The main point to remember is that the probability of detecting the electron in a region depends crucially on the wavefunction $\psi(r, \theta, \phi)$ that describes its state. The term $|\psi(r, \theta, \phi)|^2$ is called the *probability density*; this is the probability per unit volume of finding the electron at (r, θ, ϕ). The function $|\psi|^2 4\pi r^2$ is called the *radial probability density*; multiplying it by Δr gives, for spherically symmetric s states, the probability of detecting the electron in the volume element $4\pi r^2 \Delta r$ formed by a spherical shell, centred on the origin, of radius r and thickness Δr.

In Section 3.2, we stated that quantum mechanics allows a detailed description to be given of atomic transitions. In particular, it predicts the intensities of hydrogen's spectral lines, and it predicts that certain *selection rules* govern which transitions will take place.

The next question will help you summarize some of the main points made in Sections 2 and 3.

Question 3.11 In Table 3.3, there are six questions about the descriptions of the electron in the hydrogen atom given by Bohr's theory (Chapter 1) and by Schrödinger's theory. Fill in the answers to these questions given by each of the models. ■

Table 3.3 The Bohr model of the hydrogen atom compared with the Schrödinger model. (See Question 3.11.)

	Bohr's model	Schrödinger's model
1 What are the energy levels of the electron when the hydrogen atom is *not* subject to a net magnetic field?		
2 What are the possible values of the magnitude of the electron's orbital angular momentum?		
3 Can the electron have zero orbital angular momentum?		
4 What quantum numbers specify each possible quantum state of the electron?		
5 What is the role of the electron's principal quantum number, n?		
6 Does the theory enable the intensities of atomic hydrogen's spectral lines to be calculated?		

4 Electron spin

4.1 Introduction

Despite the impressive amount of detail that has been built up about the hydrogen atom in Sections 2 and 3, one vital ingredient has been missing. In this section, we shall make good this deficiency by introducing the concept of *electron spin.*

In 1921, an experiment had been performed by two German physicists, Otto Stern (1888–1969) and Walter Gerlach (1899–1979), the results of which could not be explained by Schrödinger's theory. We will not go into the details of the experiment here, but the results implied that for neutral silver atoms in an external magnetic field, there are two, and *only* two, possible values for the z-component of the *magnetic moment* of a silver atom in its ground state. This corresponds to only two quantized components of the angular momentum, an 'up' and a 'down' if you like. Such a finding is not in agreement with Schrödinger's theory because, although spatial quantization is predicted by this theory (Equations 3.6 and 3.7), the number of possible values for the z-component of the angular momentum vector is $2l + 1$, which is always an odd number (since l is always an integer). Furthermore, there should always be a component with $L_z = 0$ (corresponding to $m_l = 0$), which Stern and Gerlach did not observe. Clearly, a different explanation is required for these observations.

The magnetic moment of an atom (or current loop) is a measure of the strength of the magnetic dipole field associated with that atom or current loop.

4.2 An explanation — electron spin

An explanation for Stern and Gerlach's results came in 1925 from the work of two Dutch graduate students, Samuel Goudsmit (1902–1978) and George Uhlenbeck (1900–1988). They suggested that the electron itself possessed an intrinsic (*internal*) angular momentum, as well as the more obvious properties of mass and charge. This intrinsic angular momentum of the electron, or **electron spin** as it was called, is quite independent of any orbital angular momentum that the electron might possess due to its orbiting motion around the nucleus. Moreover, their arguments required that the spin angular momentum $\boldsymbol{S}$ should have only *two* possible orientations in an external magnetic field. That is to say, any measurement of the z-component of the spin angular momentum must yield one of *two* results.

Note *It may well help, at this point, to make an analogy with the angular momentum properties of the Earth. You know that, as well as orbiting around the Sun, the Earth also spins on its axis. The rotation about the Sun gives rise to an orbital angular momentum, while the spinning of the Earth about its own axis produces an intrinsic angular momentum, analogous to electron spin. In the case of the Earth, the magnitude of the spin angular momentum is negligible compared to the magnitude of the orbital angular momentum. In atoms, the opposite situation arises, and the orbital angular momentum can even be zero ($l = 0$). Of course, we should not take the analogy too far, or we shall be once more in danger of adopting macroscopic models in the atomic domain.*

With the introduction of the concept of electron spin, the results of the Stern–Gerlach experiment can be explained as follows. Each silver atom contains 47 electrons, and it turns out that the vector sum of the orbital and spin angular momentum for 46 of these electrons is zero. Moreover, the remaining electron has zero *orbital* angular momentum (i.e. $l = 0$). Hence, *the net angular momentum for a whole silver atom comes entirely from the spin angular momentum of a single electron* and the electron's spin angular momentum can have only two possible orientations in an external magnetic field.

4.3 Quantum numbers for electron spin

How does this extra property of the electron fit into the scheme of quantum numbers we have already established? This subsection completes the description of the hydrogen atom by answering that question.

You may recall that the intrinsic angular momentum of the electron was mentioned in the discussion of ferromagnetism in Chapter 4 of *Static fields and potentials*.

Clearly, we need another quantum number to determine the magnitude of the spin angular momentum. This so-called **spin angular momentum quantum number** is denoted by s and can have only one value, $\frac{1}{2}$. It is not surprising that s is single-valued, since it refers to an intrinsic property of an electron: in the same way, an electron has only one value of charge and mass.

The spin angular momentum quantum number s of an electron is related to the magnitude S of its spin angular momentum vector $\boldsymbol{S}$ by the equation

$$S = \sqrt{s(s+1)}\,\hbar\,. \tag{3.12}$$

This equation is exactly analogous to Equation 3.4, which relates the possible magnitudes L of the electron's *orbital* angular momentum vector to the possible values l of the electron's *orbital* angular momentum quantum number

$$L = \sqrt{l(l+1)}\,\hbar\,. \qquad \text{(Eqn 3.4)}$$

The orientations of $\boldsymbol{S}$, like those of $\boldsymbol{L}$, can be defined with respect to an arbitrary z-direction. The allowed values of the z-component of $\boldsymbol{S}$ are specified by the electron's **spin magnetic quantum number** m_s:

$$S_z = m_s\hbar. \tag{3.13}$$

This equation is analogous to Equation 3.6, which related the possible values of the z-component of the electron's *orbital* angular momentum to the possible values of its *orbital* magnetic quantum number m_l:

$$L_z = m_l\hbar. \qquad \text{(Eqn 3.6)}$$

The spin magnetic quantum number, m_s, can take only two possible values:

$$m_s = +\tfrac{1}{2} \quad \text{or} \quad -\tfrac{1}{2}. \tag{3.14}$$

The electron is often described as being 'spin up' ($m_s = +\frac{1}{2}$) or 'spin down' ($m_s = -\frac{1}{2}$). The combination of Equations 3.13 and 3.14 shows that S_z can have only two values ($\pm\frac{1}{2}\hbar$).

The quantum numbers associated with the spin angular momentum of the electron are shown in Box 3.3, where they are compared with the analogous quantum numbers associated with the *orbital* angular momentum of the electron in the hydrogen atom. However, you should note that the spin angular momentum properties apply to an electron *whether or not* it is confined within a hydrogen atom.

Box 3.3 Quantum numbers associated with the orbital angular momentum vector $\boldsymbol{L}$ and the spin angular momentum vector $\boldsymbol{S}$ of the electron in the hydrogen atom

orbital angular momentum $\boldsymbol{L}$

$$L = \sqrt{l(l+1)}\,\hbar$$

where the orbital angular momentum quantum number l can have the values $l = 0, 1, 2, \ldots, n - 1$.

$$L_z = m_l\hbar$$

where the orbital magnetic quantum number m_l can have the values $m_l = 0, \pm1, \pm2, \ldots, \pm l$.

This shows that $\boldsymbol{L}$ can have $(2l + 1)$ orientations with respect to an arbitrarily defined z-axis.

The allowed values of L and L_z depend on the value of n.

For example: for $n = 2$,

$L = 0$ or $\sqrt{2}\,\hbar$

$L_z = 0$ or $+\hbar$ or $-\hbar$

spin angular momentum $\boldsymbol{S}$

$$S = \sqrt{s(s+1)}\,\hbar$$

where the spin angular momentum quantum number s has the single value $s = \frac{1}{2}$.

$$S_z = m_s\hbar$$

where the spin magnetic quantum number m_s can have one of the two values $m_s = \pm\frac{1}{2}$.

This shows that $\boldsymbol{S}$ can have $(2s + 1)$ (i.e. just two) orientations with respect to an arbitrarily defined z-axis.

The value of S must be ....

The value of S_z must be either .... or ....

Question 3.12 (a) Use the expressions for S and S_z given in Box 3.3 to fill in their values in the space provided at the bottom of the box.

(b) Draw a sketch in the margin next to Box 3.3 of the possible orientations of $\boldsymbol{S}$ in space. ■

Note *Be sure to check that you have arrived at the right answers for this question before moving on. If you were wrong, copy the right answers into Box 3.3 now.*

Finally, it is important to realize that the concept of electron spin was not included in the Schrödinger equation that was discussed in Chapter 2 and in Sections 2 and 3. However, Wolfgang Pauli showed similar equations can be written that do take the spin of the electron into account, and in 1928 the English physicist Paul Dirac published a relativistic quantum theory of the electron that allowed the concept of electron spin to be understood theoretically.

4.4 A pattern in the degeneracy of the quantum states of the electron in the hydrogen atom

In Section 2, you saw that, according to an analysis based on the Schrödinger equation, any stationary state of an electron in a hydrogen atom is completely specified by the three quantum numbers n, l and m_l. Moreover, only the principal quantum number n determines the energy levels in the absence of an external magnetic field. States having different l and m_l combinations, but the same value for n, have the same energy, i.e. the states are degenerate.

In order to specify the quantum state of such an electron in a hydrogen atom *completely*, we must specify all its quantum numbers n, l, m_l and m_s. Since $m_s = +\frac{1}{2}$ or $-\frac{1}{2}$, there are *two* different quantum states available to an electron for each possible set of values of n, l and m_l. Table 3.4 lists the possible values of these four quantum numbers for the ten lowest quantum states of the electron in the hydrogen atom.

Table 3.4 The possible combinations of the four quantum numbers used to specify the quantum states of the electron in the hydrogen atom that are associated with the principal quantum numbers $n = 1$ and $n = 2$.

n	l	m_l	m_s
1	0	0	$+\frac{1}{2}$ $-\frac{1}{2}$
2	0	0	$+\frac{1}{2}$ $-\frac{1}{2}$
	1	+1	$+\frac{1}{2}$ $-\frac{1}{2}$
		0	$+\frac{1}{2}$ $-\frac{1}{2}$
		−1	$+\frac{1}{2}$ $-\frac{1}{2}$

Your answer to Question 3.6 showed that, for a given value of n, the number of degenerate states with different values of l and m_l is given by n^2. It is easy to show that this result holds quite generally. Now, following the introduction of spin, we see each of these states is two-fold degenerate. Thus, in total, there are $2n^2$ degenerate states in hydrogen corresponding to the principal quantum number n. The number of degenerate states for each of the first four values of n are, therefore, 2, 8, 18 and 32.

This neglects some differences resulting from spin and other effects, which are very small compared with the difference due to the principal quantum number.

Later in the chapter, you will see how these ideas can be used to give an understanding of the Periodic Table.

4.5 Summary of Section 4

In this section, you have seen that the electron has spin. Whether or not the electron is in a hydrogen atom, its spin angular momentum vector is specified using the two quantum numbers s and m_s (Equations 3.12, 3.13 and 3.14). When the electron is in the hydrogen atom, four quantum numbers are required to specify each of its quantum states: n, l, m_l and m_s (Table 3.4).

The concept of electron spin enabled an understanding of the results of the Stern–Gerlach experiment. However, a theoretical understanding of why the electron has spin was not possible until 1928, when Dirac published his relativistic quantum theory of the electron.

Question 3.13 State whether the following three statements about electron spin are *true* or *false*. In those case(s) in which you think the statement is false, give reasons for your answer.

(a) Since the spin angular momentum quantum number of an electron is always $\frac{1}{2}$, there are only two allowed values of the electron's z-component of spin angular momentum S_z.

(b) Since the spin magnetic quantum number m_s has two values, there are two possible values for the magnitude S of the spin angular momentum vector.

(c) The spin angular momentum quantum number of an electron is always $s = \frac{1}{2}$, whether or not the electron is confined in a hydrogen atom. ■

5 The structure of heavy atoms

5.1 Introduction

So far in this chapter, we have considered only the simplest possible atom, namely hydrogen. Since the hydrogen atom consists essentially of a single electron confined in a potential well due to its attraction to the proton, this system was an extension of the problem considered in Chapter 2, of a single particle confined in a three-dimensional square well. In this section, we consider how quantum mechanics can be applied to heavy atoms, by which we mean any atom that has more than one electron.

We will begin with some useful terms and definitions that enable a convenient record to be kept of the constituents of each atom. Since the electrons in heavy atoms are *confined*, you will not be surprised to learn that quantum mechanics predicts that they have stationary states that correspond to precise values of energy. But the problem is that it is not a simple matter to use Schrödinger's equation to *calculate* these levels, since the interactions between all the various particles in a heavy atom make the system very complex. Nevertheless, after making various approximations, it turns out to be possible to classify the stationary states that are available to the atomic electrons in a recognizable manner, and to determine which of these states are occupied when a heavy atom is in its ground state. Before we do that, however, we must familiarize ourselves with some of the basic properties of heavy atoms.

5.2 The constituents of heavy atoms

Each heavy atom consists of a tiny, positively charged nucleus around which negatively charged electrons are moving. We will assume that the quantum states occupied by these electrons are each characterized by the same set of four quantum numbers (n, l, m_l and m_s) that characterize each quantum state of the electron in the hydrogen atom. Moreover, we will take it that the restrictions that apply to the possible values of these quantum numbers of an electron in a *heavy* atom are the same as those that apply to the quantum numbers that specify each quantum state of an electron in a *hydrogen* atom.

The nucleus of the atom consists of positively charged *protons* and neutral particles of approximately the same mass, called *neutrons*. Protons and neutrons each have a mass that is nearly 2000 times that of the electron.

Two numbers are needed to specify the number of protons and neutrons in a nucleus. First, the number of protons that each nucleus contains is given by its **atomic number**, which is usually denoted by the capital letter Z. Second, the *total* number of protons and neutrons in the nucleus is given by its **mass number**, which is conventionally denoted by the capital letter A. Hence the numbers of neutrons in the nucleus is $A - Z$.

You know that the magnitude of the proton's charge is equal to that of the electron, so the hydrogen atom is neutral — it has no net charge. Using this idea, it's easy to see that *any* neutral atom must contain the same number of protons as electrons.

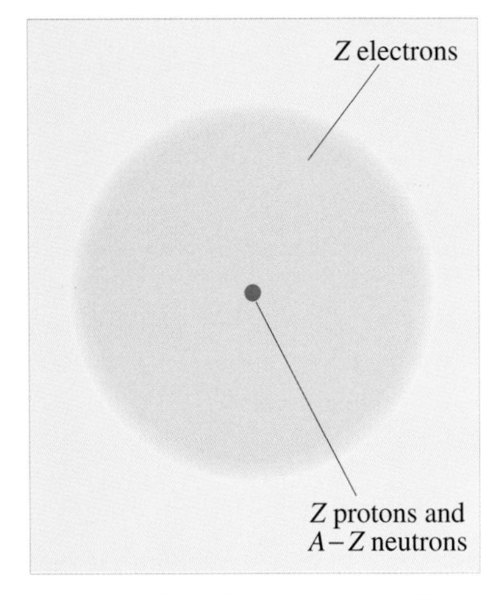

Figure 3.20 Visualization of a neutral atom whose nucleus has mass number A and atomic number Z.

Table 3.5 The first ten elements.

name	chemical symbol	Z
hydrogen	H	1
helium	He	2
lithium	Li	3
beryllium	Be	4
boron	B	5
carbon	C	6
nitrogen	N	7
oxygen	O	8
fluorine	F	9
neon	Ne	10

In other words, if the atomic number of the nucleus of a neutral atom is Z, then it must contain Z electrons (Figure 3.20). As you'll be seeing shortly, the chemical properties of an element depend on the arrangement of its constituent electrons and not on the properties of its nucleus. Table 3.5 lists the ten elements with the smallest numbers of protons and electrons. This table is for reference: you are not expected to remember it.

It's worth noting that the atoms of a particular element do not necessarily contain identical nuclei. Although the nuclei of an element contain the same number of protons (that is, they all have the same *atomic* number), they need *not* contain the same number of neutrons (that is, they need not all have the same *mass* number). Nuclei that have the *same* atomic number but different mass numbers are called the **isotopes** of a particular element, and the constituents of each isotope are noted by writing, before its chemical symbol, its atomic number (as a subscript) and its mass number (as a superscript). For example, the isotope of carbon with $A = 12$ and $Z = 6$ is written ${}^{12}_{6}\mathrm{C}$. An *ion* is an atom of a particular element which has gained or lost one or more electrons and thus carries a net electric charge.

Question 3.14 Consider a neutral neon atom that contains the nucleus ${}^{21}_{10}\mathrm{Ne}$.

(a) How many protons does it contain?

(b) How many neutrons does it contain?

(c) How many electrons does it contain?

(d) Given that the charge of the electron is -1.6×10^{-19} C, what is the charge of the nucleus?

(e) A nucleus is detected with atomic number $Z = 10$. Is it an isotope of neon? ■

5.3 Schrödinger's equation applied to electrons in heavy atoms

The electrostatic potential energy function of an electron in a heavy atom

In Chapter 2, you saw that Schrödinger's equation allows us to determine the total energy of a particle, provided that its potential energy function is known. In other words, to find the total energy of the particle using the Schrödinger equation, we must know the potential energy of the particle *at every point in space.*

In the case of the electron in the hydrogen atom, this calculation is relatively straightforward since the potential energy function of an electron moving in a Coulomb electrostatic field is simply $E_{\mathrm{el}} = -e^2/4\pi\varepsilon_0 r$, where r is the instantaneous distance of the electron from the centre of the nucleus.

But it is very much more difficult to determine the energy levels of the electrons in a *heavy* atom. Consider, for example, the argon atom, which contains eighteen electrons. If you wanted to calculate their energy levels using the Schrödinger equation, you would not only need to find the potential energy function of each electron — no mean task in itself — but you would then need to solve the corresponding Schrödinger equation — and that, to put it mildly, would be very hard indeed! You would need to take into account not only the mutual *repulsion* of the eighteen negatively charged electrons but also their *attraction* to the positively charged nucleus. One way to begin to tackle this problem is to consider the force on the electron in two extreme cases. First, picture an electron in the argon atom when it is very near the nucleus. It is reasonable to guess that, in this case, the repulsive forces on this one electron due to the seventeen others will approximately *cancel* (Figure 3.21a) and that the net force on the electron will, therefore, be due to its attraction to the nucleus. The net attractive force $\boldsymbol{F}$ will then be given by

$$\text{electron near the nucleus:} \quad \boldsymbol{F} \approx \frac{-18e^2}{4\pi\varepsilon_0 r^2}\hat{\boldsymbol{r}}, \tag{3.15}$$

where r is the distance from the electron to the centre of the nucleus and where $\hat{\boldsymbol{r}}$ is the unit vector that points from the nucleus to the electron.

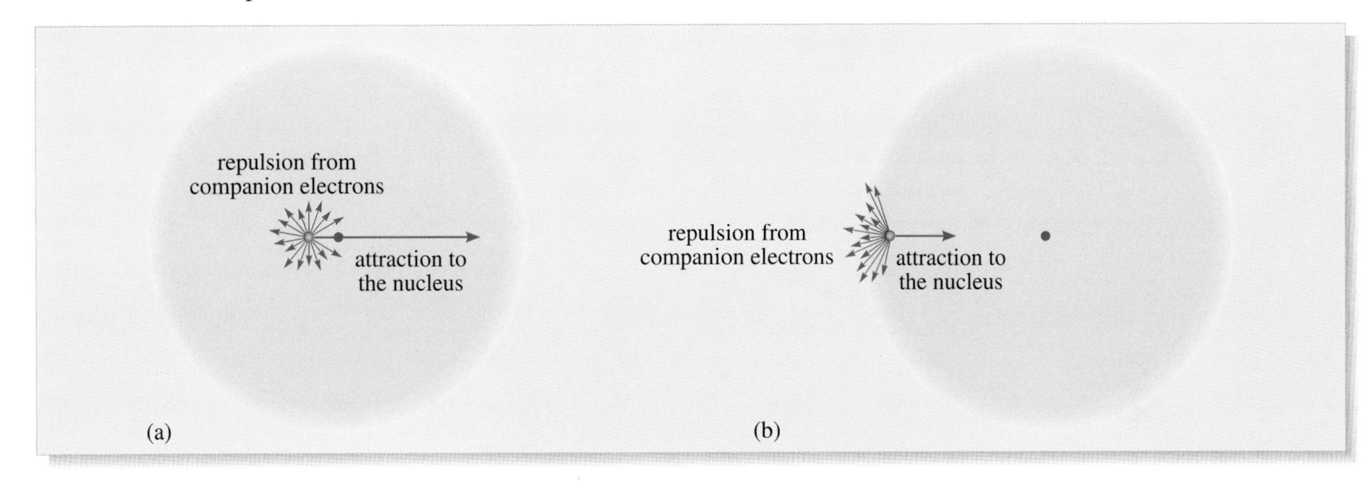

Figure 3.21 (a) When an electron is *very close* to the nucleus, the forces on it due to the *other* electrons tend to cancel. Hence the net force on the electron is mainly due to the attraction of the nucleus. (b) When an electron is *very far* from the nucleus, the repulsive forces due to the other electrons will tend to add to produce a resultant force in the radial direction, counteracting the attractive force due to the nucleus. (Notice that the components of the repulsive forces that are *perpendicular* to the radial direction tend to *cancel*.)

It is important to recognize one crucial feature of Equation 3.15: it says that when the electron is near the nucleus, the magnitude of the force acting on it depends only on r (*not* on the other variables θ and ϕ) and that the force is in a radial direction (given by the unit vector $-\hat{\boldsymbol{r}}$). As you'll be seeing later, this assumption has important implications.

Now think about the force on an electron that is *far away* from the nucleus of the argon atom. In this case, it is reasonable to expect that the repulsive forces due to the other seventeen electrons *will* be important. From Figure 3.21b, you can see that in the outward *radial* direction these forces tend to add, whereas in the perpendicular direction they tend to cancel. But the repulsive radial component is outweighed by the attractive force on the electron due to the nucleus. It turns out that the force on the electron in this case is much smaller than it is when the electron is *close* to the nucleus (Equation 3.16):

$$\text{electron far from nucleus:} \quad \boldsymbol{F} \approx \frac{-e^2}{4\pi\varepsilon_0 r^2}\hat{\boldsymbol{r}}. \tag{3.16}$$

You can see that Equation 3.16 is plausible if you regard the electron in this case as being separated from a net charge of $+e$, due to the combined effects of the nucleus (whose total charge is $+18e$) and the seventeen companion electrons (whose total charge is $-17e$).

Now, before we can apply the Schrödinger equation to an electron in the argon atom, we need to know, not the force acting on it, but rather its electrostatic potential energy E_{el}. This does not present a problem: as you saw in *Static fields and potentials*, if the Coulomb electrostatic force on a charged particle is known, then its electrostatic potential energy can be written down at once. Using Equations 3.15 and 3.16:

when the electron is near the nucleus (r small): $$\boldsymbol{F} \approx \frac{-18e^2}{4\pi\varepsilon_0 r^2}\hat{\boldsymbol{r}}; \quad E_{el} \approx \frac{-18e^2}{4\pi\varepsilon_0 r} \qquad (3.17)$$

when the electron is far from the nucleus (r large): $$\boldsymbol{F} \approx \frac{-e^2}{4\pi\varepsilon_0 r^2}\hat{\boldsymbol{r}}; \quad E_{el} \approx \frac{-e^2}{4\pi\varepsilon_0 r}. \qquad (3.18)$$

In both cases, the electrostatic potential energy of the electron is taken to be zero when it is separated from the centre of the nucleus by an infinite distance, i.e. when $r = \infty$, $E_{el} = 0$.

So far, we've discussed only the two extreme cases, in which an electron is very near to or very far from the nucleus. But what about the points in between? It is reasonable here to assume that when its distance r from the nucleus is very large or very small an electron's *actual* potential energy function will be given approximately by Equations 3.17 or 3.18, but that when r has intermediate values, the potential energy will vary smoothly between these two extremes (Figure 3.22). In technical language, we assume we can **interpolate** the electron's potential energy function between the two extremes $E_{el} \approx -18e^2/4\pi\varepsilon_0 r$ and $E_{el} \approx -e^2/4\pi\varepsilon_0 r$.

Since all the arguments so far have been general, they referred only to 'an electron', they can be taken to apply to *every* electron in the atom.

From Figure 3.22, you can see that the potential energy function of an electron is taken to depend only on r for all points (not just at small and large values of r). This assumption allows considerable insight into the quantum states that can be occupied by the electrons in the argon atom, if we also make the reasonable assumption that *each quantum state of an electron should correspond to a definite value of angular momentum*. In Section 2, you saw that this idea is borne out by the prediction of Schrödinger's equation that the electron subject to the Coulomb force in the hydrogen atom has quantum states that are each specified not only by the principal quantum number n, but also by the orbital angular momentum quantum number l and the orbital magnetic quantum number m_l.

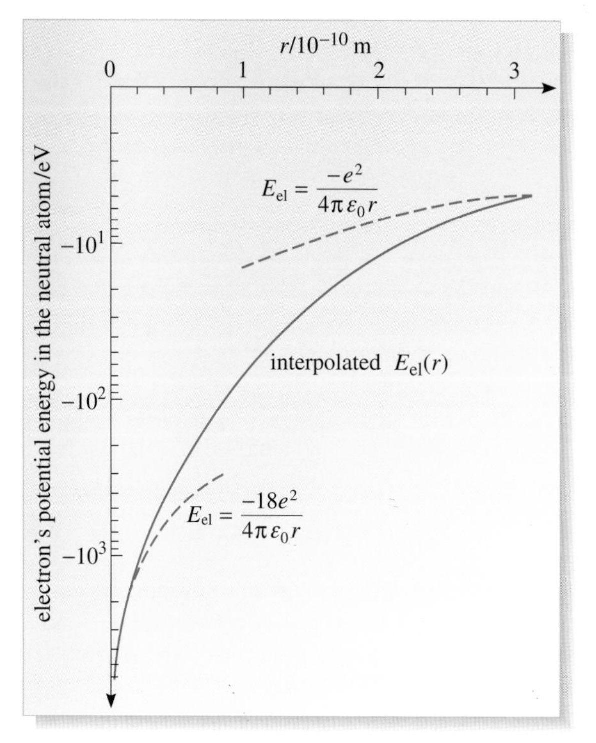

Figure 3.22 It is reasonable to suggest that the actual electrostatic potential energy function of an electron in an argon atom can be interpolated between the two extremes $E_{el} \approx -e^2/4\pi\varepsilon_0 r$ and $E_{el} \approx -18e^2/4\pi\varepsilon_0 r$.

Using these ideas, it is plausible to model the electrons in an argon atom (or in any heavy atom) as moving under the action of a force whose magnitude depends only on r and whose direction is radially inwards. This implies that the electron's quantum states (and energy levels) can also be characterized by the quantum numbers n, l and m_l (and also by m_s, since the electron has spin). *In other words, the quantum states of electrons in heavy atoms are specified by the same quantum numbers as the electron in the hydrogen atom.*

According to this model, the wavefunctions and energy levels of the quantum states characterized by these quantum numbers can be calculated from the Schrödinger equation using the potential energy function shown in Figure 3.22.

In the hydrogen atom, states with the same value of n but different values of l are degenerate, that is, they have the same energy. In heavy atoms this is not the case:

> States with the same n and different l have *different* energies, and states with lower values of l have lower energies.

This is due to the fact that the potential energy function is no longer a simple $1/r$ curve as it was for hydrogen, and this is an important fact to appreciate.

The remaining question that we need to answer is 'which of these quantum states are occupied by the electrons in the ground states of heavy atoms?', and that is the subject of the next subsection.

5.4 Pauli's exclusion principle and the electronic structures of heavy atoms

Study comment *In this subsection, chemical symbols are often used to save writing out the full names of different chemical elements. There is no need for you to remember these symbols for the purposes of this course.*

In order to determine which quantum states are occupied by the electrons in heavy atoms, it is very instructive to consider some of the properties of the elements. The properties of each element depend critically on the number of electrons in each of its atoms, as you can easily see by considering the three lightest elements, hydrogen ($Z = 1$), helium ($Z = 2$) and lithium ($Z = 3$), which contain one, two and three electrons, respectively.

At room temperature, hydrogen and helium are both colourless gases — and that is about the only property they share. Hydrogen is flammable, and it combines quite readily with other elements, for example with oxygen to form water. However, helium is non-flammable and it certainly does not combine readily with other elements: on the contrary, it is remarkably unreactive and it is said to be inert. The element lithium is different again: it is very reactive and, at room temperature, it is a *metal*!

One quantity that is particularly useful when discussing the structure of a neutral atom is the amount of energy required to remove an electron to infinity, where it no longer feels the attractive force of the nucleus. The energy needed to remove the *least* tightly bound electron from a neutral atom is called the *first ionization energy* of the neutral atom, and in Figure 3.23 it is plotted against atomic number Z for the three lightest elements. The data in this figure apply to atoms that, before their ionization, were in their electronic *ground states*, i.e. the states in which their constituent electrons occupy the lowest possible energy levels.

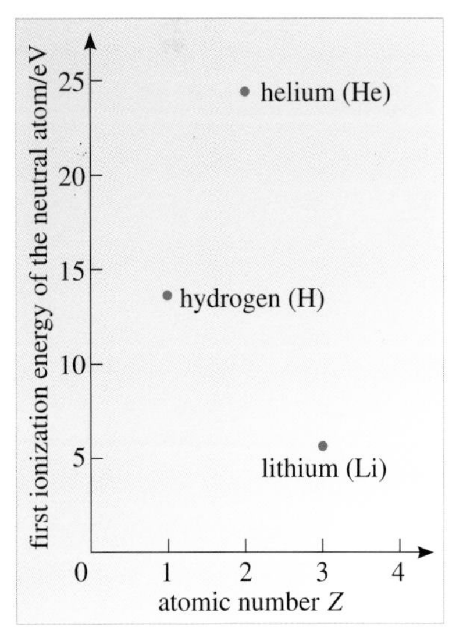

Figure 3.23 The first ionization energies (for atoms in the ground state) of the three lightest elements plotted against their atomic numbers.

You can see from Figure 3.23 that the first ionization energies do not vary smoothly as Z increases, and that helium, the *least* reactive of the three elements, has by far the *greatest* first ionization energy. In order to understand the sharp variations in these first ionization energies, we need to consider the quantum states of the electrons in heavy atoms. As you have already seen, these quantum states are each characterized by the four quantum numbers n, l, m_l and m_s, the same quantum numbers that specify the quantum states of the electron in the hydrogen atom. But the important question is 'How do we know *which* quantum states are occupied by the electrons in the ground state of a heavy atom?'

As a first guess, it would be reasonable to suggest that each electron in a heavy atom would occupy a ls quantum state (like the electron in the ground state of a hydrogen atom). But the problem is that if this were the case, then the variation of the first ionization energies with atomic number would be expected to be smoother and more gradual than it is observed to be.

The variation in the first ionization energies can, however, be explained by an extremely important principle of atomic physics that provides the crucial insight into the problem of finding *which* quantum states of the electron in an atom are occupied. This principle, which was first written down by the Austrian physicist Wolfgang Pauli (Figure 3.24), can, in the present context, be stated very simply:

Pauli's exclusion principle

No two electrons in an atom can ever occupy the same quantum state.

Wolfgang Pauli (1900–1958)

Wolfgang Pauli (Figure 3.24) was born on 25 April 1900 in Vienna, Austria. He was the son of a professor of physical chemistry. He first made his mark when he was only nineteen years old, writing an article on relativity that was admired by Einstein himself. After Pauli had finished his Ph.D., he worked with Bohr and with Born, before he became professor of theoretical physics at the Zurich Institute of Technology in 1928. During his career, he did an enormous amount of ground work in non-relativistic and relativistic quantum mechanics, and he also successfully predicted the existence of the neutrino in 1931 (twenty five years before it was discovered). He moved to the United States during the war and became an American citizen in 1946. He was awarded the Nobel Prize for physics, somewhat belatedly, in 1945. Pauli died in Zürich, Switzerland in 1958.

Figure 3.24 Wolfgang Pauli.

In *Quantum physics of matter* you will discover that all elementary particles fall into one of two categories called fermions and bosons. Only fermions obey the exclusion principle. Electrons are fermions.

A more general formulation of the exclusion principle will be presented in *Quantum physics of matter*. It follows, from the simple version given above that, since each stationary state of an electron in an atom is characterized by a particular set of quantum numbers, *no two electrons in an atom can possibly have the same set of four quantum numbers.* When the hydrogen atom was discussed in Section 2, there was no need to use the principle, since that atom contains *only one* electron, which, in its ground state, occupies a 1s ($n = 1$, $l = 0$, $m_l = 0$) quantum state with its spin either 'up' or 'down' ($m_s = +\frac{1}{2}$ or $m_s = -\frac{1}{2}$).

However, the case of the one-electron atom is the only one that *can* be studied without using the exclusion principle. When the electronic structure of a *heavy* atom is being investigated, it is always *essential* to use this principle.

In the rest of this subsection, we shall use the exclusion principle to study the structures of heavy atoms when they are in their ground states, that is, when their constituent electrons occupy the lowest possible energy levels. Let us begin with the helium atom. According to the exclusion principle, the two electrons in the ground state of this atom will *both* be 1s electrons, one with the spin magnetic quantum number $m_s = +\frac{1}{2}$, the other with $m_s = -\frac{1}{2}$ (Figure 3.25). Notice that the two 1s electrons *cannot* have the *same* spin magnetic quantum number, since that would mean them both having exactly the same set of quantum numbers.

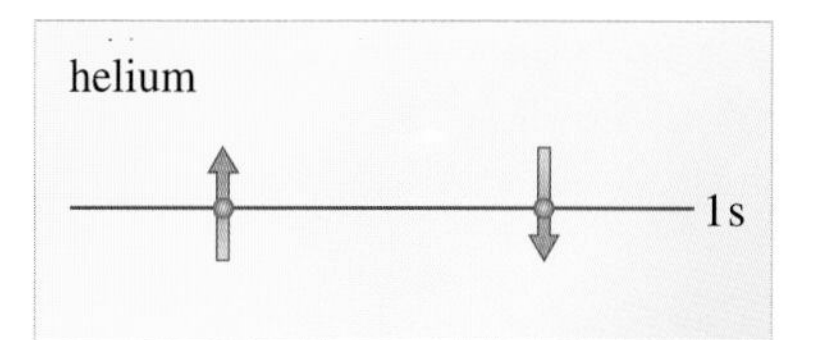

Figure 3.25 In the ground state of the helium atom, the two electrons occupy 1s quantum states. The arrows represent the two spins, $m_s = +\frac{1}{2}$ and $m_s = -\frac{1}{2}$.

The **electronic structure** of an atom is commonly written down in one of two ways. In the first, commonly referred to as the **standard notation**, the electronic structure of helium is written

He $1s^2$

where the number 1 denotes the value of the electrons' principal quantum number n, the letter s denotes, as usual, that the electrons have orbital angular momentum quantum number $l = 0$, and the superscript '2' indicates that the two electrons both occupy 1s quantum states.

The second method of writing down the electronic structure is known as the **box notation**:

He $\boxed{\uparrow\downarrow}$
1s

This method has the advantage that it shows clearly that the spins of the two electrons are opposite ($m_s = +\frac{1}{2}$ and $m_s = -\frac{1}{2}$) or, in other words, that they are *paired.*

Now think about the three electrons in the ground state of a *lithium* atom. Two of the electrons will fill the two available 1s states with $m_s = \pm\frac{1}{2}$ and according to the exclusion principle, the remaining electron must be in another quantum state. Since the lithium atom is supposed to be in its ground state, this third electron will occupy the state associated with the lowest available energy and this turns out to be a 2s state. (Remember from the previous subsection that, in heavy atoms, states with lower values of l have lower energies, so a 2s state ($n = 2$, $l = 0$) has lower energy than a 2p state ($n = 2$, $l = 1$).) This case illustrates a general rule that

lowest energy states fill first.

Notice that this electron could have the spin magnetic quantum number $m_s = +\frac{1}{2}$ or $m_s = -\frac{1}{2}$ (Figure 3.26).

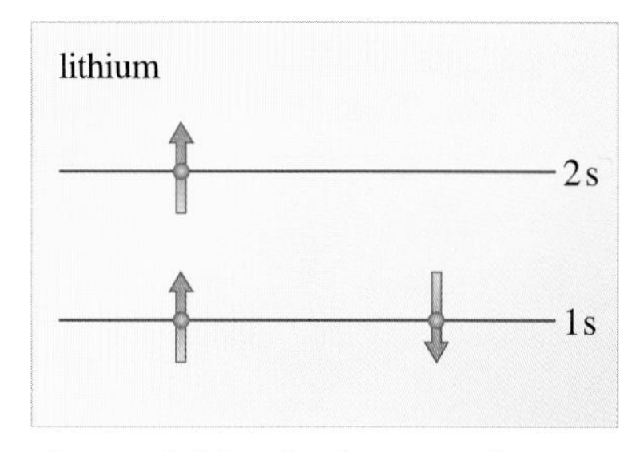

Figure 3.26 In the ground state of the lithium atom, two 'paired' electrons occupy 1s quantum states and the other electron occupies a 2s quantum state.

The electronic structure of the ground state of the lithium atom is written, in standard notation,

Li $1s^2 2s^1$

and, in box notation:

Li $\boxed{\uparrow\downarrow}$ $\boxed{\uparrow}$
1s 2s

Here the lone 2s electron is shown to be 'up' $\boxed{\uparrow}$, rather than 'down' $\boxed{\downarrow}$, but this is purely a matter of convention and not essential to the argument.

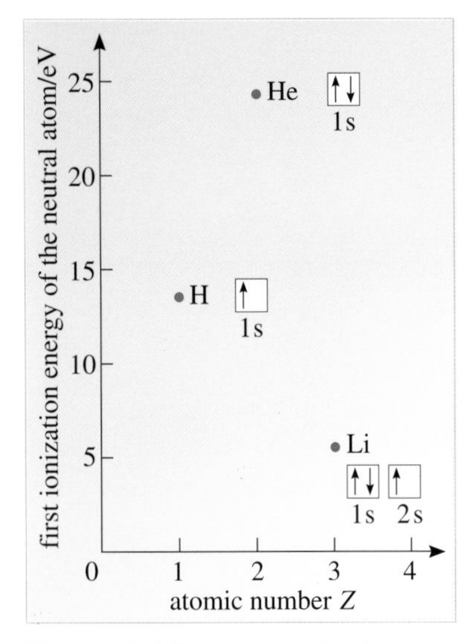

Figure 3.27 Among the three lightest elements, the helium has the largest first ionization energy.

The exclusion principle has shown that the three lightest elements — hydrogen, helium and lithium — have different electronic structures, so really it is not surprising that the first ionization energies of the elements are different. Of the three elements, helium is clearly the one with the most stable structure, since it has the highest ionization energy, as you can see in Figure 3.27.

With the electronic structure of the ground state of the lithium atom established, it is quite a simple matter to predict the structure of the ground state of the beryllium atom, which contains one more electron than lithium. Make sure you can do this by trying Question 3.15.

Question 3.15 (a) Write down what you would expect to be the electronic structure of the ground state of the beryllium atom ($Z = 4$), in standard notation and in box notation. (The chemical symbol for beryllium is Be.)

(b) Write down the quantum numbers of each of the four electrons in the ground state of the beryllium atom. ■

When we come to account for the structure of the ground state of the boron atom, which contains *five* electrons, however, a new problem appears. The first four electrons fill the 1s and 2s quantum states; which quantum state will be occupied by the remaining electron? The principle that the lowest energy states fill first suggests that it should occupy a 2p quantum state (Figure 3.28).

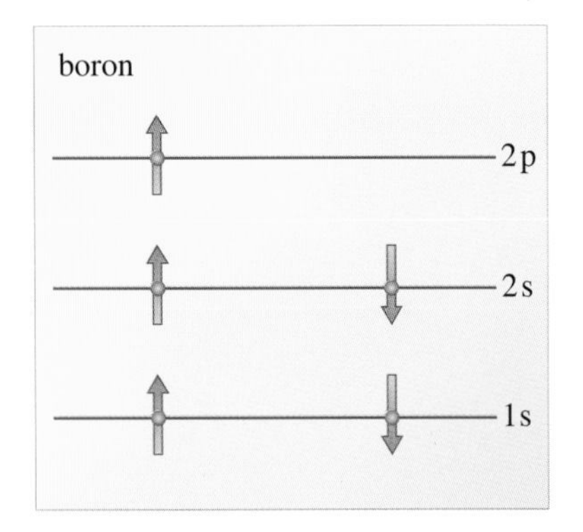

Figure 3.28 The ground state of a boron atom. (This is a schematic diagram, and you should attach no importance to the spacings of the energy levels.)

● How many 2p quantum states are available to the unpaired electron in the ground state of the boron atom?

❍ Six. The 2p quantum states are, by definition, characterized by the quantum numbers $n = 2$, $l = 1$ and $m_l = 1$ or 0 or -1. The electron can occupy any *one* of these three states with its spin 'up' or 'down' ($m_s = +\frac{1}{2}$ or $m_s = -\frac{1}{2}$) Hence, six states are accessible to it. ■

In box notation, the electronic structure of the ground state of the boron atom is written

B [↑↓] [↑↓] [↑ | |]
1s 2s 2p

Notice that there are three boxes to represent the six 2p states: the boxes correspond (in no particular order) to the pairs of $m_l = +1$, $m_l = 0$ and $m_l = -1$ states. The one arrow in the 2p box is drawn on the left only as a matter of convention: it could equally well be in one of the other two boxes.

Box notation makes it easy to see the next difficulty, which arises when we try to write down the electronic structure of the carbon atom. This contains six electrons. But which of the following is correct?

C [↑↓] [↑↓] [↑↓ | |] or [↑↓] [↑↓] [↑ | ↑ |]
1s 2s 2p 1s 2s 2p

We are really asking whether the sixth electron is paired or unpaired. The exclusion principle cannot help us here, since it is satisfied in *both* cases. The only way in which this question can be answered is to use a result of advanced quantum mechanics that is known, for historical reasons, as Hund's rule. It can be stated as follows:

Hund's rule

In the ground state of an atom, the total spin of the electrons always has its maximum possible value.

In very simple terms, Hund's rule arises because if a pair of electrons has parallel spins they are, on average, further apart. Therefore their mutual repulsion provides a smaller contribution to their potential energy, than if they have antiparallel spins.

According to this rule, the electronic structure of the ground state of the carbon atom is the one shown above on the right-hand side, since the unpaired 2p electrons in that structure have a greater total spin than the *paired* electrons in the other structure. A useful point to remember when applying this rule, is that the number of unpaired electrons in the appropriate box diagram is generally a *maximum* — this idea certainly works in the case of carbon since the actual structure of its ground state (shown on the right-hand side) has two unpaired electrons, whereas the other structure has *none*.

Question 3.16 Given that the atomic number of nitrogen is seven, write down, in box notation, the electronic structure of a nitrogen atom in its ground state. (The chemical symbol for nitrogen is N.) ■

Now let us continue to try to account for the electronic structures of other elements. The ground state of the neon ($Z = 10$) atom is written, in standard notation,

$$\text{Ne} \quad 1s^2\, 2s^2\, 2p^6$$

As you can see, the 1s, 2s and 2p states each have their full complement of electrons. This raises a question similar to ones that you've met before: in the ground state of an atom that contains eleven electrons, which quantum state will be occupied by the last electron? In order to answer this question, you need to know which quantum states will accommodate the unpaired electron in such a way that it has the lowest possible value of energy. Hopefully, you can see that this problem will recur when you try to write down the electronic structures of atoms that contain still more electrons. You need to know the order in which the quantum states are occupied: this order is shown in Figure 3.29. Now although there is no need to remember the

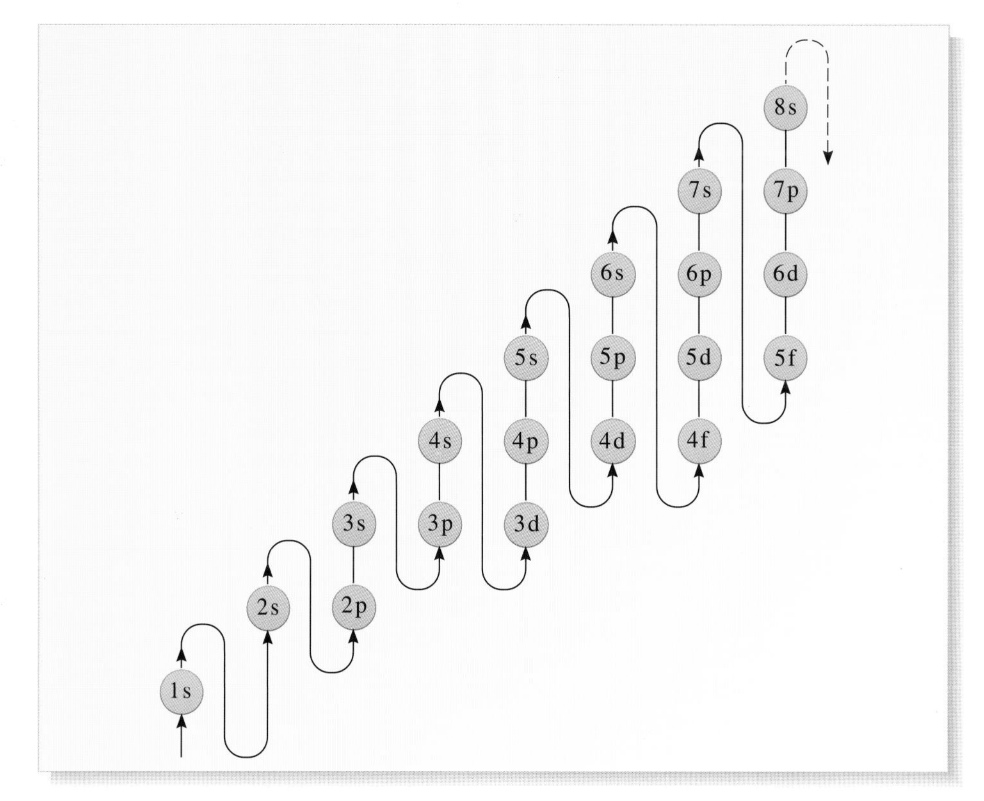

Figure 3.29 The order in which electrons fill the quantum states of the elements in their ground states. (*Note*: The order shown is nearly always correct. There are, in fact, odd exceptions, but they will not concern us in this course.)

content of this figure, you should know how to use it. But before you can do that, you need to know how many electrons can occupy quantum states associated with different values of the orbital angular momentum quantum number l. We shall discuss this next.

You have seen already that s quantum states (in which the electrons have $l = 0$) can accommodate two electrons, and that p quantum states (in which the electrons have $l = 1$) can accommodate six electrons. This information is given in Table 3.6, which is not complete since it does not give the maximum number of electrons that can occupy d and f quantum states. You can work out these maximum numbers for yourself by doing Question 3.17, after which you should enter your results in Table 3.6.

Table 3.6 The number of quantum states of an electron with the orbital angular momentum quantum number l.

orbital angular momentum quantum number, l	number of associated quantum states
0	2
1	6
2	
3	

Question 3.17 (a) Show that a set of d quantum states, in which electrons have $l = 2$, can be occupied by a maximum of 10 electrons. (b) How many electrons can occupy each set of f quantum states, in which they have $l = 3$? ■

At this point, we should perhaps point out that, so far, we have *not* been using two common (though rather misleading) terms of atomic physics. The first is the term **shell**, which is used to label the electrons in quantum states corresponding to a particular principal quantum number n. For example:

the $n = 1$ shell labels the 1s electrons;

the $n = 2$ shell labels the 2s and 2p electrons;

the $n = 3$ shell labels the 3s, 3p and 3d electrons,

and so on. The other term is **subshell**, which simply labels the electrons in a given shell that correspond to the various possible values of the orbital angular momentum quantum number l. For example, the $n = 3$ shell is said to consist of 3s, 3p and 3d subshells.

We have not used these terms so far because we think that they evoke the misleading image that the electrons in a heavy atom are in some way distributed on the surfaces of shells. This is certainly *not* correct, since the confined electrons are described by quantum-mechanical *wavefunctions,* which allow us to calculate only the *probability* of detecting the electrons in different regions within the atom. These calculations show that the electrons are certainly not distributed on surfaces within the atom! But despite these arguments, you will find the terms shell and subshell are widely used and accepted by the scientific community, and we shall resort to them in the next subsection.

Question 3.18 Write down, in *standard notation*, the electronic structure of a rubidium atom in its ground state. (The atomic number of rubidium is 37, and its chemical symbol is Rb.) ■

5.5 The Periodic Table

The orderly manner in which the quantum states are gradually occupied as atomic number increases gives us some insight into why the elements fall into groups with similar chemical properties. Figure 3.30 shows a plot of the first ionization energies against atomic number for all the elements up to radon (Rn). It has an oscillatory structure rising gradually to several sharp maxima at the noble gases. The noble gases are collectively termed Group 0 (zero) elements and they are those elements for which the occupied quantum states of highest energy are $ns^2\,np^6$ (with the exception of He, $1s^2$). The electronic structure of these elements is shown in Table 3.7. The fact that these elements all have the maximum possible number of electrons in their highest energy quantum states (i.e. 'a full outer subshell') makes them very unreactive. The Group 0 elements appear in a column down the right-hand side of the Periodic Table of the elements (Figure 3.31). The Periodic Table is arranged so that all elements with a similar outer shell electronic structure (and hence similar chemical properties) appear in vertical columns known as **groups** in which the principal quantum number of the outer shell increases as you step down the column. Thus the horizontal rows in the Periodic Table contain all the elements with the same principal quantum number in the outer shell. These rows are known as **periods** and the subshells are filling up as you move from left to right across the rows.

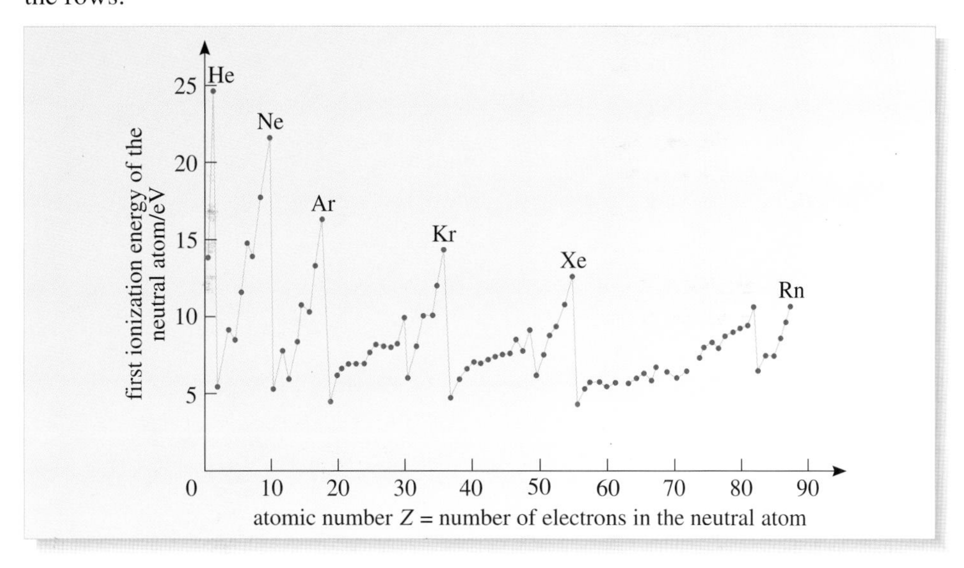

Figure 3.30 First ionization energies of neutral atoms.

Table 3.7 Ground states of Group 0 elements (noble gases). All Group 0 elements (except helium) have outer shell structures $ns^2\,np^6$.

helium	He	$1s^2$
neon	Ne	$1s^2\,2s^2\,2p^6$
argon	Ar	$1s^2\,2s^2\,2p^6\,3s^2\,3p^6$
krypton	Kr	$1s^2\,2s^2\,2p^6\,3s^2\,3p^6\,3d^{10}\,4s^2\,4p^6$
xenon	Xe	$1s^2\,2s^2\,2p^6\,3s^2\,3p^6\,3d^{10}\,4s^2\,4p^6\,4d^{10}\,5s^2\,5p^6$
radon	Rn	$1s^2\,2s^2\,2p^6\,3s^2\,3p^6\,3d^{10}\,4s^2\,4p^6\,4d^{10}\,4f^{14}\,5s^2\,5p^6\,5d^{10}\,6s^2\,6p^6$

	Group I	Group II																									Group III	Group IV	Group V	Group VI	Group VII	Group 0
Period 1	1 H																															2 He
Period 2	3 Li	4 Be																									5 B	6 C	7 N	8 O	9 F	10 Ne
Period 3	11 Na	12 Mg																									13 Al	14 Si	15 P	16 S	17 Cl	18 Ar
Period 4	19 K	20 Ca															21 Sc	22 Ti	23 V	24 Cr	25 Mn	26 Fe	27 Co	28 Ni	29 Cu	30 Zn	31 Ga	32 Ge	33 As	34 Se	35 Br	36 Kr
Period 5	37 Rb	38 Sr															39 Y	40 Zr	41 Nb	42 Mo	43 Tc	44 Ru	45 Rh	46 Pd	47 Ag	48 Cd	49 In	50 Sn	51 Sb	52 Te	53 I	54 Xe
Period 6	55 Cs	56 Ba	57 La	58 Ce	59 Pr	60 Nd	61 Pm	62 Sm	63 Eu	64 Gd	65 Tb	66 Dy	67 Ho	68 Er	69 Tm	70 Yb	71 Lu	72 Hf	73 Ta	74 W	75 Re	76 Os	77 Ir	78 Pt	79 Au	80 Hg	81 Tl	82 Pb	83 Bi	84 Po	85 At	86 Rn
Period 7	87 Fr	88 Ra	89 Ac	90 Th	91 Pa	92 U	93 Np	94 Pu	95 Am	96 Cm	97 Bk	98 Cf	99 Es	100 Fm	101 Md	102 No	103 Lr	104 Rf	105 Db	106 Sg	107 Bh	108 Hs	109 Mt	110	111	112						

lanthanides (57–70, Period 6); actinides (89–102, Period 7); transition elements (Groups between II and III)

s subshells filling (Groups I–II); f subshells filling (lanthanides and actinides); d subshells filling (transition elements); p subshells filling (Groups III–0)

Figure 3.31 The Periodic Table of the elements.

Two of the groups in the Periodic Table contain particularly reactive elements. The Group VII elements, known as halogens, all have outer shell electronic structure $ns^2\,np^5$, as you would expect since they appear one column to the left of the noble gases. Thus, all Group VII elements have one vacant space in their outer p subshell. This space is easily filled by a spare electron resulting in a negatively charged ion which is then very reactive on account of its electrostatic effects. The Group I elements on the other hand (the alkali metals) all have an outer shell structure ns^1. This outer electron is weakly bound, which is manifest in the low ionization energies of these elements (see Figure 3.30), and thus easily removed to leave a positively charged ion. Very often, of course, positively charged alkali metal ions combine with negatively charged halogen ions to form salts such as sodium chloride, our common table salt.

5.6 Summary of Section 5

In this section, you have seen that quantum mechanics can be used to investigate the behaviour of electrons in heavy atoms. The quantum states that are available to the electrons are characterized by the same set of quantum numbers (n, l, m_l, m_s) as the electron in the hydrogen atom, though the energies of these quantum states will be different in different elements and, unlike in hydrogen, states with the same value of n but different values of l are not degenerate. States of lower l have lower energy than those of higher l and so are filled first as atomic number increases.

The electronic structure of the ground state of each heavy atom is determined using three principles:

- states of the lowest energy will be filled first,
- the Pauli exclusion principle,
- Hund's rule.

The similarities between the electronic structures of several groups of chemical elements enable us to understand the Periodic Table.

Question 3.19 (a) What, in general, are the outer shell structures of noble gas atoms?

(b) Using your answer to part (a), write down the outer shell structure of the ground state of krypton (Kr), which is in Period 4 of the Periodic Table.

(c) Given that krypton has atomic number 36, write down the outer shell structure of the ground state of a bromine (Br) atom, which has atomic number 35. In what period and group of the Periodic Table would you expect bromine to lie? ■

6 Light from atoms and lasers

6.1 Introduction

In Section 3.2 we described the energy-level diagram for the ground state and excited states of the hydrogen atom and the selection rules for the allowed radiative transitions between those states, as predicted by Schrödinger's equation. These energy levels and the allowed transitions between them account for the Balmer series (Figure 1.33) and all the other series of line spectra emitted by hydrogen.

The hydrogen atom, consisting of a single proton as the nucleus and a single orbital electron, is of course the simplest of all atoms. What about more complex atoms? Sodium street lamps, for example, emit a strong yellow spectral line; the low pressure mercury vapour in fluorescent strip lights emits a strong line in the ultraviolet which is converted into visible light by the fluorescent coating on the inside of the tube; helium–neon lasers emit light on a red spectral line. What determines these emissions?

Section 5 gave you some idea of how Schrödinger's equation can be used to determine the quantum states and energy levels of atoms with many electrons. You saw, for example, that in these heavy atoms, states with the same value of the quantum number n but different values of the quantum number l have different energies, unlike in hydrogen where all levels with the same n are degenerate. This is one of the key features of the energy-level structures of complex atoms, but there are many other complications as well. In fact, the energy-level structures of many-electron atoms and the details of the selection rules governing the emission and absorption of radiation are extremely complex, and there is much current theoretical, computational and laboratory research devoted to this subject.

In this section we give a flavour of what's involved by going just two steps of complexity beyond hydrogen. First, we will look at the case of an alkali atom, such as sodium. The alkali atoms are characterized by a single electron outside closed (filled) subshells of electrons. The radiation emitted and absorbed by sodium is largely due to transitions involving this single active electron, moving in the potential energy field due to the nucleus and the two inner closed shells of electrons. Because only a single electron is actively involved, there are some similarities with hydrogen, but there are significant differences too. Secondly, we will look at the helium atom. This is the simplest case of an atom where more than one active electron is involved in the emission and absorption of light.

These two examples will give you a taste of the complexity involved in the subject of atomic spectra, and a background for studying a variety of radiative processes. One of the most exciting of these is the *stimulated emission* of light and laser action, the final topic of this chapter.

6.2 Line spectra of sodium

The ground state electronic structure of sodium ($Z = 11$) is written in standard notation as

$$\text{Na} \quad 1s^2\,2s^2\,2p^6\,3s^1$$

Showing that the $n = 1$ and $n = 2$ shells are fully occupied with two and eight electrons, respectively. (Remember that the maximum number of electrons in a shell with principal quantum number n is $2n^2$.) These ten closed-shell electrons are tightly bound, have no net angular momentum and take no active part in the emission or absorption of light by the atom. However, the unpaired outer electron, normally in the 3s ground state, can be excited into higher energy levels by electrical discharges or other means, and the spectral lines of sodium can be understood entirely in terms of the allowed transitions of this single outer electron. The ten closed-shell electrons do, however, play a passive role: they partially shield the active electron from the positive charge of the nucleus.

This shielding is quite effective when the active electron is in states of largest orbital angular momentum quantum number l, that is, when $l = n - 1$. In these high angular momentum states the active electron spends most of its time in the region outside the closed shells. Figure 3.32 shows the radial probability density for the three lowest possible states for the outer electron in sodium. Also shown is the radial probability density due to the ten inner-shell electrons. The 3s and 3p states penetrate considerably into the region occupied by the inner electrons. In the 3d state however, the electron spends nearly all of its time beyond the closed shells. In this region, the electric field due to the eleven protons in the nucleus is reduced by the electric fields of the ten closed-shell electrons. Hence, the active electron sees an 'effective' nuclear charge almost equivalent to that seen by the electron in the hydrogen atom. Not surprisingly therefore these high l states of the active electron have energies very similar to those of hydrogen. You can see this in Figure 3.33 where the energy levels of sodium are shown next to the corresponding hydrogen energy levels. All of the states with $l \geq 2$ (only nd and nf states are shown) have energies almost exactly the same as the hydrogen energy levels with the same value of n.

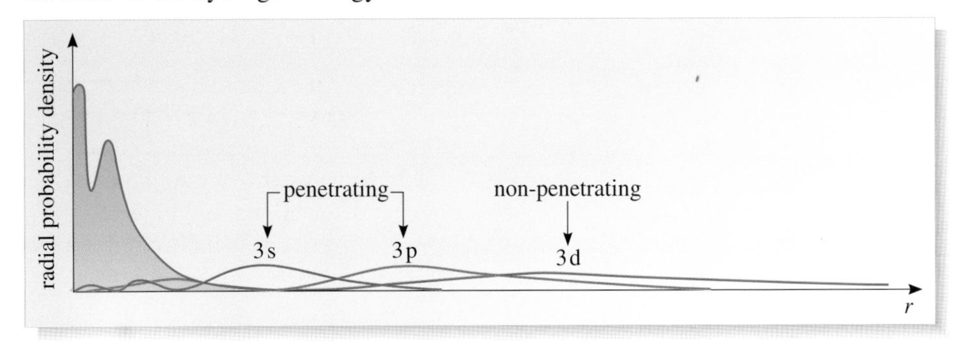

Figure 3.32 The radial probability density for the 3s, 3p and 3d states in sodium. Also shown (as the blue shaded region at low values of r) is the radial probability density for the ten inner closed-shell electrons.

When the active electron of sodium is in a state of lower angular momentum, it comes closer to the nucleus, penetrating the closed-shell region (see Figure 3.32). The shielding is then less effective and the electron sees a higher effective nuclear charge, which lowers its energy. You can see this in Figure 3.33. For example, the sodium 4p level ($n = 4$ and $l = 1$) is about 0.5 eV lower than the corresponding $n = 4$ level in hydrogen.

In the zero angular momentum case ($l = 0$ states or s states), the penetration of the active electron into the closed-shell region is largest and so the energies of the s states are much lower than the hydrogen levels with the same value of n. The sodium ground state involves, of course, the one labelled 3s. You can see that this state has an energy of about −5.13 eV. This is more than 3 eV lower than that of the corresponding $n = 3$ state of hydrogen.

The selection rules for the allowed transitions of the active electron are the same as those for hydrogen. These are given by Equations 3.10 and 3.11 in Section 3.2. The rule $\Delta l = \pm 1$, for example, means that an electron in the 4p state can make a downward transition to the 3d state ($\Delta l = 1$) or the 4s state ($\Delta l = -1$). However, a transition from 4p to 3p is forbidden since $l = 1$ for both p states, which would give $\Delta l = 0$. When an allowed transition takes place, a photon is emitted with energy $\Delta E = hf$, where ΔE is the energy of the upper state minus the energy of the lower state, and f is the frequency of the emitted radiation.

Question 3.20 The strong yellow line in the spectrum of sodium arises from a transition from the excited 3p level to the 3s ground state level. By estimating the energies of these levels from Figure 3.33, confirm that the emitted photon has a wavelength of about 590 nm (yellow). This is known as the sodium D line. ■

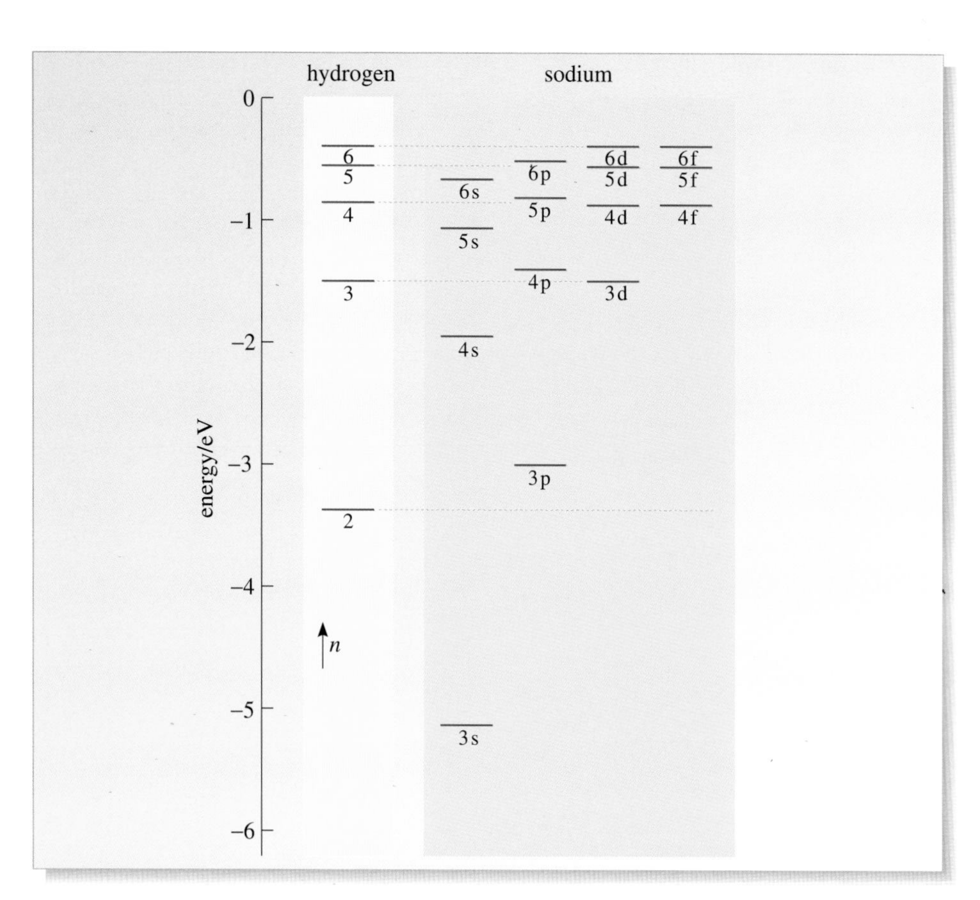

Figure 3.33 Energy levels of hydrogen are shown on the left. (*Note* The ground state at −13.6 eV is not shown.) On the right are the energy levels of the active electron in sodium. Each sodium energy level is labelled by its principal quantum number, n, and the spectroscopic symbol representing the angular momentum quantum number l (i.e. the symbol s means $l = 0$, p means $l = 1$, etc., see Table 3.1 of Section 2.6). The states with high values of the angular momentum quantum number l, have energies very close to the hydrogen energy levels with the same value of n. States with lower values of l have considerably lower energies than the corresponding hydrogen levels.

Other features of the sodium spectrum include the fact that each of the energy levels in Figure 3.33, apart from the s state levels, is, in fact, a doublet, that is, each level consists of two closely spaced levels, as shown in Figure 3.34.

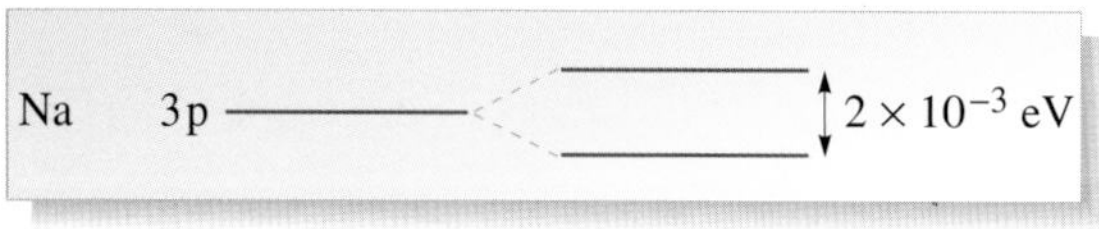

Na 3p 2×10^{-3} eV

Figure 3.34 Most of the sodium energy levels (all but the $l = 0$ states) are doublets. The 3p level, for example, consists of two closely separated levels due to the two possible alignments of the electron's spin.

This is an effect of the spin of the active electron. You have seen that both the spin and orbital angular momentum of the electron can be aligned 'up' or 'down' with respect to any specified direction in space. Quantum theory allows the electron's spin angular momentum to be aligned either parallel or antiparallel (i.e. opposite) to its orbital angular momentum. Now remember that the angular momentum of an electrically charged body has an associated magnetic moment. There is therefore a potential energy associated with the orbital and spin magnetic moments. The potential energy is positive when the two angular momenta are parallel, and negative when they are antiparallel. Hence, the total energy of the electron is different in the parallel and antiparallel states. This effect is said to arise from the **spin–orbit interaction** of the electrons. The interaction is in some ways analogous to that of two magnets lying next to one another. They have a higher potential energy when like poles are together than when unlike poles are together. The spin–orbit energy difference is quite small, only about 2×10^{-3} eV in the sodium 3p level (which is invisible on the scale of Figure 3.33), but its effect can easily be observed when the spectral lines are seen through a spectrometer. For example, the sodium D line is seen as a doublet, as a result of the excited 3p state being a doublet. The wavelengths of the two lines in the doublet are 589.6 nm and 589.0 nm.

Question 3.21 Why are the s states not doublets? ■

The doublet structure is found in the energy levels of all the alkali atoms (but not in the s levels) and is an example of the so-called **fine structure** of spectral lines. Fine structure effects are also seen in hydrogen, but they are very small and involve relativistic effects and other quantum effects as well as the effects of electron spin.

6.3 Line spectra of helium

The helium atom has two electrons. In the ground state both electrons are in the 1s state but with opposite spins. This ground state configuration is written as $1s^2$. The spectral lines emitted by helium can be understood in terms of there being two distinct sets of excited state energy levels. We call these two sets singlets and triplets for reasons that will become clear shortly.

The two sets of energy levels, together with the hydrogen levels for comparison, are shown in Figure 3.35. (You need not be concerned with the labelling of the levels.) Spectral lines are observed corresponding to transitions within each set of levels. No spectral lines are observed corresponding to transitions from one set of levels to the other. Why is this?

In fact, it was once thought that helium consisted of two distinct elements which were called orthohelium and parahelium.

Well, the excited states of helium involve one electron remaining in the 1s state with the other electron excited into one of the unoccupied states lying above the 1s ground state, i.e. 2s, 2p, 3s, etc. However, the 1s electron isn't altogether passive. The energies of the excited states depend crucially on how the two spin angular

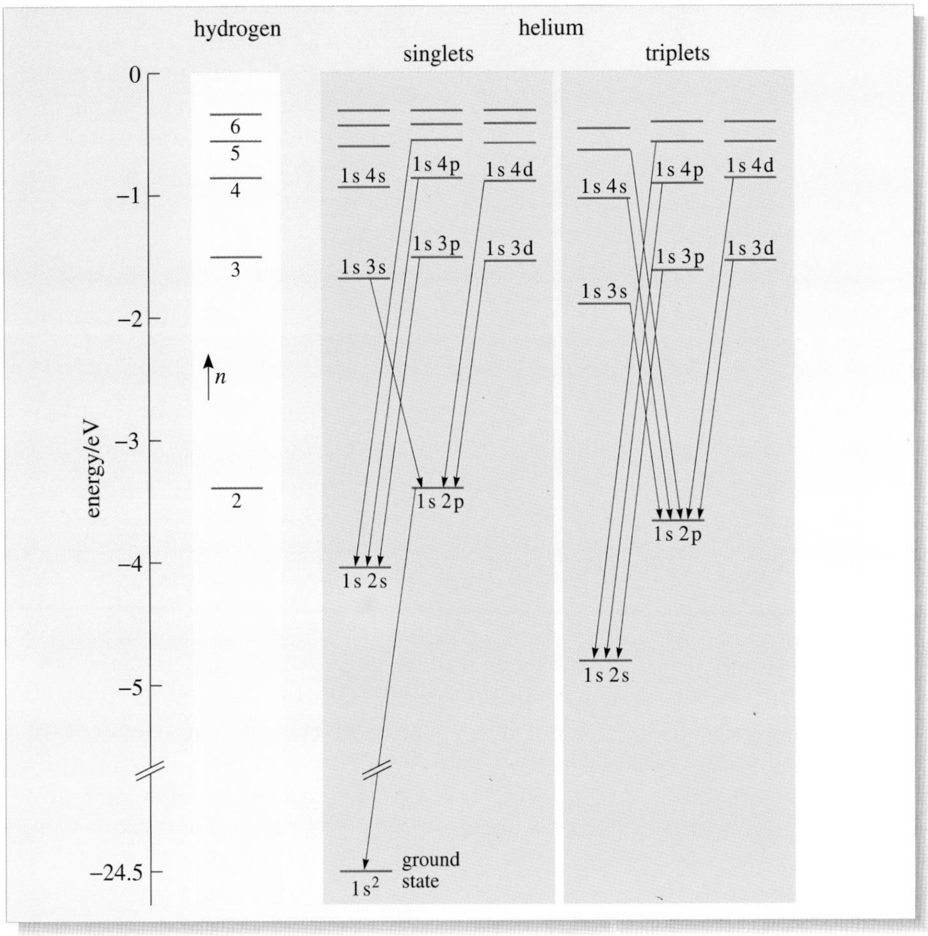

Figure 3.35 Energy levels of the helium atom, showing some allowed transitions. The levels are labelled using a standard spectroscopic convention for two-electron systems.

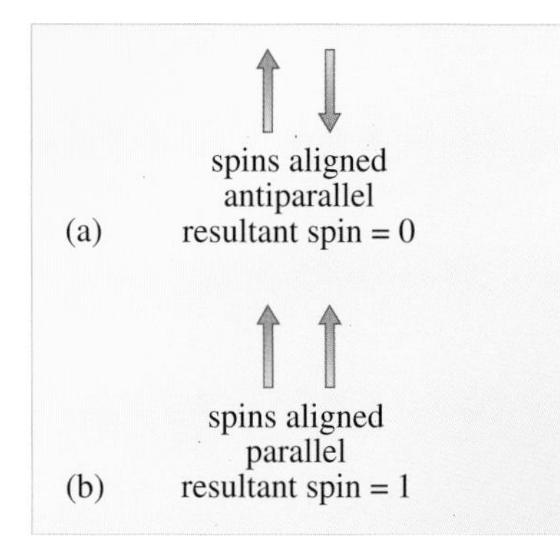

Figure 3.36 The two arrows represent the spin angular momentum vectors of the two electrons in helium. Quantum theory requires that they be aligned either (a) antiparallel or (b) parallel.

momenta of the electrons are aligned with respect to one another. As you might expect, quantum mechanics imposes restrictions. The two spins may be aligned either parallel to one another or antiparallel, as illustrated in Figure 3.36.

In the antiparallel case, the two spins cancel one another and so the net spin of the electrons in the helium atom is zero. This is the key to the *singlet* energy levels shown in Figure 3.35. The overall structure of these energy levels is similar to that of sodium. The unexcited 1s electron provides some shielding which lowers the energy when the excited electron is in a state of low orbital angular momentum. For example, the excited state of lowest energy is when the excited electron is in the 2s state ($l = 0$) which penetrates into the region close to the nucleus. You can see that this level is about 0.7 eV lower than the next level where the excited electron is in the 2p state with higher angular momentum ($l = 1$). There is, however, an important difference from the sodium structure. There is no net electron spin and so the energy levels are not split into doublets — they are said to be singlets.

Can you guess how the triplet energy levels in helium (Figure 3.35) arise?

In the triplets the spin states of the two electrons are aligned parallel to one another (Figure 3.36b). You can see from Figure 3.35 that the overall structure of the triplet energy levels is similar to that of the singlets. For example, because of the shielding effect provided by the unexcited 1s electron, the lower angular momentum states have lower energies than higher angular momentum states. There are two significant differences however. You can see from Figure 3.35 that the triplet levels are all lower than the corresponding singlet levels. This is due to a very subtle consequence of the exclusion principle and arises because a pair of electrons with parallel spins spend, on average, more of their time at greater distances from each other than a pair of electrons with antiparallel spins. The electromagnetic potential energy of the mutual repulsion is therefore lower. The other difference, not shown in Figure 3.35, is, of course, that some of these levels are split into three, that is, they are, in general, triplets. This is because quantum mechanics allows the parallel electron spin combination (with resultant spin = 1) to have one of three alignments with respect to the orbital angular momentum of the excited electron.

The magnetic potential energy due to different alignments of the spin and orbital 'magnets' is different in the three cases as you might expect. These small energy differences cause the three-way splitting, or triplet structure, of the energy levels. This is another example of the spin–orbit effect. Of course, states where both electrons have $l = 0$ are not split since there is then no orbital angular momentum with which the spin angular momentum can interact.

Of course, selection rules operate in helium, restricting the number of possible transitions and spectral lines. They are indicated by the arrows in Figure 3.35. The most important selection rule forbids transitions between the triplet levels and the singlet levels. The origin of this selection rule is the absence of any effective mechanism in helium gas for 'flipping' the electron spins.

The energy level structures of other 'two-electron systems' show most of the same features as in helium. For example, all the Group II metals (e.g. Be, Mg, Ca, etc.) have two electrons outside closed shells and, like helium, show a singlet and triplet structure with a selection rule forbidding transitions between singlet and triplet levels. However, this selection rule is not absolute. Mercury, for example, is a two-electron system in which one of its strongest spectral lines is a so-called intercombination line at 253.7 nm in the ultraviolet.

Question 3.22 Why is there no $1s^2$ energy level in the helium triplets? ■

6.4 Light from lasers

As we pointed out in Section 1, laser light is, nowadays, found in many everyday situations. No rock concert would be complete without a 'light-show' created by laser beams; surgeons use lasers to perform delicate operations on the eye; and CD players use laser beams to scan the tracks on a disc. The operation of lasers is firmly based on quantum-mechanical principles.

The word **laser** is an acronym for **l**ight **a**mplification by the **s**timulated **e**mission of **r**adiation. Unlike many acronyms, this one conveys a very accurate impression of the process it describes. To explain how a laser light source works, it is worth considering how atoms absorb and emit light.

6.5 Absorption and emission of light

Atoms are normally in their ground state, of energy E_0. They can be excited into a state of higher energy E_1, by any mechanism that can supply the required energy $E_1 - E_0$. This can happen in a variety of ways. For example, an electrical discharge (a spark) through a gas can cause some of the atoms, through collisions with energetic electrons or ions, to gain enough energy to excite other atoms. Another way is to irradiate the atom with a beam of photons of energy $E_1 - E_0$. Such photons, carrying an energy equal to the energy difference, are called **resonant photons**. A resonant photon can *stimulate* the atom to make a transition from the ground state level to the excited energy level as shown in Figure 3.37a. This process is sometimes called **stimulated absorption**, or simply, **absorption**.

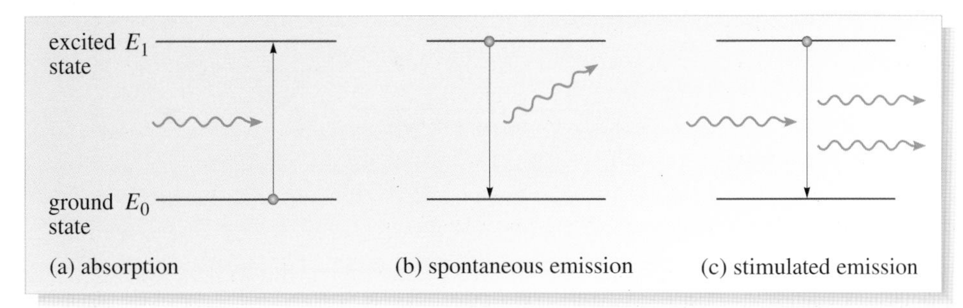

Figure 3.37 Three types of transition in an atom.

An atom will stay in an excited state for a short period of time only, before *spontaneously* returning to its ground state and emitting a photon of energy $E_1 - E_0$ (Figure 3.37b). The emitted photon has the same energy as the resonant photon that was absorbed, but the direction of the emitted photon and the time of its emission cannot be predicted. Thus, the atom decays in a random way, characterized by a certain time constant that depends on the atom and the energy levels involved. This random process is called **spontaneous emission**. The time constant for the spontaneous emission is called the **lifetime** of the excited energy level. Lifetimes are typically between a few nanoseconds and a few hundred nanoseconds. For example, the lifetime of the excited 3p level in sodium (Figure 3.33) is 16 ns (i.e. 1.6×10^{-8} s). This does not mean that all sodium atoms excited into the 3p level will decay after 16 ns. Spontaneous decay is a random process. The lifetime of 16 ns is the average time that you would measure for a very large number of excited sodium atoms. This random behaviour is typical of quantum-mechanical decay processes.

In 1916, Einstein realized that excited atoms can emit light by another mechanism, which can compete with spontaneous emission. Just as an incident photon can stimulate an atom to make an upward transition to a higher energy level, so an incident resonant photon can *stimulate* an *excited atom* to make a *downward transition* to the lower energy level. Thus, if an atom is in an excited state of energy E_1, and a beam of photons of energy $E_1 - E_0$ is shone on the atom, the atom can be stimulated to decay back to the ground state by emitting a photon of energy $E_1 - E_0$ (Figure 3.37c). This process is called **stimulated emission**, and it has special properties.

In stimulated emission, the emitted photon is highly correlated with the incident photon. That is, the emitted photon not only has the same energy, and therefore wavelength, as the incident one, but it propagates in the same direction and has the same phase. In other words, the stimulated light is *coherent* with the incident beam and hence *amplifies* it (i.e. the intensity of the light after stimulated emission has occurred is greater than the intensity before). This is in contrast with the random process of spontaneous emission where the excited atom emits the photon in an arbitrary direction and with a random phase.

Stimulated emission and spontaneous emission are competing processes. So how do we know which process will, in fact, occur when an excited atom is irradiated with resonant photons? It depends on which process is fastest. Stimulated emission occurs more rapidly when the stimulating photon beam is of high intensity. Spontaneous emission is fast when the lifetime of the excited state is short. Lifetimes of excited states vary widely but there is a systematic effect: lifetimes depend strongly on the size of the energy difference $E_1 - E_0$, and therefore on the wavelength or frequency of radiation produced by the transition. Lifetimes tend to be shorter for high-frequency transitions. Thus, spontaneous emission occurs more rapidly in high-frequency transitions than in low-frequency transitions. This means that spontaneous emission competes more strongly with stimulated emission in transitions that emit blue or ultraviolet light than for transitions that emit red or infrared radiation. Of course, this all assumes that the atom is in the excited state. If the atom is in the ground state it can only absorb light.

It follows from the above paragraph that stimulated emission can be used to produce intense, highly directional beams of light. Of course, in order to do so, we need to have a population of atoms maintained in an excited state and illuminated by an *intense* beam of resonant photons that will encourage stimulated emission, and the transition should be chosen to have a *long lifetime* against spontaneous emission.

Figure 3.38 The American physicist Charles H. Townes (1915–) conceived the idea of a microwave laser in 1951. In 1958, together with his brother-in-law, he showed theoretically how lasers could be made to work in the optical part of the spectrum. He was awarded the Nobel Prize for physics in 1964.

For many years, the concept of stimulated emission remained little more than a scientific curiosity, but in the 1950s Charles Townes (Figure 3.38) realized that stimulated emission could form the basis for producing very intense beams of very monochromatic radiation. He constructed a device that used stimulated emission from excited ammonia molecules to generate intense beams of monochromatic microwave radiation. Microwaves have relatively long wavelengths and therefore low frequencies, and so Townes was able to exploit the fact that spontaneous emission is a relatively slow process at low frequencies. However, these ideas were extended into the visible part of the spectrum by 1960, when the first laser was built. Even today, however, it is much more difficult to make lasers that operate in the high-frequency ultraviolet region of the spectrum than in the visible and infrared regions.

To achieve laser action it is necessary to arrange for a large number of atoms to be maintained in the upper level of a suitable transition. When there are more atoms in the upper level than in the lower level, a **population inversion** is said to exist. Under these conditions, an incident beam of resonant photons will be amplified by stimulated emission rather than attenuated by absorption. Of course, the difficult part is to maintain the population inversion against the loss of excited state atoms by stimulated emission and, more importantly, by the competing process of spontaneous emission. This can be achieved by a variety of mechanisms known collectively as

pumping. The pumping mechanism used in the helium–neon laser, for example, is indicated in Figure 3.39. The idea is to maintain a population inversion on the transition between the two excited neon levels indicated. This is done by setting up an electrical discharge in a tube containing a mixture of helium and neon gases. This discharge excites some helium atoms into excited states that cannot normally make transitions to the helium ground state because the selection rules forbid it. However, by coincidence, there are excited neon levels at almost the same energy as the excited helium level, and collisions between helium and neon atoms can allow the excited helium atom to transfer its energy to a ground state neon atom. (Collisional transitions are not subject to the same selection rules that govern radiative transitions.) Thus the excited helium atom returns to its ground state and the neon atom goes into one of the excited states. This process can 'pump' a large number of neon atoms into the upper of the two excited levels shown in Figure 3.39, so that a population inversion exists between the excited neon level and the almost empty lower level. Thus, stimulated emission can occur on this neon transition. Of course, the lower neon level will tend to fill and reduce the population inversion. However, this effect is reduced by the fact that the lower neon level has a very short lifetime and so it empties rapidly and the population inversion can be maintained so long as the discharge is maintained.

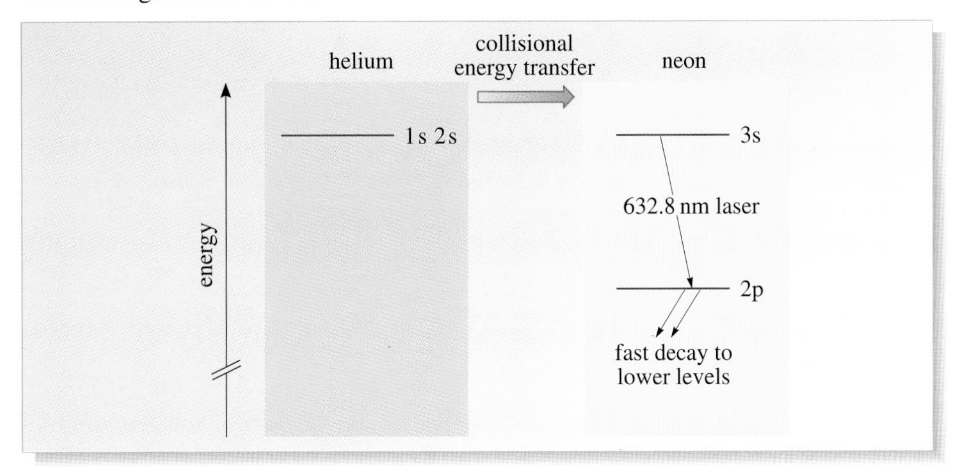

Figure 3.39 The energy levels involved in the operation of the helium–neon laser.

There is a huge variety of lasers available today using many different types of pumping mechanisms. The essential point is that, in all lasers, some pumping mechanism is used to maintain a population inversion so that stimulated emission can occur at a greater rate than absorption. The amplification by stimulated emission is aided by a pair of mirrors (Figure 3.40) so that the emitted photons make many passes to and fro through the lasing material. This 'feedback' allows a very high intensity of light to build up inside the mirrors, thereby making stimulated emission on the lasing transition more likely than spontaneous emission. One of the mirrors is only partially reflecting, so that a beam of light — the laser beam — can escape.

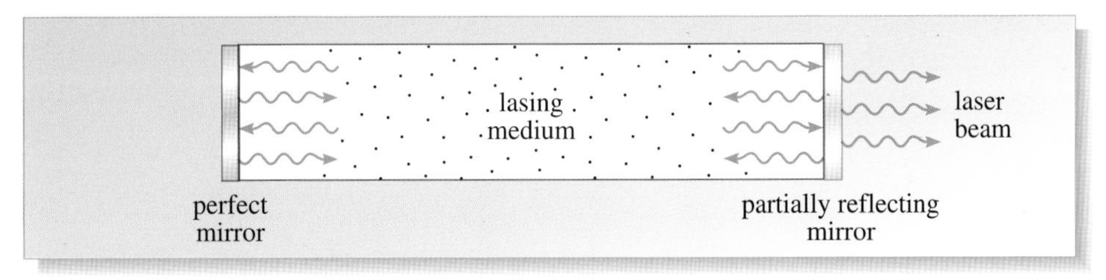

Figure 3.40 A laser is constructed by placing the lasing medium between two mirrors, one of which is only partially reflecting, thus allowing the laser beam to emerge.

It is important to emphasize that there are three features of laser light that distinguish it from ordinary light produced by spontaneous emission:

- The laser beam is intense because the light beam between the mirrors passes many times through the amplifying medium.
- The laser beam is monochromatic and is described by an almost pure sine wave. This is because the beam, moving to and fro between the mirrors, forms a standing wave. Thus an integral number of half-wavelengths must fit exactly between the mirrors. No other wavelengths can be amplified. This makes laser light much more monochromatic than light emitted by spontaneous emission.
- Laser light forms a narrow, well-directed beam, which spreads at large distances by diffraction only. The spread by diffraction is through an angle of only about λ/d where d is the beam diameter. This angle is normally only a few milliradians. The high directionality is explained by the fact that all the photons produced by stimulated emission travel in the same direction. Again, this is unlike the radiation produced by spontaneous emission, which is emitted in all directions and spreads out in spherical waves from the source. A sodium street lamp, for example, emits the yellow D line radiation as spontaneous emission.

Question 3.23 A laser emits a beam of light with a circular cross-section of diameter 1 mm (and wavelength $\lambda = 633$ nm). How big is the spot 100 m away? ■

Box 3.3 Applications of laser light

The special features of laser light have led to several important applications. In a famous scene from a James Bond film, a laser appears to cut through a metal plate (and threatens to do the same to our hero). At the time the film was made, this was not possible: early lasers were far too weak to perform such feats. Nowadays, however, high-power lasers are indeed used for cutting, and for welding sheets of metal together (see Figure 3.1 at the beginning of this chapter).

One of the more spectacular applications of high-power lasers is in the attempt to generate useful energy by controlled nuclear fusion. Several different approaches are currently being investigated and it is not yet clear which, if any, will eventually succeed. In the approach based on laser light, a tiny pellet of solid material is surrounded by powerful lasers so that light is shone on the pellet symmetrically from many different directions. The lasers all fire simultaneously, delivering 200 kJ energy to the pellet in less than a nanosecond (10^{-9} s). As a consequence, the pellet heats up very rapidly and violently implodes. The aim is to make the implosion so rapid that individual nuclei in the pellet are driven into contact with one another. This requires great force because the nuclei have to overcome their mutual Coulomb repulsion, but, if it can be made to work, nuclear fusion will ensue and immense power will be liberated — one would hope in a controlled way. It is a testament to the power of modern lasers that such an approach is even being considered!

The fact that laser light forms a narrow beam with a very well-defined direction can be used to measure distances to extreme accuracy. For example, the *Apollo 14* astronauts left behind a special reflector on the Moon's surface. The distance from an Earth-bound observatory to the Moon was then determined by firing a laser pulse through a telescope and measuring the time

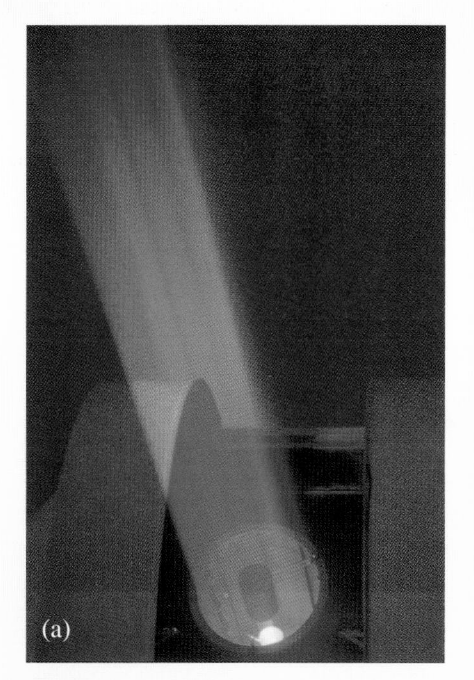

Figure 3.41 (a) A pulse of laser light fired at the Moon from the observatory near Grasse, France. The pulse is of high power and spreads by only a few kilometres. (b) A special mirror, left on the moon by *Apollo 14* astronauts, reflects the pulse back to Earth. By measuring the round trip time of transit, the distance of the mirror is found.

of flight of its round trip to and from the detector (Figure 3.41a, b). In this way, the distance between the telescope and the reflector was measured to within a few centimetres, and a series of measurements has confirmed that the Moon is receding from us at the rate of a few centimetres per year.

The fact that laser light is highly monochromatic ensures that any interference patterns produced will be very clear and easy to interpret. This has an application in compact disc (CD) players, where an interference pattern is produced between a reference laser beam and a laser beam that is reflected from steps cut in the playing surface of the disc. Another famous application of interference of laser light is the production of *holograms* or 'three-dimensional photographs'. In order to prepare the hologram of an object, light from a laser is split into two beams by a half-silvered mirror. One beam illuminates the object, is scattered from it, and then falls onto a photographic plate. The other beam proceeds directly to the photographic plate without encountering the object at all. The photographic plate therefore records the *interference pattern* between these two beams. At first sight, the developed image looks rather unpromising with no clear pattern evident. However, if the image is illuminated with laser light from the original laser, a *three-dimensional* image of the object is revealed. If you change your viewpoint, you will observe the effects of parallax in the image. In particular, objects that are hidden behind others from the initial viewpoint will come into view from other angles.

Question 3.24 The scheme suggested in Figure 3.37 involves stimulated emission from an excited level down to the ground state, while that shown in Figure 3.39 involves stimulated emission between two excited levels. For an operational laser, the second scheme is likely to be more successful than the first. Can you explain why?

Question 3.25 In laser fusion, light of wavelength 1.1×10^{-6} m is emitted in a 200 kJ pulse. How many excited atoms have undergone stimulated emission in forming this pulse? ■

7 Closing items

7.1 Chapter summary

1 The electron in the hydrogen atom is subject to the Coulomb interaction with the proton in the nucleus and it is the electrostatic potential energy associated with this interaction ($-e^2/4\pi\varepsilon_0 r$) that is inserted into the Schrödinger equation for the electron in hydrogen.

2 Because of the spherical symmetry of the hydrogen atom, the wavefunctions are most conveniently expressed in spherical polar coordinates (r, θ, ϕ). Each stationary state wavefunction $\psi(r, \theta, \phi)$ can be written as the product of three wavefunctions that each depend on only one of the coordinates (r, θ, ϕ). That is

$$\psi(r, \theta, \phi) = \psi_1(r) \times \psi_2(\theta) \times \psi_3(\phi). \tag{3.2}$$

3 When the electron is confined within the hydrogen atom (i.e. when it has $E_{\text{tot}} < 0$) the solutions to Schrödinger's equation are stationary state wavefunctions each of which has a definite energy.

4 The energy levels of the hydrogen atom are given by

$$E_{\text{tot}} = -\frac{1}{n^2}\left\{\frac{m_e e^4}{8h^2\varepsilon_0^2}\right\} = -\frac{13.6}{n^2}\ \text{eV} \quad \text{(where } n = 1, 2, 3, \ldots \text{ etc).}$$

There is also a continuum of positive energy states available to the electron. These correspond to the states of an ionized hydrogen atom.

5 Three quantum numbers n, l and m_l are required to specify the wavefunction of the electron in hydrogen according to Schrödinger's equation.

6 The principal quantum number n is chiefly responsible for determining the energy of the electron. It can take any integer value.

7 The orbital angular momentum quantum number l determines the magnitude of the electron's orbital angular momentum L according to

$$L = \sqrt{l(l+1)}\,\hbar \quad l = 0, 1, 2, \ldots, n-1. \tag{3.4 and 3.5}$$

8 The orbital magnetic quantum number m_l determines the component of the electron angular momentum along an arbitrarily chosen z-axis, according to

$$L_z = m_l\hbar \tag{3.6}$$

where m_l can take the values

$$m_l = 0, \pm1, \pm2, \ldots, \pm l. \tag{3.7b}$$

9 In the absence of a magnetic field, the energy levels of the electron in hydrogen are degenerate. That is, states with the same value of n but different values of l and m_l have the same energy.

10 When a magnetic field is applied, the energy levels of hydrogen are split due to a magnetic energy term E_{mag} that depends on m_l, namely

$$E_{mag} = m_l \left(\frac{e\hbar}{2m_e} \right) B_{ext}. \tag{3.8c}$$

11 Standard spectroscopic notation is used to characterize the wavefunctions of electrons in atoms.

12 The spatial distribution of the electron in hydrogen is often illustrated using the electron cloud picture, which represents the probability density $|\psi(r, \theta, \phi)|^2$ by the density of a pattern of dots.

13 All states with $l = 0$ are spherically symmetric, whilst those with $l \geq 1$ have distributions that depend on θ. However, $|\psi_3(\phi)|^2$ is constant for all states.

14 For spherically symmetric s states the radial probability density $4\pi r^2 |\psi|^2$ is the probability per unit radial distance of finding the particle at radius r.

15 Atoms can make transitions between energy levels by absorbing or emitting photons of light of an appropriate energy. The frequency of the light is given by $f = (E_2 - E_1)/h$, where E_1 and E_2 are the energies of the lower and upper levels, respectively. These radiative transitions are governed by selection rules, in particular, they must obey

$$\Delta l = \pm 1 \quad \text{and} \quad \Delta m_l = 0 \text{ or } \pm 1. \tag{3.10 and 3.11}$$

Collisional transitions are not subject to the same selection rules.

16 The electron has an intrinsic angular momentum called spin which has magnitude S, and which is determined by its spin angular momentum quantum number s, such that

$$S = \sqrt{s(s+1)}\,\hbar. \tag{3.12}$$

The quantum number s can take only one value: $\frac{1}{2}$.

17 The z-component of the electron's spin relative to an arbitrarily defined z-axis is determined by the spin magnetic quantum number m_s such that

$$S_z = m_s\hbar. \tag{3.13}$$

The quantum number m_s has only two possible values: $+\frac{1}{2}$ and $-\frac{1}{2}$.

18 Schrödinger's equation can be applied to many-electron atoms but it is much more difficult to find the potential energy function. Nevertheless, the stationary states in heavy atoms can be specified by the same quantum numbers (n, l, m_l and m_s) as for the electron in a hydrogen atom.

19 Electrons are fermions and obey the Pauli exclusion principle. This means that no two electrons in an atom can have the same set of four quantum numbers. This means that they must occupy the available quantum states, filling from the lowest energy upwards and obeying Hund's rule. This states that, in the ground state of an atom, the total spin of the electrons always has its maximum possible value. This ordering in the filling up of available states (see Figure 3.29) gives rise to the Periodic Table of the elements.

20 Heavy atoms have much more complicated energy-level diagrams than hydrogen, and transitions between these energy levels gives rise to complicated spectra. In particular, in hydrogen-like alkali metal atoms, the penetration of closed shells lowers the energy of s and p states. Also, in many atoms, including sodium and helium, spin–orbit interaction gives rise to doublets and higher multiplets of lines.

21 Laser light can be produced by maintaining a population inversion on an atomic transition and utilizing the process of stimulated emission on that transition.

7.2 Achievements

After studying this chapter, you should be able to:

A1 Explain the meaning of all the newly introduced (emboldened) terms in this chapter.

A2 Understand and use the notation associated with a spherical polar coordinate system.

A3 Describe qualitatively how the Schrödinger equation is used to study the hydrogen atom.

A4 Recognize the allowed combinations of l and m_l for a given principal quantum number n of the electron in the hydrogen atom. Calculate the magnitude and the z-component of the orbital angular momentum of the electron in a hydrogen atom, given the values of l and m_l.

A5 Discuss qualitatively the effects of an external magnetic field on the energy levels of atomic hydrogen (neglecting the effects of electron spin).

A6 Use spectroscopic notation to specify different quantum states.

A7 Make a critical comparison between the predictions of the model of the hydrogen atom based on Schrödinger's equation and the corresponding predictions of Bohr's model.

A8 State whether a given radiative transition is allowed or forbidden using the appropriate selection rules.

A9 Explain how the spin angular momentum of an electron is specified by its spin angular momentum quantum number s, and by its spin magnetic quantum number, m_s.

A10 Deduce the constituents of a neutral atom, given the atomic number and mass number of its nucleus.

A11 Write down, in box notation and in standard notation, the electronic structure of an element of a given atomic number, using Pauli's exclusion principle, Hund's rule and the information given in Figure 3.29.

A12 Write down, in standard notation, the structure of the outer shell of a typical noble gas atom in its ground state, and deduce the structures of the outer shells of the elements in Groups I–VII.

A13 Describe qualitatively the features of the energy levels of heavy atoms including the origin of doublet and triplet structures and the lowering of s and p state energies.

A14 Describe the nature of radiative transitions between atomic energy levels and how this can lead to the production of laser light.

7.3 End-of-chapter questions

Question 3.26 (a) How many quantum numbers are required to completely specify the wavefunction of an electron in an atom?

(b) An electron is in the $n = 4$ energy level in hydrogen. How many states are there with this energy, and what are the quantum numbers associated with each state?

Question 3.27 A hydrogen atom is placed in a magnetic field of magnitude 10^{-2} T. What is the resulting energy difference between the $m_l = \pm 1$ states and the $m_l = 0$ state in the 2p level?

Question 3.28 Figure 3.42 shows some of the energy levels available to a hypothetical alkali-type element. How many different radiative transitions are possible, connecting the levels shown on the diagram? Draw the allowed transitions on the figure.

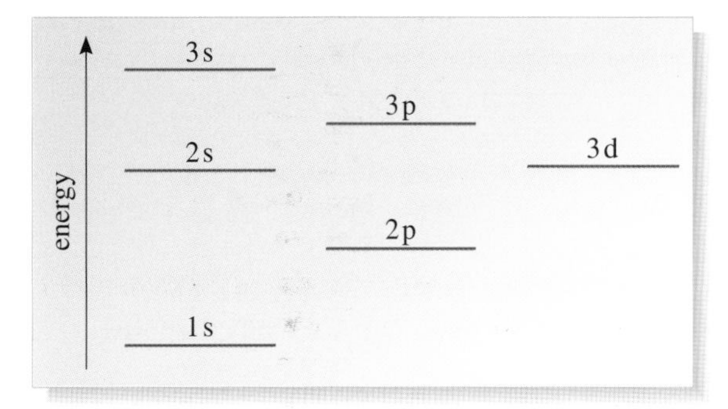

Figure 3.42 For use with Question 3.28.

Question 3.29 The atomic number of oxygen (symbol O) is 8. Write down, in both box notation and standard notation, the electronic structure of oxygen in its ground state.

Question 3.30 According to Einstein, the probability of stimulated emission between two atomic energy levels is the same as the probability of absorption between the same two levels (see Figure 3.37a and c, for example). Use this fact to show that, for a laser to produce amplification, it is necessary to have more atoms in the upper level than in the lower level of the laser transition (i.e. a population inversion). ■

Chapter 4 The interpretation of quantum mechanics

1 Formalism vs. interpretation

Is quantum mechanics a successful scientific theory? What you have learnt in the first three chapters of this book clearly shows that the answer is "Yes". Quantum mechanics permits the accurate calculation of many quantities that cannot be predicted in any other way. The frequencies and relative intensities of spectral lines, for example, were utterly mysterious when they were first observed, yet they can now be predicted with great accuracy — thanks to quantum mechanics. Some of the most accurate predictions in the whole of science have been made using quantum mechanics.

This enormous success has, however, been achieved at a price. Quantum-mechanical predictions generally involve complicated mathematical procedures, centred on the use of Schrödinger's equation to determine wavefunctions, and the detailed investigation of those wavefunctions. These mathematical aspects of quantum mechanics constitute the **formalism** of the subject: the equations and protocols that allow quantum mechanics to make its many successful predictions. By almost any standards, the development of the formalism of quantum mechanics has been one of the great success stories of modern science, but the resulting theory is highly mathematical and rather abstract. Some of its most important elements, such as wavefunctions, seem rather distant from common experience and everyday language, even though they appear to be essential to the success of the theory.

Because of the gulf that exists between the mathematical formalism of quantum mechanics and the everyday world of things and phenomena, there is a widely felt need for an **interpretation** of quantum mechanics — a detailed account of the physical significance of the different parts of the formalism. There is no such need in classical mechanics, because there the 'interpretation' is obvious and self-evident. The interpretation of quantum mechanics is far more challenging, and much more unsettling. You encountered an example of this in Chapter 2, when Max Born's interpretation of the wavefunction was introduced. According to Born the quantity $|\Psi(\boldsymbol{r}, t)|^2 \Delta V$ should be identified as the probability, at time t, of finding a particle whose wavefunction is $\Psi(\boldsymbol{r}, t)$ in a small volume ΔV centred on the point with position vector $\boldsymbol{r}$. The introduction of this particular element of interpretation was quite revolutionary, since it marked the recognition of the key role of probability in quantum mechanics. However, it certainly did not settle all of the questions that surround the interpretation of quantum mechanics. Born's interpretation of the wavefunction marked the beginning of a debate, not the end of one.

Today, over seventy years after the creation of quantum mechanics, there is a widespread feeling that physicists have learned to use the theory without really understanding it. The formalism has been mastered, but not the interpretation. The purpose of this chapter is to highlight those features of quantum mechanics that present the greatest challenges to simple interpretation, and to describe some of the main ideas that have emerged from the ongoing effort to understand what quantum mechanics is really about.

2 Description and prediction in quantum mechanics

This section is mainly concerned with the way in which quantum mechanics is used to describe the state of a physical system, and to predict measurable properties associated with that state. That is to say, it is concerned mainly with formalism. It starts by reviewing and elaborating those parts of the formalism that were introduced in Chapters 2 and 3. It then extends the formalism by introducing the related concepts of *eigenstate* and *eigenvalue*, and by explaining the importance of *linear superposition* and *coherence* in quantum mechanics. The section ends with a first look at what is generally regarded as the 'conventional' interpretation of quantum mechanics and at some of the issues that it raises.

2.1 Quantum systems

The starting point of any application of quantum mechanics is the identification and specification of the system to which the formalism will be applied. Such a system is often referred to as a **quantum system**, though its specification generally involves purely classical concepts. You have already met many examples of quantum systems. The simplest, considered in Chapter 2, consisted of a particle of mass m that was free to move in one dimension. In that chapter you were also introduced to a variety of other one-dimensional systems in which the potential energy of the particle was described by a function $E_{\text{pot}}(x)$ that was either a well, a barrier or a step. Some of these potential energy functions are illustrated in Figure 4.1.

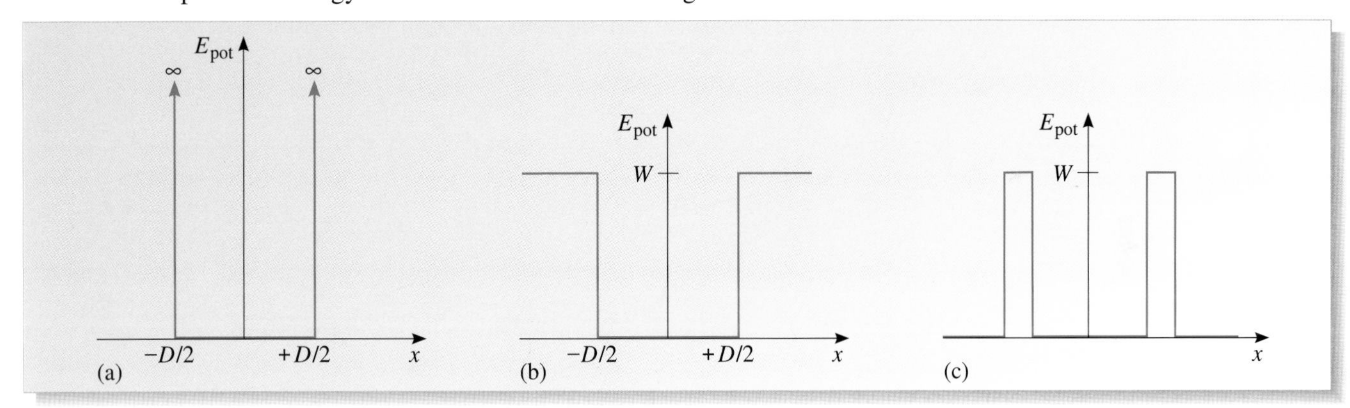

Figure 4.1 Some of the one-dimensional potential energy functions that were considered in the discussion of quantum mechanics in Chapter 2.

A more complicated quantum system formed the main subject of Chapter 3 — the hydrogen atom. In that case, the system consisted of an electron (a particle of mass m_e, charge $-e$, etc.) moving in three dimensions under the influence of a positively charged nucleus. The details of the nucleus were ignored; all of its effects on the electron were accounted for by assuming that the electron's potential energy was described by the function $E_{\text{pot}}(r) = -e^2/(4\pi\varepsilon_0 r)$, with the radial coordinate r being measured from the centre of the nucleus. This potential energy function is shown in Figure 4.2.

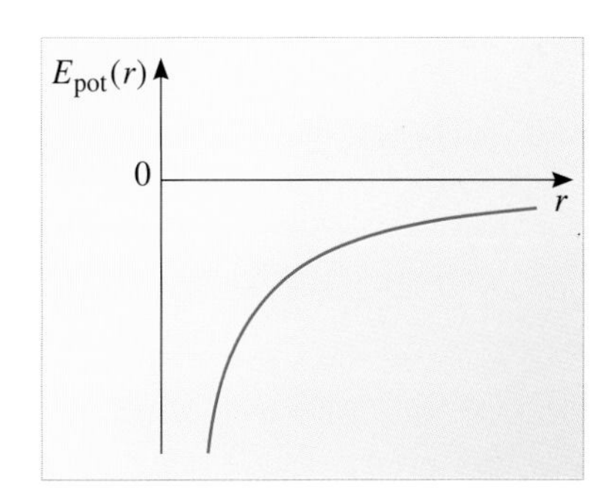

Figure 4.2 The electrostatic potential energy function for the electron in the hydrogen atom in Chapter 3.

Another example of a quantum system is provided by the idealized two-slit electron diffraction set-up shown in Figure 4.3. In this case, electrons (i.e. particles of mass m_e and charge $-e$) travelling from the electron gun G to the fluorescent screen S, encounter a diaphragm D that is impenetrable apart from two narrow slits, each of width w, the centres of which are separated by a distance d. Although this system was introduced in Chapter 2, it was not treated there in anything like the detail of the systems described above. Had it been subjected to such detailed analysis we would have found it useful to introduce a coordinate system of the kind shown in Figure 4.3, and we would have described the influence of the diaphragm in terms of some potential energy function $E_{pot}(x, y, z)$ that would have involved the parameters w and d amongst others. This would have been a crucial part of our specification of the system.

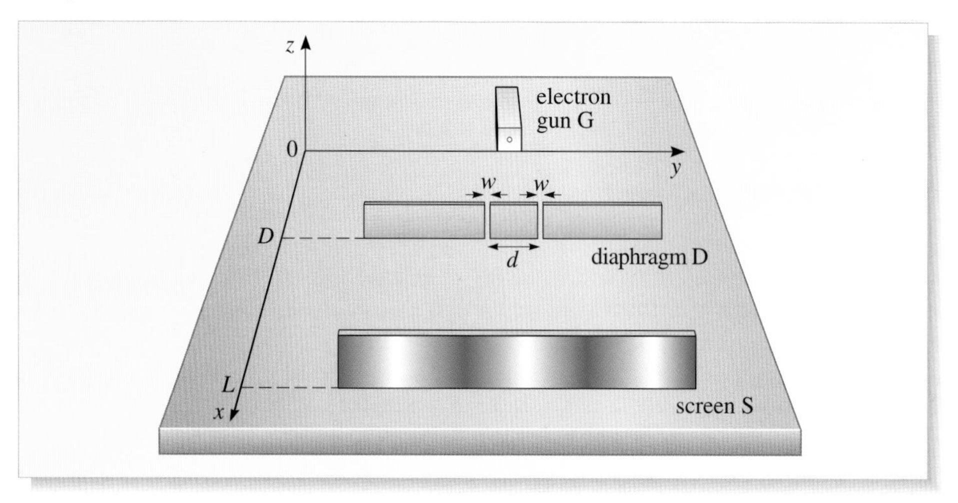

Figure 4.3 A two-slit electron diffraction set-up. Electrons from an electron gun G, travelling towards a fluorescent screen S, encounter a diaphragm D that contains two narrow slits. For the purposes of quantum-mechanical analysis, the effect of the diaphragm will generally be represented by a potential energy function $E_{pot}(x, y, z)$ that involves the parameters w and d. The screen shows the kind of interference pattern that would be observed when an intense steady beam of electrons flows through the apparatus. The bright bands correspond to places where large numbers of electrons arrive, and the dark bands are places where few, if any, arrive. (Note that the slit width and separation are shown greatly exaggerated.)

The main point to take away from this discussion is the following:

> The specification of a quantum system typically involves a number of parameters (such as m_e and $-e$) and a potential energy function $E_{pot}(x, y, z)$.

As you will see in the next subsection, these items of information, that are specific to a particular system, can be used as inputs when the time comes to apply the general formalism of quantum mechanics to the system they describe.

2.2 States and observables

Having specified a system of interest, the next step in applying the formalism of quantum mechanics is to investigate the possible states of that system. The **state** of a system refers to the condition of the system, and implies a sufficiently detailed description of that condition to distinguish it from any other condition that would cause the system to exhibit different behaviour.

Both classical mechanics and quantum mechanics make use of the concept of state, and both kinds of mechanics relate the concept of a state to measurable properties such as position, energy and momentum. These measurable properties of a physical system are referred to as the **observables** of the system, and it is the values of a system's observables that are generally the focus of experimental interest. However, where classical mechanics and quantum mechanics differ is in the way they describe the state of a system, and in the way they relate the state to the values of the system's observables. These differences are profound and important, so we shall discuss them in some detail.

In classical mechanics, specifying the state of a system involves assigning values to sufficient of its observables to uniquely determine the state. For instance, to specify the state of a particular golf ball, you might give the simultaneous values of its position, its velocity, its acceleration and its angular momentum. These values would determine the condition of the golf ball at some particular time, and could be used as the starting point for an equally detailed prediction of its condition at some future time.

Contrast this with the situation in quantum mechanics. There, as you saw in Chapter 2, specifying the state of a system generally implies specifying the time-dependent wavefunction Ψ that corresponds to the state. The wavefunction is found by solving an appropriately formulated version of the time-dependent Schrödinger equation — one that takes full account of any system-specific parameters such as masses and charges, as well as the relevant potential energy function. Not until the Schrödinger equation has been solved and the appropriate wavefunction determined is it possible to make detailed statements regarding the values of observables such as position or momentum. Even then, the sort of statements that quantum mechanics permits are quite different from those allowed in classical mechanics. In the quantum-mechanical case, Heisenberg's uncertainty principle will limit the extent to which the values of certain pairs of observables can be simultaneously determined, and for any given observable the most that can generally be predicted *in advance* of a measurement are the *possible* outcomes of that measurement and the *probability* of each of those possible outcomes. The following example will help to illustrate this.

Consider again the two-slit electron diffraction experiment illustrated in Figure 4.3, but suppose now that the intensity of the electron beam is reduced to such an extent that electrons pass through the diffraction apparatus one at a time. What is observed under these circumstances was discussed in Chapter 2. The arrival of a single electron at the screen S will create a single spot of light on the screen, thus revealing the particle-like nature of the electron (Figure 4.4a). However, if, over a period of time, the spots created by the arrival of a large number of individual electrons are recorded, then it will be found that they form an interference pattern similar to that

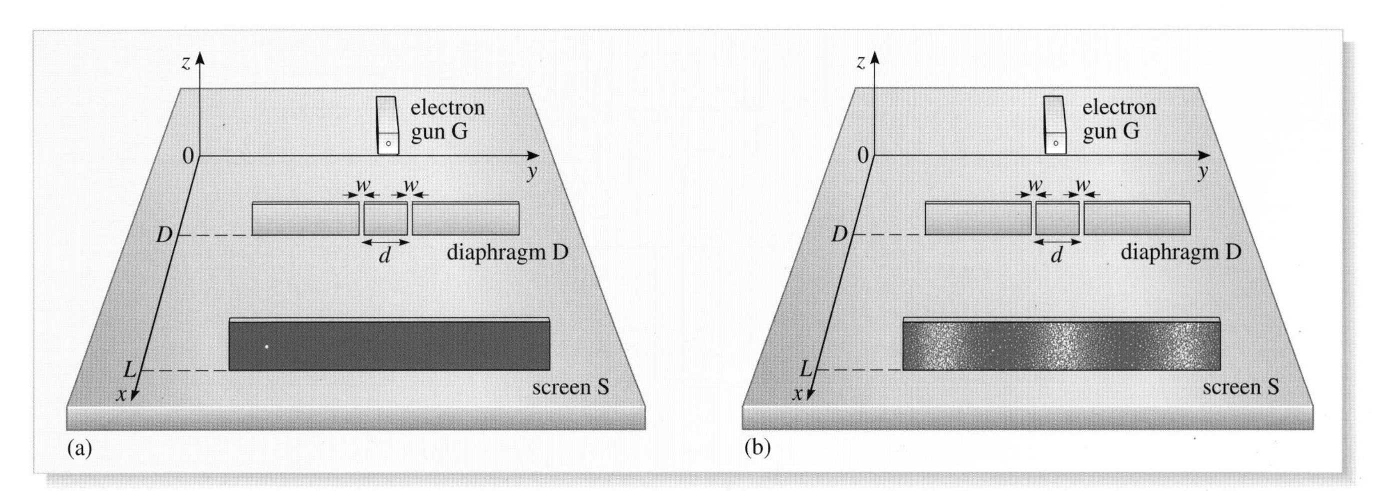

Figure 4.4 (a) In a two-slit electron diffraction experiment, the arrival of a single electron at the screen creates a single spot of illumination. (b) A record of the spots created by many such individual electrons will reproduce the interference pattern produced by an intense beam of electrons.

created by a steady intense beam (see Figure 4.4b). This is a clear indication of the wave-like nature of the electron. It is the task of quantum mechanics to provide a coherent account of the behaviour of a system that exhibits this classically contradictory wave–particle duality. This is what Schrödinger's wave mechanics achieves, by using a wavefunction to describe the state of the system.

The state of an electron that sets out from the electron gun, at $x = 0$, at some time close to $t = 0$, and travels towards the screen, can be described by a wavefunction $\Psi(x, y, z, t)$ that takes the form of a (complex) travelling wave packet. At time $t = 0$, when the electron is just starting out on its journey to the screen, its wavefunction will be concentrated near the electron gun. As t increases, the wave packet will evolve, moving towards the screen and spreading out as it does so, as indicated in Figure 4.5. (The increasing spread of the wavefunction is a consequence of the unavoidable uncertainty in the initial momentum of the localized electron.) The encounter with the slits will divide the wave packet into two parts, but they are both parts of the same wave packet and still describe the state of a *single* electron. As time continues to pass, the two parts of the evolving wavefunction will overlap and interfere. At time $t = T$, when the electron might be observed to arrive at the screen, the value of the wavefunction at a point on the screen with coordinates (L, y, z) will be given by $\Psi(L, y, z, T)$. The range of y and z values for which $\Psi(L, y, z, T)$ is not zero determines the *possible* outcomes of a measurement of the electron's arrival point on the screen, while the function $|\Psi(L, y, z, T)|^2$ determines the relative *probability* of each of those possible outcomes. The fact that this latter function has the form of a two-slit interference pattern ensures that the detection of many individual electrons, each described by a similar wavefunction, will eventually create the observed interference pattern.

More will be said about the nature and significance of measurement in quantum mechanics in the next section.

Observing a spot on a fluorescent screen at $x = L$, constitutes a *measurement* of the electron's position (specifically, of its y- and z-coordinates). The process of measurement is highly significant in quantum mechanics, since it is measurement that provides information about the values of observables at a particular time. Indeed, in quantum mechanics, statements about the value of an observable only become meaningful in the context of a measurement. In the absence of an actual measurement, the most we can say about the value of an observable is what we can learn from the wavefunction that describes the state of the system, and that, as we have already stressed, is limited to the *possible* outcomes of a measurement and the *probability* of each of those possible outcomes.

Nobody who has studied classical physics would have much difficulty in understanding a statement such as, 'An electron moving with constant velocity $\boldsymbol{v} = (1, 0, 0)\,\mathrm{m\,s^{-1}}$, and which passes through the origin at time $t = 0$, will reach the point $\boldsymbol{r} = (1, 0, 0)\,\mathrm{m}$ at time $t = 1\,\mathrm{s}$.' However, this is just the kind of statement that quantum mechanics does not allow us to make about an electron. There are two general features of quantum mechanics that limit what can be said about measurements of observables.

1. **Indeterminacy**. This term describes the impossibility of measuring, with total precision, the values of all the observables of a quantum system simultaneously. It is always possible to find pairs of observables that cannot be precisely measured at the same time. Heisenberg's uncertainty principle for position and momentum provides the best known example of this, but indeterminacy is a general feature of quantum mechanics, the uncertainty principle for position and momentum simply brought it to light.
2. **Indeterminism**. This term describes the fact that, in quantum mechanics, determining the precise values of some observables at a particular time will not

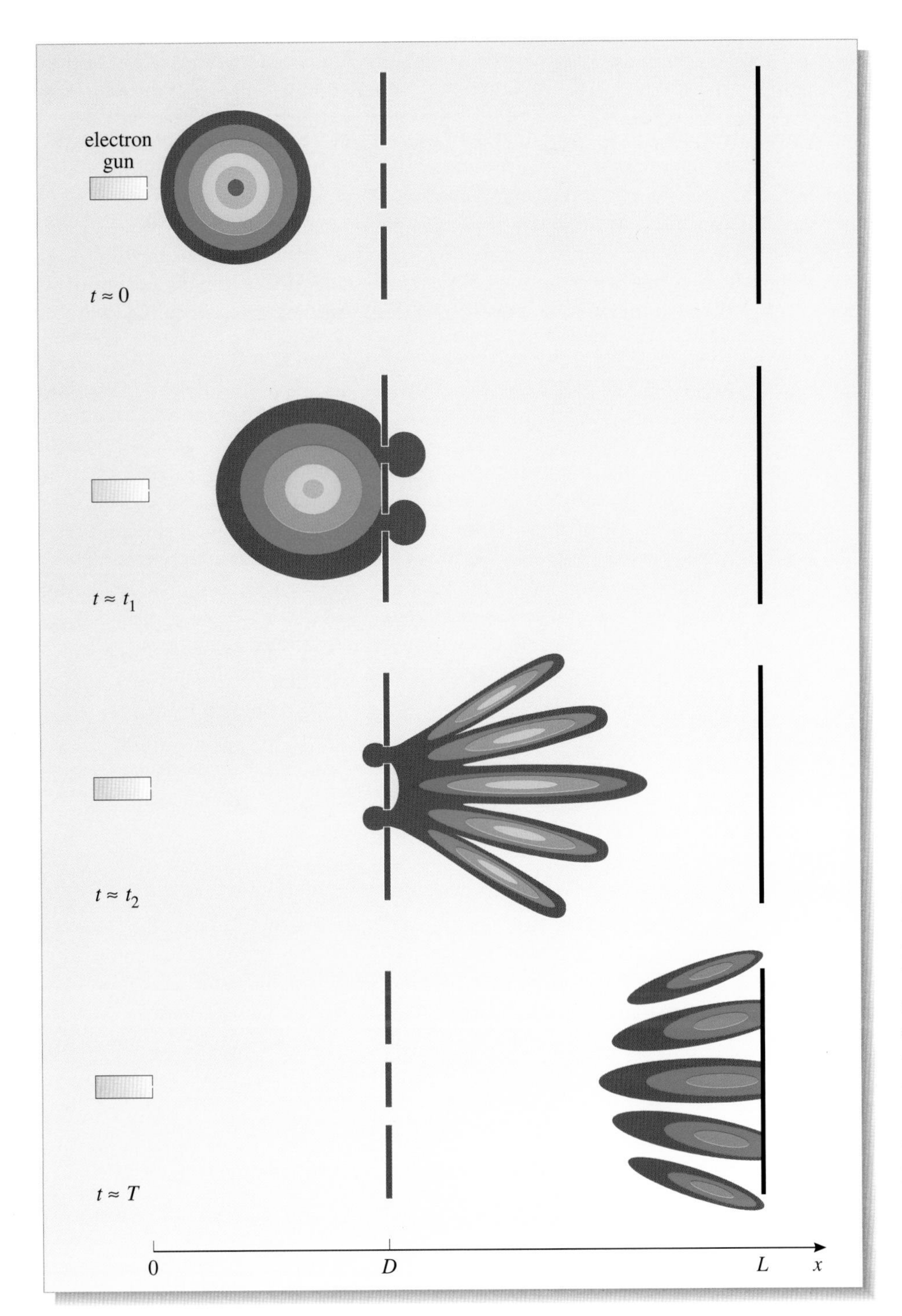

Figure 4.5 The changing wavefunction of a single electron in a two-slit diffraction experiment. The wavefunction itself is a complex quantity, so it cannot be easily illustrated; for this reason the diagram actually shows the value of the (real) positive quantity $|\Psi|^2$ at various times, and it displays these values as contour plots. Even so, the figure is highly schematic and makes no attempt to accurately represent the detailed wavefunctions that can be computed for this situation.

generally determine the precise values of those observables at later times. This is in sharp contrast to classical mechanics, which is fully deterministic, implying that measurements of certain observables at one time do determine the values of those observables at later times.

It is important to realize that the indeterminacy and indeterminism of quantum mechanics are both matters of principle. They are consequences of the mathematical structure of the formalism and have nothing to do with practical considerations about the difficulty of taking accurate experimental readings or anything of that nature.

When thinking about the states and observables of a quantum system it is always useful to keep indeterminacy and indeterminism in mind. This will help you to avoid using concepts that cannot be justified. In the case of two-slit electron diffraction, for example, it is tempting to suppose that the electron follows a definite but unknown path through the apparatus, and that the use of wavefunctions simply provides a way of dealing with our ignorance of the details of the motion. This, however, is not the case. According to the conventional interpretation of quantum mechanics, the electron's position only has a value when it is measured. In the absence of a measurement, the electron's position is not just unknown, but *indeterminate*. Prior to the measurement, the electron is not at some unknown position, rather, it is not at any position at all.

In the absence of a measurement, the formalism of quantum mechanics neither provides nor requires any information about the precise position of the electron at any time. The coordinates x, y and z that appear in the argument of the wavefunction $\Psi(x, y, z, t)$ simply label points where the electron *might* be found if a measurement is performed — they are *not* the instantaneous position coordinates of the electron at time t. By analysing the wavefunction it is possible to determine all the possible outcomes of a measurement of position (or, in fact, of any other observable) at time t, and to predict the probability of each of those possible outcomes. However, the formalism of quantum mechanics does not give any general procedure for predicting the precise outcome of any particular measurement. The prediction of possible outcomes and their respective probabilities is as far as quantum mechanics goes. This point was made at the start of our discussion of quantum mechanics, in Chapter 2, but it is of such importance that it deserves to be highlighted again.

In quantum mechanics the state of a system is represented by a wavefunction. By analysing the wavefunction, the possible outcomes of any measurement of a given observable can be predicted, and so can the probability of each of those possible outcomes. This, however, is as far as quantum mechanics goes. It is not generally possible to predict the precise outcome of any particular measurement.

Although there are general techniques for carrying out the analysis of wavefunctions mentioned above, we shall not pursue them here. Instead, in the next two subsections, we shall investigate some special types of state in which the prediction of possible measurement outcomes and their probabilities is especially simple. First, however, we need to recall another point about wavefunctions and probabilities that was first introduced in Chapter 2; the process of *normalization*.

A probability value of 1 is conventionally taken as an indication of certainty. A tossed coin has a probability of 0.5 of coming down as heads and a probability of 0.5 of coming down tails, but it is certain to come down as one or the other — the probability of that is 1. In a similar way, there is a probability of 1 of finding, at some location, a particle that is known to be somewhere. In order that the wavefunction describing the state of such a particle should predict a probability of 1 for finding the particle somewhere, it is necessary that the wavefunction should satisfy a mathematical requirement known as a *normalization condition*. For a wavefunction $\Psi(x, y, z, t)$, this amounts to dividing the region where the particle might be found into many small regions, each of volume ΔV, calculating the probability of finding the particle in each of those regions, $|\Psi(x, y, z, t)|^2 \Delta V$, and then demanding that the sum of all those separate probabilities should be 1. (In practice, due to the continuous nature of space, this 'sum' is usually more properly represented by a definite integral.)

Wavefunctions that meet this condition are said to be *normalized*. It is generally desirable that wavefunctions should be normalized, since this simplifies their interpretation in terms of probability. However, it is not unusual to encounter functions that fail to satisfy the normalization condition, even though they are perfectly acceptable wavefunctions in all other respects. Fortunately, it is generally possible to normalize such functions simply by dividing them by an appropriate number. This process is called *normalization*.

You can generally assume that the wavefunctions you meet in this chapter have been normalized. In the few cases where this assumption is not valid, either steps will be taken to normalize them, or you will be asked to normalize them yourself as part of a question.

Question 4.1 At the beginning of this subsection it was said that specifying the state of a system involves providing 'a sufficiently detailed description of its condition to distinguish it from any other condition that would cause the system to exhibit different behaviour'. But you have since seen that identical measurements performed on the same state of a quantum system may lead to different outcomes. How might the words 'different behaviour' in the first sentence be interpreted in the context of quantum mechanics if that first sentence is to remain true? ■

2.3 Eigenstates and eigenvalues

Of course, it is always possible that the state of a quantum system will be such that a measurement of some selected observable has only one possible outcome. In other words that there is a possible outcome that is predicted to occur with probability 1. When this is the case, the state of the system is said to be an **eigenstate** of the selected observable, and the one value that will inevitably be obtained for the observable when the measurement is made is said to be the **eigenvalue** of the observable that corresponds to that particular eigenstate. (The word 'eigen' comes from the German for 'characteristic', and is a reminder that many of the founders of quantum mechanics were German speaking.)

Many of the wavefunctions that were discussed in Chapters 2 and 3 represented eigenstates of energy, since many of them corresponded to a single value of the energy observable. In those particular cases the states were also *stationary states*, since it was possible to separate the space and time dependence of Ψ, and write the wavefunction as a product of the form

$$\Psi(x, t) = \psi(x)\,\phi(t). \tag{4.1}$$

As Chapter 2 explained, the function of x represented by the lower case Greek letter psi, $\psi(x)$, describes the *profile* of the wavefunction at some particular time, and is said to be the 'time-independent wavefunction' of the state. Such functions are found by solving the time-independent Schrödinger equation

$$\frac{d^2\psi}{dx^2} + \frac{2m}{\hbar^2}(E_{\text{tot}} - E_{\text{pot}}(x))\psi = 0. \tag{4.2}$$

Each of the time-independent wavefunctions that satisfies this equation corresponds to some particular value of the total energy E_{tot}, which represents the eigenvalue of energy associated with that solution.

Figure 4.6 shows the time-independent wavefunctions of some of the energy eigenstates for a particle of mass m confined in an infinite square well of width D. These particular eigenstates correspond to different values of a *quantum number*, n, that has been used to label the different time-independent wavefunctions. As you saw in Chapter 2, this quantum number also appears in the expression for the corresponding energy eigenvalues:

$$E_n = \frac{n^2h^2}{8mD^2}. \tag{4.3}$$

The energy eigenvalues for $n = 1, 2, 3, 4$, and the potential energy function $E_{\text{pot}}(x)$ are also illustrated in the figure.

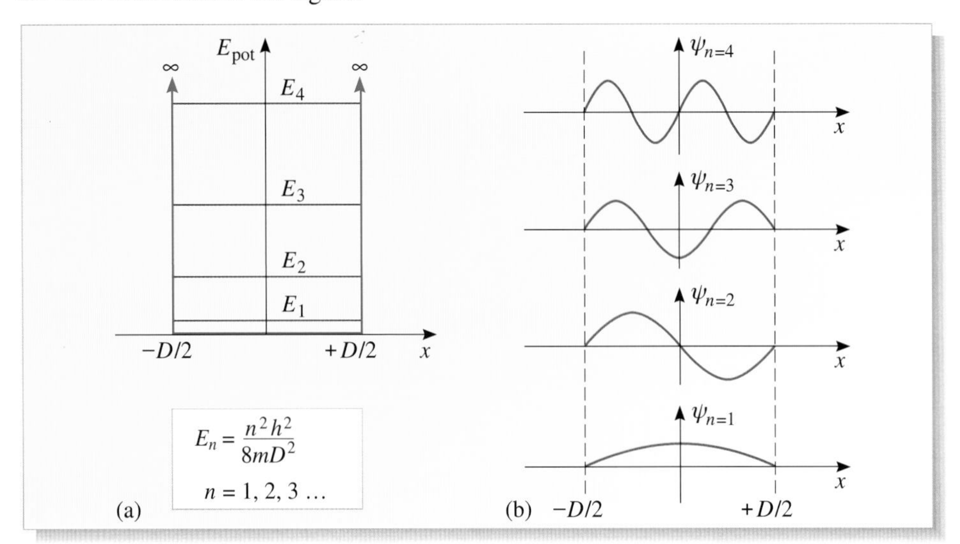

Figure 4.6 The time-independent wavefunctions (b) associated with some of the eigenstates of energy of a particle confined in an infinite square well, (a). The square well itself, and the energy eigenvalues that correspond to the eigenstates are also shown.

Question 4.2 What is the eigenvalue of energy associated with the (a) the 1s state of the hydrogen atom and (b) the 2p state? ■

The fact that the energy eigenstates of confined systems are also stationary states gives them a special significance and makes them especially easy to discuss. However, it is important to note that the concepts of eigenstate and eigenvalue are quite general and may be applied to any observable, not just to energy.

Eigenstates and eigenvalues play an important part in quantum mechanics. Historically, the series of scientific papers in which Schrödinger introduced his formulation of quantum mechanics (published in the German journal *Annalen der Physik*, in 1926) were collectively entitled 'Quantization as an Eigenvalue Problem'. The fact of the matter is, that according to quantum mechanics, the result yielded by a measurement of any observable must be an eigenvalue corresponding to some particular eigenstate of that observable. This is true, whatever the state of the system prior to the measurement. If we know the state prior to the measurement, then quantum mechanics enables us to predict which of the various eigenvalues is a possible result. It also allows us to calculate the probability of each of those possible results. But even if we don't know the state of the system, we can still be sure that when we measure some observable, the result will be an eigenvalue corresponding to *some* eigenstate of that observable. We shall have more to say about the significance of this later. For the moment, here is a reminder of the key points.

> The eigenvalues of a given observable are the only possible outcomes of any measurement of that observable. The state of the system (as described by its wavefunction) determines the relative probability of measuring each of those eigenvalues. A state in which a particular eigenvalue is predicted with probability 1 is said to be an eigenstate of the given observable and may be said to correspond to the relevant eigenvalue.

2.4 Superposition states

Many of the peculiarities of quantum mechanics can be traced to one fundamental feature of the mathematical formalism: the fact that any two solutions of a system's time-dependent Schrödinger equation can be combined to produce a third solution. We can use the wavefunctions of the particle in the infinite square well to illustrate this important fact.

Suppose that $\Psi_1(x, t)$ and $\Psi_2(x, t)$ are the normalized time-dependent wavefunctions that correspond to the time-independent wavefunctions $\psi_{n=1}(x)$ and $\psi_{n=2}(x)$ shown in Figure 4.6. Since $\Psi_1(x, t)$ and $\Psi_2(x, t)$ are both solutions to the time-dependent Schrödinger equation for a particle in an infinite square well, it follows from the mathematical nature of the Schrödinger equation that their sum $(\Psi_1(x, t) + \Psi_2(x, t))$ will also be a solution and will therefore represent a possible wavefunction of the system. This sum would not represent a *normalized* wavefunction, but we can easily scale it to produce the normalized wavefunction

$$\Psi(x, t) = \frac{1}{\sqrt{2}}[\Psi_1(x, t) + \Psi_2(x, t)]. \tag{4.4}$$

Such a wavefunction is said to be a normalized **linear superposition** of the wavefunctions Ψ_1 and Ψ_2, and the state described by that linear superposition may be referred to as a *superposition state*.

The reason why a superposition state of this kind is interesting, and so very characteristic of quantum mechanics, is that it combines features that would appear self-contradictory in a classical state. In the case of Equation 4.4, for example, the wavefunction consists of equal contributions from eigenfunctions corresponding to two different energy eigenvalues; $E_1 = h^2/8mD^2$ and $E_2 = 4h^2/8mD^2$. This means that if things are arranged in such a way that the particle in the infinite well is actually in the state described by Equation 4.4, then a measurement of the energy of that particle has *two* possible outcomes E_1 or E_2, and the probability of obtaining either result is 0.5. This implies that in, say, a thousand measurements of energy, performed on the same system, prepared each time in the same superposed state, the result will be E_1 on about 500 occasions, and E_2 on about 500 occasions. The fact that identical measurements repeatedly carried out on an identically prepared system can lead to different outcomes is typical of quantum mechanics, but it is totally at odds with classical mechanics.

The process of forming linear superpositions is not restricted to eigenfunctions of energy, nor is it restricted to equal combinations of two solutions of Schrödinger's equation. Any solutions of the relevant time-dependent Schrödinger equation may be linearly superposed, and we can combine as many of those solutions as we wish. The general rule for forming such combinations is known as the *principle of superposition*, and may be expressed as follows.

The principle of superposition

If $\Psi_1, \Psi_2, \Psi_3, \ldots \Psi_n$, are n solutions of the time-dependent Schrödinger equation for a given system, then that equation will also be satisfied by any linear superposition of the form

$$\Psi = c_1\Psi_1 + c_2\Psi_2 + c_3\Psi_3 + \ldots + c_n\Psi_n \tag{4.5}$$

where $c_1, c_2, c_3 \ldots c_n$ are numbers that are, in general, complex.

It should be noted that solutions to the time-independent Schrödinger equation cannot generally be superposed in this way.

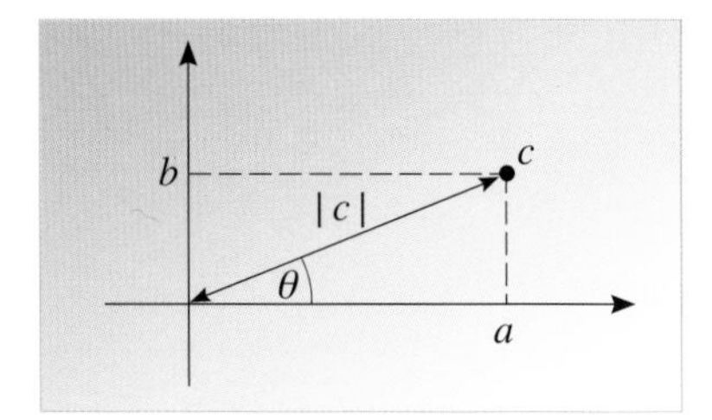

Figure 4.7 Complex numbers may be thought of as generalizations of ordinary real numbers to two dimensions. Any given complex number c may be specified in terms of its modulus $|c|$ and a polar angle θ. Alternatively, c can be specified in terms of its components along the horizontal and vertical axes (usually called the real and imaginary axes in this context): such specifications are usually written in the form $c = a + ib$, where i denotes an algebraic quantity with the property $i^2 = -1$.

As explained in Chapter 2, a complex number c involves *two* real numbers, typically, its modulus, $|c|$, and an angle, θ, (see Figure 4.7). A 'visualization' of the process of linear superposition is shown in Figure 4.8. As you can see, the process is similar to that of adding component vectors to produce a resultant vector, though in the case of linear superpositions there may well be more than three independent 'directions' and the coefficients c_1, c_2, etc. may be complex. We shall not pursue the analogy between linear superposition and vector addition here, but the relationship is a deep one.

● Write down the values of c_1 and c_2 in the linear superposition

$$\Psi = \frac{\Psi_1 + \Psi_2}{\sqrt{2}}.$$

❍ $c_1 = 1/\sqrt{2}$ and $c_2 = 1/\sqrt{2}$. (Real numbers, such as $1/\sqrt{2}$, are special cases of the more general class of complex numbers.) ■

The general linear superposition described by Equation 4.5 is not necessarily normalized. However, it will be normalized if two additional conditions are also satisfied:

1 The wavefunctions $\Psi_1, \Psi_2, \Psi_3, \ldots \Psi_n$ are individually normalized.

2 The numbers c_1, c_2, c_3, etc. are such that

$$|c_1|^2 + |c_2|^2 + |c_3|^2 + \ldots + |c_n|^2 = 1. \tag{4.6}$$

Provided *both* these conditions are satisfied, Ψ will be a normalized wavefunction, and the particle it describes will therefore have a probability of 1 of being found somewhere when its position is measured.

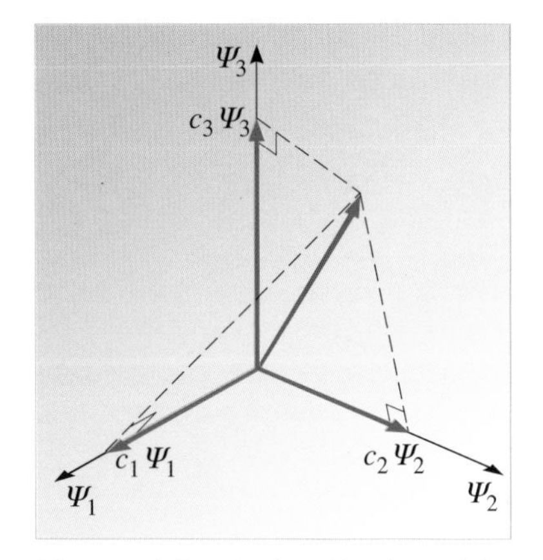

Figure 4.8 A visualization of the process of linear superposition. This figure is not supposed to represent any particular linear superposition, but rather to suggest an analogy between the linear superposition of wavefunctions (with complex coefficients) and the summation of vectors (with real coefficients).

● If Ψ_1 and Ψ_2 are individually normalized, is the linear superposition $\Psi = (\Psi_1 + \Psi_2)/\sqrt{2}$ also normalized?

❍ Yes, because $|c_1|^2 = \frac{1}{2}$ and $|c_2|^2 = \frac{1}{2}$, so $|c_1|^2 + |c_2|^2 = 1$. ■

The great advantage of dealing with *normalized* linear superpositions is that they lead to simple predictions concerning the probabilities of predicted measurement outcomes. This point is most easily made by considering an example. Suppose Ψ_1, Ψ_2 and Ψ_3 are three eigenstates of energy for some quantum system, corresponding to the energy eigenvalues E_1, E_2 and E_3. If the state of the system is represented by the linear superposition

$$\Psi = c_1\Psi_1 + c_2\Psi_2 + c_3\Psi_3 \tag{4.7}$$

then the only possible outcomes of a measurement of the energy will be E_1 or E_2 or E_3. Quantum mechanics does not allow us to say which of the three possible values will result from any particular measurement performed on the system while it is in the state described by Equation 4.7, but, provided the linear superposition has been normalized, quantum mechanics does predict the probability of each of the possible outcomes. If we denote the three probabilities by $P(E_1)$, $P(E_2)$ and $P(E_3)$, quantum mechanics predicts that:

$$P(E_1) = |c_1|^2$$

$$P(E_2) = |c_2|^2$$

$$P(E_3) = |c_3|^2.$$

Note that, because we are dealing with a *normalized* superposition, it is certain to be the case that $|c_1|^2 + |c_2|^2 + |c_3|^2 = 1$. This implies that, as long as the system remains in the superposed state, the probability that a measurement of the system's energy will result in one of the three possible outcomes is 1, i.e. it is certain.

This simple relationship between linear superpositions and predicted probabilities can be generalized to any linear superposition of eigenstates for any observable. In general, if we consider some observable A (not necessarily the energy) with eigenvalues $a_1, a_2, a_3, \ldots a_n$, then we can say the following.

Given a system in a state represented by the normalized wavefunction

$$\Psi = c_1\Psi_1 + c_2\Psi_2 + c_3\Psi_3 + \ldots + c_n\Psi_n,$$

where the n normalized wavefunctions $\Psi_1, \Psi_2, \Psi_3, \ldots \Psi_n$ represent eigenstates of some observable A, and correspond, respectively, to eigenvalues $a_1, a_2, a_3, \ldots a_n$, then the only possible outcome of any measurement of that observable is one of those eigenvalues, and the probability that it will be a_i is given by

$$P(a_i) = |c_i|^2 \quad (\text{for } i = 1, 2, 3, \ldots n). \tag{4.8}$$

It's worth stressing again that the wavefunctions Ψ_i referred to here, which represent eigenstates of the observable A, are not necessarily eigenstates of energy. The general indeterminacy of quantum mechanics that prevents particles from simultaneously having precisely determined positions and momenta, may actually prevent the eigenstates of the observable A from also being eigenstates of energy.

Question 4.3 A quantum system has been prepared in such a way that its state is described by the wavefunction $\Psi = (\Psi_1 + 2\Psi_2 - 2\Psi_3 + \Psi_4)/\sqrt{10}$, where the normalized wavefunction Ψ_k represents an eigenstate of an observable L, with Ψ_k corresponding to the eigenvalue $L_k = k\hbar$, for $k = 1, 2, 3, 4$. Confirm that Ψ is a normalized wavefunction, and write down the probabilities of obtaining the following results in a measurement of the observable: $\hbar$, $3\hbar$, $5\hbar$. ■

It is instructive to apply the idea of a superposition state to the two-slit electron diffraction experiment we discussed earlier. Figure 4.9 shows a modification of the original arrangement in which the slits are labelled 1 and 2, and one of the slits is blocked. If slit 2 is blocked, each electron must travel through slit 1, and the state of each of those electrons can be described by a wavefunction Ψ_1 that reflects this. Recording the arrival of a large number of those electrons on the screen S will produce the single-slit diffraction pattern shown in Figure 4.9, *not* the two-slit

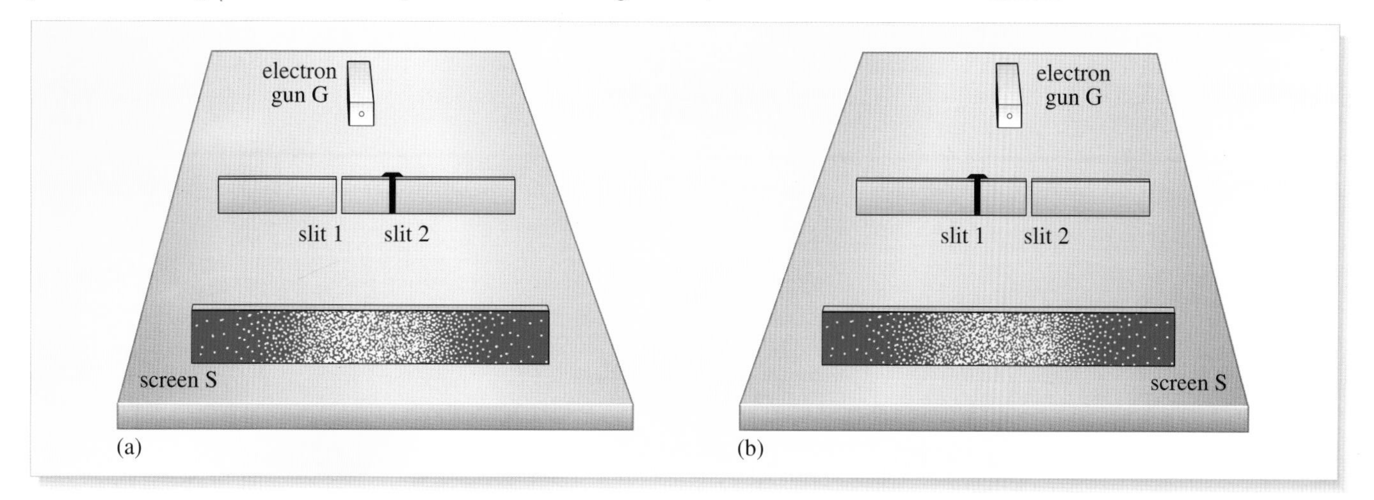

Figure 4.9 (a) A two-slit electron diffraction set-up in which slit 2 has been blocked. The screen shows a single-slit diffraction pattern rather than a two-slit interference pattern. (b) Opening slit 2 and closing slit 1 changes the wavefunction of electrons passing through the system, but hardly affects the pattern recorded on the screen.

Figure 4.10 The two-slit interference pattern recorded when both slits are open.

interference pattern of Figure 4.10. If slit 2 is opened and slit 1 closed, the state of each electron will be described by a somewhat different wavefunction, Ψ_2, but the pattern recorded on the screen will be practically unchanged — it will again be a single-slit diffraction pattern. However, if both slits are open, the state of each electron can be represented by the linear superposition

$$\Psi = \frac{1}{\sqrt{2}}(\Psi_1 + \Psi_2)$$

and the pattern recorded on the screen will be the two-slit interference pattern of Figure 4.10.

When only one slit is open, the intensity pattern will be proportional to either $|\Psi_1|^2$ or $|\Psi_2|^2$, but when both slits are open the intensity is proportional to

$$|\Psi|^2 = \left|\frac{1}{\sqrt{2}}(\Psi_1 + \Psi_2)\right|^2 = \frac{1}{2}|\Psi_1 + \Psi_2|^2 . \tag{4.9}$$

It is this latter intensity distribution that accounts for the interference effects seen when both slits are open. As a result, when both slits are open there will be points on the fluorescent screen where electrons will never be observed, even though electrons may be seen at those points when either slit 1 or slit 2 is open on its own.

The important point to note about this is that the function represented by $|\Psi_1 + \Psi_2|^2$ is quite different from the function represented by $|\Psi_1|^2$ or $|\Psi_2|^2$ or even $|\Psi_1|^2 + |\Psi_2|^2$. To say that the wavefunction of a system is $\Psi = (\Psi_1 + \Psi_2)/\sqrt{2}$ is *not* the same as saying that the wavefunction is *either* Ψ_1 *or* Ψ_2, even with probabilities attached to the two alternatives. The wavefunctions Ψ_1 and Ψ_2 individually correspond to classically comprehensible statements about the electron — 'it went through slit 1' or 'it went through slit 2' — but the linear superposition $(\Psi_1 + \Psi_2)/\sqrt{2}$ does not. In particular, it does not correspond to the statement 'either the particle went through slit 1 or it went through slit 2'. Rather, it corresponds to the non-classical statement that 'the particle was free to go through both of the slits in an indeterminate way'.

As long as the route of each electron is indeterminate the contributions represented by Ψ_1 and Ψ_2 are said to add **coherently**, implying that they can interfere (like coherent beams of light) and it is then that their joint effect results in an intensity distribution that is proportional to $|\Psi_1 + \Psi_2|^2$. Modifying the experiment so that it becomes possible to say which slit each electron went through would remove the indeterminacy and destroy the coherence. As a result, the intensity distribution would be proportional to an 'incoherent' sum of the form $|\Psi_1|^2 + |\Psi_2|^2$, and the interference effects would vanish. This will be true irrespective of *how* the path of the electron was determined.

Coherence is a subject to which we shall return later. For the present, however, we conclude this section by taking a first look at the interpretation of quantum mechanics.

2.5 A first look at interpretation

The issue of its physical interpretation has troubled quantum mechanics ever since its birth. Schrödinger himself initially misunderstood the physical significance of the wavefunction. However, because the earliest predictions of quantum mechanics concerned eigenvalues and dealt with quantities such as atomic energy levels and the frequencies of spectral lines, the subject was able to achieve spectacular successes even before Born introduced his probabilistic interpretation of the wavefunction in 1926. Nonetheless, after Born's seminal contribution there was a clear need for a comprehensive physical interpretation of the new formalism. Inevitably, this was influenced by the prevailing philosophical climate of the time.

In the mid-1920s, intellectual debate about the nature of truth and meaning was dominated by a tough-minded attitude that goes under the broad name of **positivism**. The fundamental tenet of this philosophy, in one of its more extreme forms, is that no sentence is really meaningful unless we can verify it directly with our senses. This attitude can be used to justify the assertion that: a scientific statement can only be about the results of experiments. Related to this is the doctrine of **instrumentalism**, which asserts that the purpose of a scientific theory can only be to predict the results of experiments. A consequence of these views is that a scientific term is meaningful only if there is a precisely defined experimental procedure for observing or measuring it. A further step leads to **operationalism**, the view that only such observationally defined quantities should be allowed at any stage, to enter in to a theory. This view was inspired by the way that Einstein set up the special theory of relativity, where he replaced Newton's unexamined assumption that every event had a unique time and a set of spatial coordinates by precise prescriptions for measuring these quantities.

This brief caricature of positivism fails to do justice to a complex philosophical issue that was viewed differently at different times by the founders of quantum mechanics. Interested readers should look elsewhere for further details.

In Copenhagen, in the 1920s, Niels Bohr led a school of thought in physics that was rooted in positivism, and it was Bohr, together with Heisenberg and others, who came to provide what was long regarded (and still is by many) as the 'conventional' interpretation of quantum mechanics. According to this **Copenhagen interpretation**, the only meaningful statements made by quantum mechanics, concern the setting-up of experiments and the results of experiments. All other statements, for example those about wavefunctions, are simply subordinate parts of a quantum-mechanical calculation, whose only meaning lies in the experimental propositions they link. The intermediate stages in the calculation have no significance of their own outside the final statement about experimental apparatus.

The Copenhagen interpretation accepted the mathematical formalism developed by Heisenberg, Schrödinger, Dirac and others; it adopted Born's probabilistic interpretation of the wavefunction, and it also encompassed a statement of Bohr's known as the **principle of complementarity**. This reflects the 'wave–particle duality' that is implicit in quantum mechanics, and asserts that it is acceptable to use classically incompatible forms of language (such as those concerning waves and particles) when discussing quantum phenomena, though not to use them simultaneously. Bohr himself felt that this was a new philosophical principle of fundamental importance, but many later commentators have felt it to be rather vague, and modern treatments of quantum mechanics often ignore it entirely, even though they accept the general thrust of the Copenhagen interpretation.

One of the many questions to which the Copenhagen interpretation provides a clear (though not uncontested) answer concerns the meaning of the probabilities that quantum mechanics predicts. Do probabilities say as much as can be said about an individual quantum system in which the value of an observable is truly indeterminate until it is measured? Or are the probabilistic statements of quantum mechanics of a more statistical nature, along the lines of 'well, if you had a large number of identically prepared copies of the system, the predicted probability of a certain outcome represents the fraction of those systems that would already possess that property'? This is a subtle distinction, but an important one. The former 'individual system' approach is essentially saying that quantum mechanics is a complete theory, but inevitably limited to using probability in an inherent and unavoidable way. The alternative 'statistical' approach could be applied even if individual systems actually possessed fully determined properties and our current need for probability in quantum mechanics was a result of our present inability to determine all those properties. The statistical approach therefore tends to be favoured by those who think that quantum mechanics is incomplete — just a part of some more comprehensive theory, the discovery of which will remove the need to use probability in any 'essential' way.

The view taken by Bohr and his colleagues was that the formalism of quantum mechanics was essentially complete and therefore provides the fullest possible description of individual quantum systems that nature allows. According to the Copenhagen interpretation then, probability is an inherent part of the quantum mechanical description of the world and cannot be avoided. This implies that the process of making a measurement is of profound importance, since it somehow converts a 'potentiality', a possible outcome with an associated probability, into an 'actuality', a real measured result. We shall discuss the nature of the measurement process, and its role in the interpretation of quantum mechanics, in detail in Section 3 of this chapter.

The Copenhagen interpretation has important implications for our whole conception of reality, and for what it means to say that microscopic entities such as electrons actually exist. For instance, if we use it to interpret what happens when we turn on a television set, it tells us that we should *not* think that there is a stream of very small particles actually moving from the electron gun to the screen and causing the screen to produce an image. Rather, there is an electron gun and a screen, but, linking the observable behaviour of these macroscopic entities, there is a calculation that results in a verifiable prediction of the scene on the screen, but which involves meaningless statements about unobserved microscopic electrons in its intermediate stages. As Bohr himself put it:

> There is no quantum world; there are only quantum calculations of the behaviour of macroscopic apparatus.

Bohr is not saying that the electrons do not exist, but he is stressing that the success of quantum-mechanical calculations does not justify any claim that they do. The Copenhagen doctrine allows one to think of a quantum world of particles in doing calculations, but only as a useful fiction. Although this was for many decades the 'official line' on quantum mechanics, few physicists can have succeeded in maintaining such an austere discipline in their thought. It is particularly unsatisfactory to those who think that the purpose of physical science is to find out about a real physical world, and that calculating the results of experiments is a secondary aim, not a primary one, only interesting as a check on the truth of a theory. We shall return to the issue of quantum mechanics and reality in Section 4 of this chapter.

Physicists have always differed in their reaction to the Copenhagen interpretation. Some have taken the view that any discussion of interpretation is 'just philosophy' and have preferred to get on with the business of using the formalism of quantum mechanics to make yet more successful predictions. Others have taken the challenge of interpretation more seriously and have looked for a different interpretation of the formalism, or even a modified version of the formalism that preserves the existing successes but which has a simpler or more palatable interpretation. Neither approach has resulted in any undisputed victory, but a number of recent developments have caused the interpretation of quantum mechanics to become a very active field of research again, after several decades of relative quiescence.

2.6 Summary of Section 2

Here is a brief summary of the key points to emerge from Section 2.

The specification of a quantum system involves a number of parameters (such as masses and charges) and, typically, a potential energy function. This information can be used in the formulation of a version of the time-dependent Schrödinger equation that is specific to the system concerned. The states of the system are described by time-dependent wavefunctions, usually denoted Ψ, that satisfy this differential equation. (In situations where the state of the system is a stationary state, the essential information contained in the wavefunction may be expressed in terms of a time-independent wavefunction, usually denoted ψ, that satisfies an appropriate version of the time-independent Schrödinger equation.)

The measurable properties of a system are called observables. By analysing the wavefunction that describes the state of a system, it is possible to predict the possible outcomes of any measurement of a given observable, and the probability of each of those possible outcomes. This, however, is as far as quantum mechanics goes. The indeterminacy of quantum mechanics means that, whatever the state of the system, it will not generally be possible to assign values to all the observables simultaneously. Also, the indeterminism of quantum mechanics means that, even if the value of an observable is precisely determined at some particular time, it will not generally be possible to predict the precise value of that observable at later times.

The possible outcomes of a measurement of a given observable are known as the eigenvalues of that observable. A state in which a particular measurement outcome is predicted to occur with probability 1 is called an eigenstate, and may be said to correspond to the relevant eigenvalue.

Given a state described by the normalized linear superposition

$$\Psi = c_1\Psi_1 + c_2\Psi_2 + c_3\Psi_3 + \ldots + c_n\Psi_n, \qquad \text{(Eqn 4.5)}$$

where Ψ_1, Ψ_2, Ψ_3, ... Ψ_n represent eigenstates of some observable, and correspond to eigenvalues a_1, a_2, a_3, ... a_n, then the only possible outcome of a measurement of that observable is one of those eigenvalues, and the probability that it will be a_i is given by

$$P(a_i) = |c_i|^2 \quad (\text{for } i = 1, 2, 3, \ldots n) \qquad \text{(Eqn 4.8)}$$

where c_i is a (complex) number and $|c_i|$ is its modulus.

The 'conventional' interpretation of quantum mechanics is known as the Copenhagen interpretation. According to this, a wavefunction provides the fullest possible description of the state of an individual quantum system. This implies that probabilities are an inherent part of the description of nature. It also implies that systems do not generally 'possess' values of observables, but that such values may be measured. The Copenhagen interpretation thereby gives great significance to the process of measurement, and calls into question naive views concerning the 'reality' of microscopic entities.

3 Measurement in quantum mechanics

It should be clear from the summary of the last section that measurement plays a crucial role in quantum mechanics. A measurement has the effect of turning a potentiality (a possible outcome with an associated probability) into an actuality (a measured result). This puts measurement at the heart of many of the debates about the interpretation of quantum mechanics. It is for this reason that we devote this section to the measurement process and to some of its implications.

3.1 The nature and effect of a measurement

Although the term 'measurement' has already been used several times it has not so far been properly defined. In quantum mechanics, **measurement** is a technical term with a precise meaning. It describes a physical interaction between a quantum system and a measuring device that results in the measuring device being left in some state that represents the outcome of the measurement. The final state of the measuring device might, for example, correspond to a reading on a scale, or an orientation of a pointer, or just some combination of indicator lights being on rather than off.

The Copenhagen interpretation puts particular emphasis on measurement since it assumes that measuring devices are essentially classical pieces of apparatus (certainly, the readings they provide are 'classical'). So measurement provides the interface between a comprehensible world of classical objects and results, and a much less comprehensible world of quantum systems and states that can be known only through the measurement results that it provides (Figure 4.11).

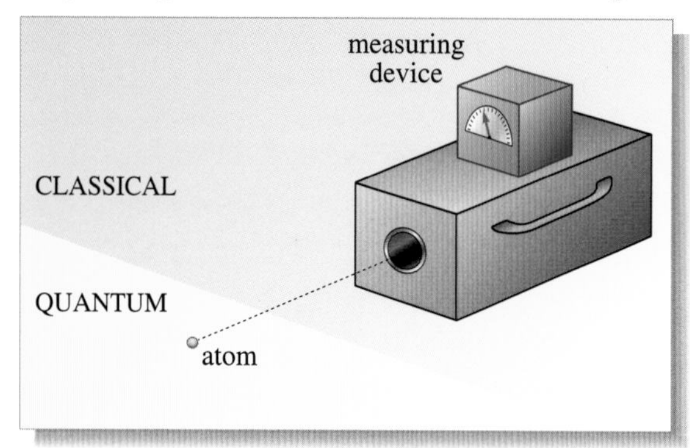

Figure 4.11 According to the Copenhagen interpretation, measurements carried out on quantum systems must be performed by devices that are essentially classical. The interpretation therefore implies a 'boundary' that separates the classical world of everyday objects from the quantum world that underpins it.

Now, a measurement performed on a quantum system has no point unless it provides information about the state of that system. What, then, does a measurement of a certain observable A, that yields the result a, tell us about the state of the measured system? The conventional answer to this question is this:

> Immediately after a measurement of an observable A, the state of the measured system will be the eigenstate of A that corresponds to the measured eigenvalue a.

So, according to this conventional view, the process of making a measurement will force the system into the eigenstate that corresponds to the measured result. This eigenstate may not have been the state before the measurement, but it will be the state immediately after the measurement. This can be confirmed by making another measurement of the same observable, A, immediately after the first. If the second measurement really is performed *immediately* after the first, its outcome is certain to be the same as the first. But the only state that guarantees a definite outcome of this kind is one that is an eigenstate of the measured observable.

Now consider the implications of this for a system that, prior to a measurement of observable A, is in a state described by the linear superposition

$$\Psi = c_1\Psi_1 + c_2\Psi_2 + c_3\Psi_3 + \ldots + c_n\Psi_n$$

where Ψ_1, Ψ_2, Ψ_3, ... Ψ_n are n normalized wavefunctions that represent eigenstates of the observable A, corresponding to eigenvalues a_1, a_2, a_3, ... a_n. If A is measured, and the result a_i is obtained, then immediately after that measurement the state of the system will be represented by the wavefunction Ψ_i. This abrupt change in the system's wavefunction, from Ψ to Ψ_i, is described by saying that the measurement causes the **collapse of the wavefunction**.

The collapse of a wavefunction can be illustrated by looking again at the case of single electrons passing through the two-slit diffraction apparatus of Figure 4.4. An electron passing through the two-slit apparatus has a time-dependent wavefunction that consists of two parts, Ψ_1 and Ψ_2, corresponding respectively to passage through slit 1 or slit 2. Now, as long as no steps are taken to determine which slit an individual electron passes through, its route is indeterminate and the two contributions may be added to produce a wavefunction $\Psi = (\Psi_1 + \Psi_2)/\sqrt{2}$. The two terms add coherently, so they can produce interference effects. Hence the intensity distribution created by a large number of similar electrons arriving at the screen will be described by $(|\Psi_1 + \Psi_2|^2)/2$, i.e. the familiar two-slit interference pattern of Figure 4.10.

However, if the indeterminacy in each electron's route is removed by carrying out some kind of measurement that reveals which slit it passed through, the measurement will cause each electron's wavefunction to collapse to either Ψ_1 or Ψ_2. The probability distributions corresponding to these two distinct wavefunctions are $|\Psi_1|^2$ and $|\Psi_2|^2$, respectively. The result of observing the arrival at the screen of a large number of electrons that have had their wavefunctions collapsed in this way (by an earlier measurement) will be described by an incoherent sum of the form $(|\Psi_1|^2 + |\Psi_2|^2)/2$ — the sum of two single-slit diffraction patterns, of the kind shown in Figure 4.9.

The process of wavefunction collapse is illustrated in a highly schematic way in Figure 4.12.

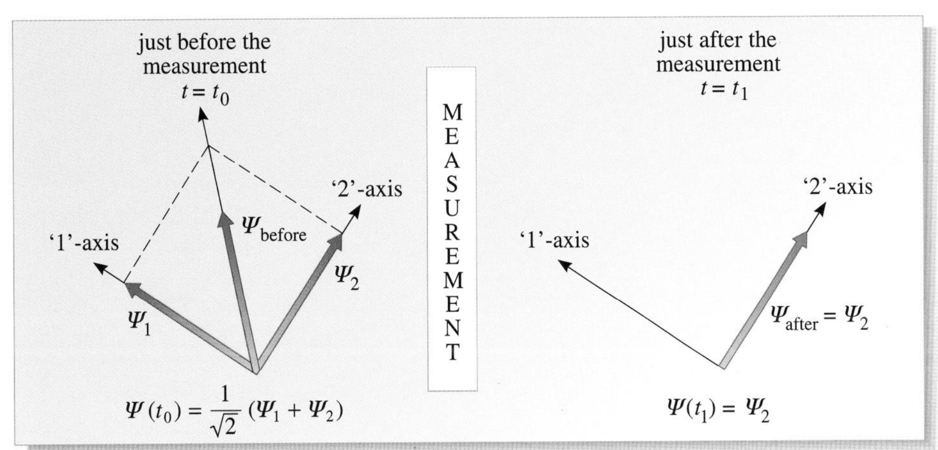

Figure 4.12 A highly schematic view of the process of wavefunction collapse. Prior to the collapse the wavefunction Ψ consists of two parts that describe different eigenstates of a given observable. These are indicated by perpendicular vectors, corresponding to Ψ_1 and Ψ_2. After the measurement, the wavefunction has collapsed onto one of those two eigenstates. Overall normalization is preserved throughout.

It was said earlier that quantum mechanics predicts the possible outcomes of a measurement and the probability of each of those possible outcomes. You should now be able to see that an alternative way of expressing this is to say that quantum mechanics predicts the *possible* states to which the wavefunction of a system might collapse, and the *probability* that it will collapse to each of those possible states. However, just as quantum mechanics does not generally predict the precise result of a particular measurement, so it does not predict the precise eigenstate to which the wavefunction will collapse as a result of a particular measurement. To that extent the collapse of the wavefunction is unpredictable as well as abrupt.

Question 4.4 An atom is initially in a state described by the wavefunction $\Psi = (2\Psi_1 + \Psi_2)/\sqrt{5}$, where Ψ_1 and Ψ_2 are normalized wavefunctions describing eigenstates of energy that correspond to eigenvalues E_1 and E_2, respectively. If a measurement shows that the energy of the atom is E_1, what is the wavefunction of the atom immediately after the measurement? ■

3.2 Measurement and time evolution

If we now try to summarize the fundamental principles governing the change of a quantum-mechanical wavefunction with time, we find that we have to make a special statement to cover the effect of a measurement on a system. Thus:

1 As long as a system is not observed, its wavefunction Ψ changes continuously and predictably according to the time-dependent Schrödinger equation.

2 When a system is subjected to a measurement, its wavefunction changes discontinuously and unpredictably to that describing an eigenstate of the measured observable whose eigenvalue a is the actual result of the measurement.

Unpredictable, discontinuous changes, as described in the second of these statements, are a familiar feature of the subatomic world under the name of *quantum jumps*. Historically, the notion that an electron 'jumps' from one allowed orbit to another, emitting radiation as it does so, was an essential part of Bohr's semi-classical theory of the atom. Today, it is almost a part of everyday life that Geiger counters click as radioactive nuclei decay, making a quantum jump from one nuclear species (e.g. radium) to another (e.g. radon plus an α-particle). That such a jump is possible can be predicted, so can the probability that the jump will occur in any given time interval, but the precise time at which a *particular* atom or nucleus will make the quantum jump is *not* predictable.

Actually, the term, 'quantum jump' is not an essential part of quantum mechanics. At a fundamental level, the only discontinuous processes in quantum mechanics are those stimulated by a measurement. Spontaneous jumps, such as those that appear to occur in the emission of radiation from an excited level of an atom, may be described entirely in terms of *measurements* of the energy of the atom at some time t. Given an atom that is initially in some excited state, it is generally possible to predict the probability that a measurement of its energy at some later time t will produce a result that corresponds to some state of lower energy. This probability may be taken to represent the likelihood that a quantum jump to the state of lower energy has occurred by time t.

It is a very mysterious (and, many would add, unsatisfactory) feature of quantum mechanics that it requires a special rule to describe the effect of a measurement. The Copenhagen view of the measurement process is that it involves an unanalysable interaction between a quantum system and a classical measuring apparatus. But it is generally accepted that the operation of the measuring device is actually underpinned by quantum physics — all measuring devices are ultimately made up of atoms and molecules that are themselves described by quantum mechanics. For the purposes of the Copenhagen interpretation the measuring device must be treated classically, but this is a logical rather than a physical requirement.

Many physicists have been concerned by this distinct separation between classical and quantum descriptions. It raises many questions. Is there a real physical separation, so that some systems, such as atoms, can only be meaningfully described by quantum physics, while others, such as voltmeters and ammeters, can only be meaningfully described in terms of classical physics? If not, how is the classical world of everyday experience to be explained? Why is it that everyday objects such as tables and chairs can be said to possess properties such as position and momentum, when the atoms and molecules of which the objects are composed only acquire properties as a result of measurement, and then only fleetingly? Also, is the measurement process really something separate and distinct from the normal evolution of a quantum system? If the measuring device itself could be treated quantum mechanically (even if a classical result has to somehow emerge), might not the evolution of the whole system, including the measuring device, be described by a single wavefunction that satisfies some appropriate version of the Schrödinger equation? As you will see in the next two subsections, attempts to answer these quite reasonable questions have led to some extraordinary and controversial proposals concerning quantum mechanics and its interpretation.

3.3 Schrödinger's cat and Wigner's friend

Schrödinger's cat's a mystery cat, he illustrates the laws;
The complicated things he does have no apparent cause;
He baffles the determinist, and drives him to despair
For when they try to pin him down — the quantum cat's not there!

Schrödinger's cat's a mystery cat, he's given to random decisions;
His mass is slightly altered by a cloud of virtual kittens;
The vacuum fluctuations print his traces in the air
But if you try to find him, the quantum cat's not there!

Schrödinger's cat's a mystery cat, he's very small and light,
And if you try to pen him in, he tunnels out of sight;
So when the cruel scientist confined him in a box
With poison-capsules, triggered by bizarre atomic clocks,
He wasn't alive, he wasn't dead, or half of each; I swear
That when they fixed his eigenstate — he simply wasn't there!

Amongst the many physicists to experience disquiet over the interpretation of quantum mechanics, one of the most prominent must surely have been Schrödinger himself. In 1935, while living in Oxford, in exile from the rising tide of Nazism in his native Austria, Schrödinger published an article that highlighted some of his concerns. In the article he cleverly emphasized the problem of making sense of measurement in quantum mechanics with his famous story of a cat in a box (Figure 4.13). This poor beast is now known as **Schrödinger's cat**.

Figure 4.13 Schrödinger's cat, shut in a box with a radioactive nucleus and a lethal contraption that will be triggered if and when the nucleus decays. What determines the vitality of the cat? The triggering of the device or the observation of the state of the cat?

Imagine, he said, a cat placed in a closed box containing a 'hellish device' consisting of a single radioactive radium nucleus, a radiation detector and a phial of poison gas, connected up so that when the nucleus decays the phial will be smashed, releasing the gas and killing the cat. Initially the nucleus is undecayed and the cat is alive. But, due to the possibility of radioactive decay, the wavefunction of the nucleus must be a linear superposition of two parts, one representing the undecayed radium nucleus, the other representing the decayed nucleus. As time passes, the first part will decrease and the second part will increase, as the likelihood of decay having occurred increases. This may sound harmless enough for a quantum system such as a nucleus, but think what it implies for the wavefunction of the cat. Treating the cat as a quantum system, its wavefunction will also contain two terms, one representing a live cat, the other a dead cat. What does it mean, especially from the cat's point of view, to say that it is in a superposition state with the wavefunction

$$\Psi_{\text{cat}}(t) = \Psi_{\text{alive}}(t) + \Psi_{\text{dead}}(t)? \tag{4.10}$$

If you take the view that the wavefunction is not the concern of the cat, since it only has meaning in the intermediate stages of a quantum calculation, the problem of its interpretation becomes no less worrying. In order to put the cat into a definite state of life or death it will be necessary to perform a measurement of its vitality. (Actually we should be a little more careful; it is not the cat but the whole box, including the cat and the nucleus, that is in a superposition state. However, the vitality of the cat can be regarded as one of the observables of the whole system.) The measurement could be performed simply by an observer opening the box and seeing if the cat is alive or dead. At this point the wavefunction collapses to an eigenstate in which the cat is either alive or dead. But this means that if the cat is found to be dead when the box is examined, it only became dead at the moment when the box was opened. Therefore it was the cat-loving observer who actually killed the cat!

The serious point behind this is that although superposition states may be fully acceptable when dealing with quantum systems, they present difficult problems of interpretation when applied to macroscopic systems, especially in the context of measurement. Schrödinger's cat is essentially a metaphor, useful for highlighting the problems associated with superpositions of macroscopic states. What, if anything, do such states represent? And at what stage in the measurement process do they collapse to the state that represents the measured result? Schrödinger's cat is well behaved as long as we are prepared to treat it as a classical measuring device, that is at all times either alive or dead (the Copenhagen view), but as soon as we regard it as a quantum system, with its own wavefunction, it becomes very capricious indeed.

One way of overcoming the difficulties presented by Schrödinger's cat and answering the question 'when does the wavefunction collapse?', has been proposed by the British mathematical physicist Sir Roger Penrose. He has suggested that when quantum systems become sufficiently 'large', some new physical process, possibly determined by gravity, comes into play that forces the system's wavefunction to collapse, compelling the system to behave classically. The logical distinction between quantum system and classical measuring device required by the Copenhagen interpretation would then become a real physical distinction that prevents macroscopic objects such as cats from ever existing in superposed states, even though they are composed of atoms and molecules that can be in such states.

This is an appealing view, but it faces many difficulties and has not won widespread support. One problem arises directly from the improvement of experimental techniques which continue to increase the range and size of quantum phenomena that can be observed and measured. For example, it is now possible to perform diffraction experiments with large molecules, not just single electrons (see Figure 4.14). Many physicists feel that a physical separation of the classical and quantum domains based simply on size or mass, or any similar parameters, is becoming harder and harder to maintain. The supposedly non-existent quantum world is in danger of growing in size until it approaches the dimensions of the clearly existent macroscopic world.

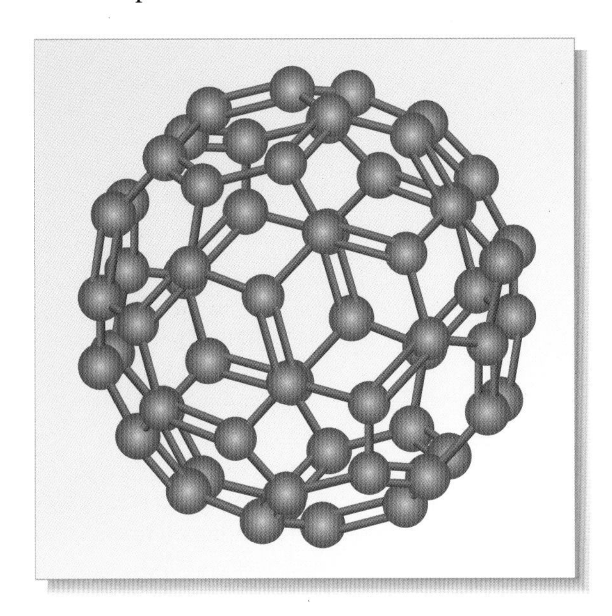

Figure 4.14 The soccer-ball shaped molecule of Buckminsterfullerine consists of 60 carbon atoms and is commonly known as a Buckyball. At the time of writing it represents the largest system to have been definitively shown to exhibit the kind of quantum interference effects that are also demonstrated by electrons.

Another attempt at explaining how a classical everyday world manages to emerge from an underlying quantum reality makes use of the so-called **decoherence effect**. As we have seen, one of the clearest manifestations of quantum physics is the interference pattern seen in a two-slit electron diffraction experiment. This is a consequence of the coherence of different parts of the electron's wavefunction. Destroy that coherence — by determining which slit each electron went through, for instance — and the interference pattern will disappear. In the absence of coherence the electron behaves much more like a classical particle.

The decoherence effect refers to the way in which interactions between a system and its environment tend to quickly destroy the coherence of different parts of the system's wavefunction and thus cause it to behave in a classical way. Perhaps, for macroscopic systems such as cats, chairs, tables and people, interaction with the environment is such an ever-present part of reality that the system is, in effect, constantly being 'measured' so that its wavefunction is permanently collapsed. In this approach the wavefunction of a macroscopic system never undergoes any real collapse, but it generally behaves as though such a collapse has already taken place. This idea is very popular at present and seems certain to play a role in future interpretations of quantum mechanics. Whether decoherence alone is sufficient to account for the emergence of the classical world is, however, still not clear.

An entirely different solution to the question 'when does the wavefunction of a measured system collapse?' can be arrived at by pursuing an idea due to Eugene Wigner (1902–1995), a Nobel Prize-winner who made many contributions to quantum mechanics. Suppose Wigner wants to find out what has happened to Schrödinger's cat, but instead of looking into the box himself he asks a friend to do so. He could regard the room, the box, the cat and his friend as one big quantum system, which is in a superposition of states until he makes a measurement by asking his friend whether the cat is alive or dead. He might refer everything to his own consciousness, and say that the wavefunction of the room collapses only when he hears his friend's answer. But this, argued Wigner, would be arrogantly egocentric: if he asked his friend what she thought before he asked his question, she would say that she already knew the state of the cat. Wigner argued that her knowledge shows that the wavefunction had already collapsed, and concluded that the agency of the collapse must be human consciousness. Most physicists, however, feel that to bring consciousness into the theory is unsatisfactory and unnecessary; we will now see what can be done without it.

3.4 A second look at interpretation

In this subsection we briefly examine the famous (some would say infamous) **many worlds interpretation** of quantum mechanics. This interpretation manages to avoid the conventional picture of a wavefunction that collapses at the moment of measurement. The measuring devices that it involves can be treated as instruments that obey the rules of quantum physics, rather than having to maintain the Copenhagen pretence that they are (logically at least) classical devices. The interpretation also offers ways of avoiding the difficulties that arise when trying to make sense of superpositions of macroscopically different states, of the kind highlighted by Schrödinger's cat.

The many worlds interpretation was first proposed by an American, Hugh Everett, in the late 1950s under the more sober sounding title of 'relative state interpretation'. It is widely regarded as the most extravagant of all the major interpretations since, in its most extreme form, it literally requires you to accept the existence of a very large, quite possibly infinite, number of parallel universes that coexist along side our own. Many physicists feel that accepting any version of the many worlds interpretation is too high a price to pay, even to answer the riddles of quantum physics, and discount this particular interpretation on those grounds alone. Nonetheless, the many worlds interpretation has its merits and its ardent supporters.

The crux of the many worlds interpretation is the idea that no system can ever be regarded as being in a single well-defined quantum state that is independent of the rest of the Universe. Whatever the state of the system, the rest of the Universe must be in some unique *relative state*.

By considering measurements of a given observable, using a measuring device that retained a memory of the results of previous measurements of that observable, Everett was able to show that the evolving wavefunction of the Universe would consist of many branches, each corresponding to a particular sequence of outcomes for the value of the measured observable. This branching wavefunction is indicated schematically in Figure 4.15.

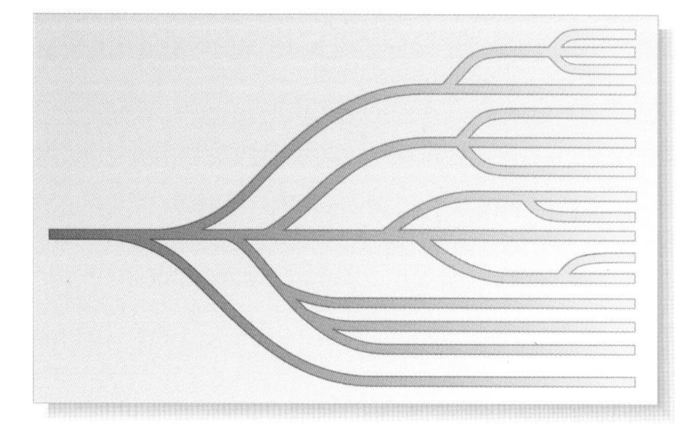

Figure 4.15 The branching, but not collapsing, wavefunction of the many worlds interpretation.

The contribution that any one of these branches makes to the overall wavefunction is determined by the conventional formalism of quantum mechanics and therefore corresponds to what would normally be described as the probability of obtaining those results in a series of measurements. However, from the point of view of the many worlds interpretation, each branch of the universal wavefunction corresponds to a parallel universe that actually exists and the 'probability' of obtaining a given sequence of results reflects the fraction of those parallel universes in which that particular sequence of results actually occurred.

Because it utilizes the conventional formalism of quantum mechanics, the many worlds interpretation is certain to share the calculational successes of more conventional interpretations. On the other hand it avoids the troublesome collapse of the wavefunction by claiming that the full wavefunction always evolves smoothly and continuously in accordance with the deterministic time-dependent Schrödinger equation. It dispenses with the parts of the wavefunction that do not correspond to the past history of our Universe by associating them with parallel universes that are in some sense orthogonal to our own. Schrödinger's cat is both alive and dead, but the universes in which it persists are different from those in which it perishes.

Since the 1980s, one of the most vocal supporters of the many worlds interpretation has been the British physicist David Deutsch, a major contributor to the relatively new field of quantum computation. You can find out more about Deutsch's own version of the many worlds interpretation and its possible relevance to quantum computers in his popular book *The Fabric of Reality*, (Penguin Books, 1997).

4 Non-locality and realism in quantum mechanics

Quantum mechanics successfully accounts for a wide range of *correlations* between classical measurements, but, according to the Copenhagen interpretation, this does not provide evidence for the *reality* of the quantum entities that are used to account for those correlations. For example, a measurable signal supplied to a TV set from an antenna can be correlated with the picture seen on the screen by means of quantum calculations involving electrons. However, the Copenhagen interpretation, at least in its most extreme form, refuses to accept this as evidence that the electrons actually exist. The Copenhagen view specifically denies quantum entities the sort of reality that classical systems acquire from the properties they possess independently of any measurements. In quantum mechanics the values of observables may be measured, but the results do not generally indicate values that were 'possessed' prior to the measurement and those results are, in any case, limited by the sort of quantum indeterminacy represented by Heisenberg's uncertainty principle.

This section is very much concerned with the reality of the quantum world. We start by considering some arguments, by Einstein and others, against the quantum indeterminacy that seems to stand in the way of the simplest kind of property-based 'reality' of the quantum world. We then go on to examine the work of John Bell who investigated the extent to which such simple-minded reality conflicts with testable predictions of quantum mechanics. Bell's results lead us to another interpretation of quantum mechanics that combines simple existence with quantum mechanics, but does so at a high price.

4.1 The Bohr–Einstein debate

From its inception, the Copenhagen interpretation found a resolute opponent in Albert Einstein. At conferences in 1927 (Figure 4.16) and in following years, Einstein presented a series of thought experiments designed to show that variables such as position and momentum, *could* be measured simultaneously even though the uncertainty principle said that it was impossible. Bohr, the principal architect and defender of the Copenhagen interpretation, found flaws in each of these proposals in turn. We shall consider just one of Einstein's proposals to provide a taste of the kind of debate that took place between these two giants of theoretical physics.

One of the simplest manifestations of uncertainty is the wave-like behaviour of a particle after it has passed through a small hole in a diaphragm. Knowing that it has gone through the hole, we have information about the position of the particle. After passing through a pinhole, however, the particle's wavefunction spreads out in all directions as shown in Figure 4.17. The interpretation of this in particle terms is that the direction of the particle's momentum (more precisely, its momentum component in the plane of the diaphragm) is uncertain. If the particle is observed by watching for the flash it makes on a fluorescent screen placed beyond the diaphragm, we cannot precisely predict where the flash will be observed. This is all in accord with the uncertainty principle. But, said Einstein, we don't have to accept this uncertainty.

Figure 4.16 The participants at the 1927 Solvay Conference in Brussels, one of the first occasions on which Einstein was able to challenge Bohr over the interpretation of quantum mechanics. The gathering included many of the leading physicists of the day. Einstein is seated fifth from the left in the front row, to the left of Einstein are Lorentz, Madam Curie and Planck. In the middle row, from right to left are Bohr, Born, de Broglie, Compton and Dirac. In the back row, Heisenberg (third from the right) stands behind Born, with Pauli to the left. Schrödinger (sixth from the right), is standing behind Einstein, Dirac and Compton.

If we know the momentum of the particle *before* it went through the hole, and if we measure the change in the momentum of the diaphragm as a result of the particle's passage through the hole, then, by using the principle of momentum conservation, we can determine the momentum of the particle *after* it has passed through the hole without having to disturb the particle itself. Knowing the position at which the particle passes through the diaphragm, *and* the momentum of the particle as it emerges from the hole, we can, said Einstein, predict precisely where it will appear on the screen.

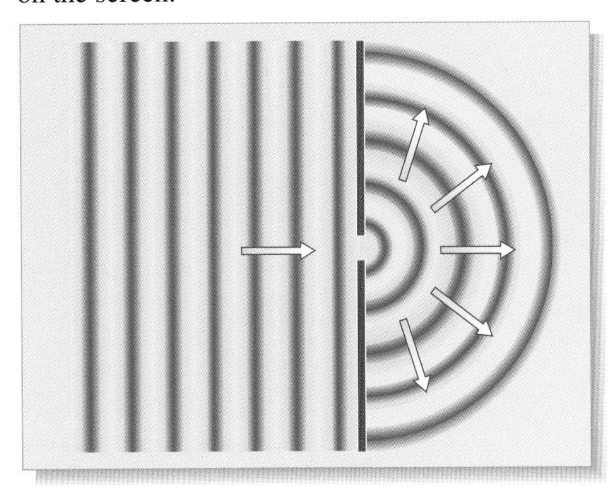

Figure 4.17 The behaviour of a wave after passing through a pinhole. The wave spreads out, illustrating the phenomenon of diffraction.

Applying this to a two-slit diffraction experiment, Einstein argued that it gave a way of determining which slit the particle goes through, contradicting the quantum-mechanical view that the existence of the interference pattern on the fluorescent screen (see Figure 4.18) makes this impossible. Suppose the particle hits the screen at a point C opposite the source G. If it had reached this point by passing through the upper slit A, it would have had to move upward from G to A and then downward from A to C. It would therefore have acquired downward momentum from the diaphragm and would have given the diaphragm an equal upward kick. But if it had passed through the lower slit B, it would have given the diaphragm a kick downwards. So, by measuring the change in the momentum of the diaphragm as the particle passes through, we can distinguish between the two possible paths GAC and GBC, without affecting the interference pattern on the fluorescent screen.

Bohr replied that one must also apply the uncertainty principle to the diaphragm that contains the slits. Einstein's proposal requires precise knowledge of the momentum of the diaphragm; its position therefore becomes uncertain. But in order to obtain an

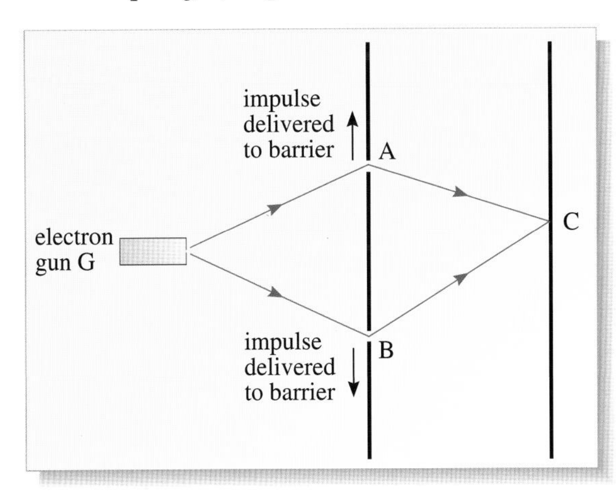

Figure 4.18 Momentum in the two-slit experiment.

interference pattern on the screen, it is necessary that the pair of slits should have a precise position relative to the screen (it is their coordinate parallel to the screen that is important, and this is just the direction in which Einstein needs to know the component of momentum). If the position of the slits is uncertain, then the position of the dark and light bands in the pattern will also be ill-defined, so they will cancel each other out and there will be no interference. In keeping with the principle of complementarity, Bohr insisted on the importance of the experimental arrangement: if this is such as to bring out the wave-like interference pattern, then it will make it impossible to determine the path taken by particle-like electrons.

The Bohr–Einstein debate continued over many years and included one celebrated occasion on which Bohr's rebuttal of one of Einstein's proposals made clever use of Einstein's own general theory of relativity. Following that particular encounter things quietened down for a while, but in 1935 Einstein presented a new challenge to quantum mechanics — one that is still much discussed today, more than 60 years later. This new challenge is the subject of the next subsection.

4.2 The Einstein–Podolsky–Rosen argument

In 1935, Einstein, together with Boris Podolsky (1896–1966) and Nathan Rosen (1909–1995), produced another argument showing that the momentum and position of a particle must both exist simultaneously as 'elements of reality' and that quantum mechanics, as conventionally formulated and interpreted, must therefore provide only an incomplete description of the physical world.

The original **Einstein–Podolsky–Rosen argument** is rather complicated, so we shall present it in a modified version devised by the physicist David Bohm, though we shall continue to associate it with the names of Einstein, Podolsky and Rosen, or EPR for short. This reformulation of the EPR argument concentrates on measurements of an electron's spin, so the basic facts about such measurements are reviewed in Box 4.1.

Box 4.1 The quantum mechanics of electron spin measurements

Every electron is said to have spin $\frac{1}{2}$. This means that it is possible to associate with any electron an 'intrinsic' angular momentum $\boldsymbol{S}$, and a spin quantum number $s = \frac{1}{2}$, that are related by the equation

$$S = \sqrt{s(s+1)}\,\hbar.$$

According to quantum physics, any measurement of a particular component of an electron's spin along some chosen direction can only have two possible outcomes: the result must be either $+\hbar/2$ or $-\hbar/2$. (These are the only eigenvalues for this kind of measurement.) Consequently, if we arbitrarily choose some particular direction to be the z-direction, and then measure the component of an electron's spin in that direction we will inevitably find that

$$\text{either} \quad S_z = +\hbar/2 \quad \text{or} \quad S_z = -\hbar/2.$$

Similar statements can be made about measurements of the component of an electron's spin along any other direction. So, $+\hbar/2$ and $-\hbar/2$ are the only possible outcomes for a measurement of S_x or S_y, or indeed for a measurement of $S_{\boldsymbol{a}}$ made in some completely arbitrary direction defined by a vector $\boldsymbol{a}$.

Now measurements of spin components, such as S_x and S_y, are subject to an indeterminacy relation similar to the uncertainty principle for position and momentum. In the case of position and momentum, the uncertainty principle implies that the precise measurement of any one of the position coordinates of a particle (x or y or z) is incompatible with a simultaneous and precise measurement of the corresponding momentum component (p_x or p_y or p_z). In the case of spin components it can be shown that quantum indeterminacy implies that a measurement of any one of the spin components, S_z say, is incompatible with a simultaneous measurement of the perpendicular components S_x and S_y.

The EPR argument concerns a pair of electrons somehow produced together and in such a way that the total spin of the pair is zero. The two electrons, let's call them A and B, may be supposed to travel in opposite directions towards well separated measuring devices that are each capable of determining some chosen component of an electron's spin, S_z say (see Figure 4.19). As explained in Box 4.1, the only result that such measuring devices can yield for S_z is either $+\hbar/2$ or $-\hbar/2$. However, because each pair of electrons has a total spin of zero, it must be the case that whatever value of S_z is measured for A, the opposite result must be measured for B.

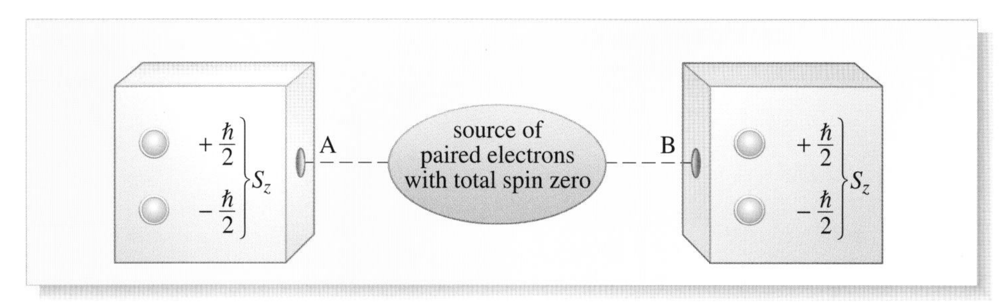

Figure 4.19 An EPR experiment. The result of a spin component measurement at either detector is indicated by the flash of a light.

The EPR argument is specifically concerned with a pair of electrons that have been produced in a state with the following wavefunction

$$\Psi = \frac{1}{\sqrt{2}}[\Psi_+(\mathrm{A})\Psi_-(\mathrm{B}) - \Psi_-(\mathrm{A})\Psi_+(\mathrm{B})] \qquad (4.11)$$

Note the minus sign in Equation 4.11 between the two terms.

where $\Psi_+(\mathrm{A})$ is the eigenstate of the S_z observable that corresponds to the eigenvalue $+\hbar/2$, and describes the situation in which a measurement of S_z performed on electron A will certainly yield the result $+\hbar/2$, while $\Psi_-(\mathrm{A})$ describes a state in which a measurement performed on A yields the result $-\hbar/2$. (Similarly for $\Psi_+(\mathrm{B})$ and $\Psi_-(\mathrm{B})$.) It is clear that the wavefunction Ψ defined in Equation 4.11 describes the state of two electrons since each of its terms is a product of wavefunctions that separately describe electrons A and B, even though those electrons might be found to be far apart. What is less clear, but also true, is that this particular wavefunction implies that any measurement of the spin component S_z for one of the electrons will always be *correlated* with the result of a similar measurement performed on the other electron, wherever that electron may be. This follows from the way in which measurement causes the wavefunction of a quantum system to abruptly collapse and comes about as follows.

According to conventional quantum mechanics, Equation 4.11 implies that an initial measurement of S_z, carried out at one of the measuring devices, is equally likely to yield the result $S_z = +\hbar/2$ or $S_z = -\hbar/2$. However, once that initial measurement has been made, the wavefunction of the pair will immediately collapse to either $\Psi_+(\mathrm{A})\,\Psi_-(\mathrm{B})$ or $\Psi_-(\mathrm{A})\,\Psi_+(\mathrm{B})$, according to the measured result. (For instance, measuring $S_z = -\hbar/2$ for A will cause the wavefunction of the pair to collapse to $\Psi_-(\mathrm{A})\,\Psi_+(\mathrm{B})$.) A subsequent measurement of S_z at the other detector will then inevitably lead to the opposite result to that obtained in the first measurement. So, although the spin components of the two electrons are indeterminate prior to any measurement, as soon as a given component of spin is measured for one electron, the result of a similar measurement performed on the other electron can be confidently predicted to yield the opposite result. The spin components of the two electrons are indeterminate prior to any measurement, but nonetheless certain to produce results that are completely correlated when they are measured.

Question 4.5 Suppose that the pair of electrons, A and B, were described by a wavefunction of the form

$$\Psi = \frac{1}{\sqrt{4}}[\Psi_+(\mathrm{A})\Psi_-(\mathrm{B}) + \Psi_+(\mathrm{A})\Psi_+(\mathrm{B}) + \Psi_-(\mathrm{A})\Psi_+(\mathrm{B}) + \Psi_-(\mathrm{A})\Psi_-(\mathrm{B})].$$

List all the possible outcomes of the measurement of S_z for the two electrons in this case. Use your results to explain why it would be true to say that this wavefunction describes a state in which the spin components are indeterminate prior to the first measurement, but *not* true to say that, when both measurements have been performed, their results are correlated. ■

The state described by the wavefunction of Equation 4.11 is a particular example of what is generally referred to as an **entangled state**. States of this kind always describe systems consisting of two or more parts (such as the electrons A and B), and often imply correlations between the outcomes of measurements performed separately on the different parts of the system. Any wavefunction that describes such an entangled state will have the property that it *cannot* be written as a simple product of wavefunctions describing the different parts of the system. Equation 4.11 is of this kind; it *cannot* be written as a simple product, even though it is the sum of such products.

Question 4.6 Confirm that the wavefunction considered in the previous question does *not* describe an entangled state by showing that it can be written as a simple product of the form $\Psi = \Psi(\mathrm{A})\,\Psi(\mathrm{B})$. Do this by writing down expressions for the wavefunction of each electron $\Psi(\mathrm{A})$ and $\Psi(\mathrm{B})$ in terms of the normalized wavefunctions $\Psi_+(\mathrm{A})$, $\Psi_-(\mathrm{A})$, $\Psi_+(\mathrm{B})$ and $\Psi_-(\mathrm{B})$, paying attention to the overall normalization of the full wavefunction Ψ. ■

Now, according to conventional quantum mechanics, the EPR state described by Equation 4.11, does not associate any particular value of S_z with either electron A or electron B prior to the first measurement. Yet, because Equation 4.11 describes an entangled state, as soon as a measurement of S_z is performed on one electron, the outcome of that measurement will instantly determine the result of a similar measurement subsequently performed on the other electron, even though it might be far away. This raises a number of issues. For instance, if A and B are far apart and the measurements are performed in quick succession, then any signal conveying information about the outcome of the first measurement might well have to travel faster than light if it is to get to the site of the second measurement in time to influence its outcome. How can the information that the wavefunction has collapsed travel so quickly (possibly instantly) from one measuring device to the other? Einstein in particular, the discoverer of special relativity, must have found the suggestion that quantum mechanics required faster than light connections between different places highly unpalatable. (He described such connections as 'spooky'.) On the basis of this, and other arguments, EPR concluded that properties such as spin components really were 'possessed' by particles and that measurements simply revealed the values of those possessed properties. There was no need, according to EPR, to 'communicate' the result of a spin measurement from one measuring device to the other since both electrons already 'possessed' definite spins that were just waiting to be revealed by a measurement. The fact that this ran counter to the conventional understanding of quantum mechanics (in which, prior to a measurement, the spins were *indeterminate* rather than simply unknown) was, in the view of EPR, simply an indication that conventional quantum mechanics was incomplete.

The feature of conventional quantum mechanics that the EPR argument brought to light, the ability of measurements performed at one point to instantly determine the outcomes of measurements performed elsewhere, is now usually referred to as **non-locality**.

Bohr was quick to recognize the significance of the EPR attack on the Copenhagen interpretation. After careful thought he produced a response that made use of the principle of complementarity to refute the EPR argument. The main point of Bohr's refutation was that the kind of locality that Einstein, Podolsky and Rosen were assuming was simply inconsistent with quantum mechanics. While the two electrons remained in an entangled state they simply couldn't be regarded as truly separate no matter how far apart they were.

The Copenhagen response did not satisfy Einstein, who continued to argue that physics should combine **locality** (no faster than light connections) with **realism** (microscopic entities possess all their properties at all times, independently of measurements of their values). The issue remained a point of disagreement between Bohr and Einstein until the time of the latter's death in 1955 (Bohr died in 1962.) However, that was not the end of the story.

4.3 Bell's theorem

There was no essential development in the EPR stalemate until 1964, when John Bell (Figure 4.20), a physicist at the European Centre for Particle Physics (CERN), pointed out a totally unexpected feature of the EPR argument. He proved that in a more elaborate form of the spin component experiment, the kind of local realism that EPR advocated led to experimental predictions that conflicted with the predictions of conventional quantum mechanics. Clearly, if such experiments could be performed they would provide a new test of the adequacy of quantum mechanics.

Figure 4.20 John S. Bell (1928–1990).

Figure 4.21 shows an idealized version of the sort of modified EPR apparatus that could be used to investigate Bell's result. In this particular version of the experiment the measuring devices are no longer restricted to measuring the z-component of spin. Instead, each has a 'component selector' that may be used to select any direction in the plane of the page. Once the selector of a measuring device has been set, that measuring device will always measure spin components in that particular direction until the setting is changed. Note, however, that no matter what direction has been selected, the result of any individual measurement can still only be $+\hbar/2$ or $-\hbar/2$, since these are the only possible outcomes for *any* measurement of an electron's spin component, irrespective of the direction along which the spin component is measured.

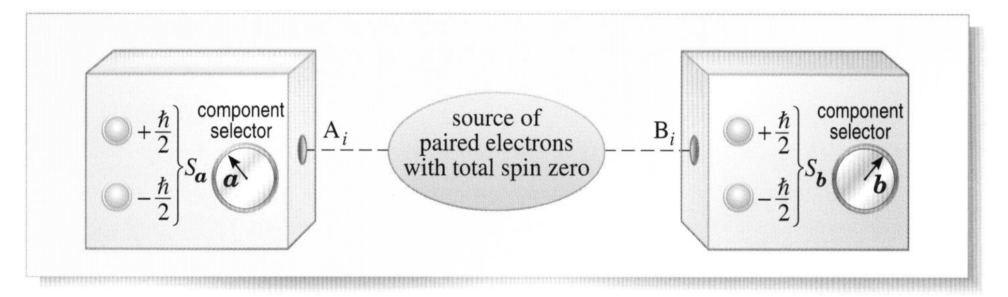

Figure 4.21 A modified EPR experiment in which the measured spin components of electrons A_i and B_i are $S_a(A_i)$ and $S_b(B_i)$, in directions $\boldsymbol{a}$ and $\boldsymbol{b}$, respectively.

Now, imagine setting the selectors on the two detectors to the directions represented by the vectors $\boldsymbol{a}$ and $\boldsymbol{b}$ in Figure 4.21. If the spin components are measured for a pair of electrons A and B in the usual EPR entangled state of Equation 4.11, then the detectors will indicate the measured values of $S_a(\mathrm{A})$ and $S_b(\mathrm{B})$. Since each of these results must be either $+\hbar/2$ or $-\hbar/2$, the product of the two results, $S_a(\mathrm{A}) \times S_b(\mathrm{B})$, must be either $+\hbar^2/4$ or $-\hbar^2/4$.

Both results are now allowed for both electrons as the vectors $\boldsymbol{a}$ and $\boldsymbol{b}$ are not parallel to each other.

If similar measurements are made for a large number of entangled pairs (A_1, B_1), (A_2, B_2), (A_3, B_3), ... (A_N, B_N), then for the ith pair the quantity $S_{\boldsymbol{a}}(A_i) \times S_{\boldsymbol{b}}(B_i)$, will turn out to be either $+\hbar^2/4$ or $-\hbar^2/4$. (Different pairs will generally give different results for the product, but each of those paired results must be *either* $+\hbar^2/4$ *or* $-\hbar^2/4$.) Consequently, if you take the average of all those paired results and multiply it by $4/\hbar^2$, you will obtain a single number that lies between +1 and −1. This number, which depends on the chosen directions $\boldsymbol{a}$ and $\boldsymbol{b}$, will be denoted by the symbol $C(\boldsymbol{a}, \boldsymbol{b})$; it signifies the extent to which the measured spin components are correlated. For example, if $\boldsymbol{a}$ and $\boldsymbol{b}$ are parallel then the requirement of zero total spin for the pair means that $S_{\boldsymbol{a}}(A_i) = -S_{\boldsymbol{b}}(B_i)$ for every pair, so that $C(\boldsymbol{a}, \boldsymbol{b}) = -1$. Whereas, if $\boldsymbol{a}$ and $\boldsymbol{b}$ are antiparallel then it will always be the case that $S_{\boldsymbol{a}}(A_i) = S_{\boldsymbol{b}}(B_i)$, in which case $C(\boldsymbol{a}, \boldsymbol{b}) = 1$. Other choices for $\boldsymbol{a}$ and $\boldsymbol{b}$, including that shown in Figure 4.21, will result in different values for $C(\boldsymbol{a}, \boldsymbol{b})$. Since $C(\boldsymbol{a}, \boldsymbol{b})$ is a measure of spin component correlation, and since it is also a function of the vectors $\boldsymbol{a}$ and $\boldsymbol{b}$, it is generally referred to as a *spin component correlation function*.

Bell discovered that, by combining spin component correlation functions in an appropriate way, it is possible to construct quantities for which the quantum-mechanical predictions conflict with the consequences of the EPR assumption of locality and realism. One such quantity is

$$|C(\boldsymbol{b}, \boldsymbol{b}) + C(\boldsymbol{b}, \boldsymbol{c}) + C(\boldsymbol{a}, \boldsymbol{b}) - C(\boldsymbol{a}, \boldsymbol{c})|$$

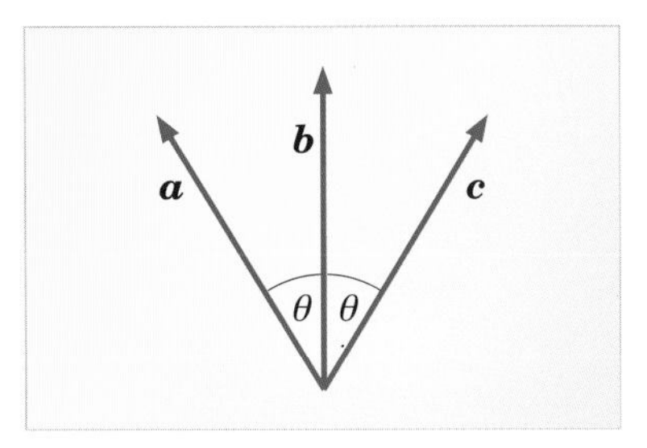

Figure 4.22 Three directions, determined by the vectors $\boldsymbol{a}$, $\boldsymbol{b}$ and $\boldsymbol{c}$. Note that the angle between $\boldsymbol{a}$ and $\boldsymbol{b}$ is θ, as is the angle between $\boldsymbol{b}$ and $\boldsymbol{c}$.

where the three vectors, $\boldsymbol{a}$, $\boldsymbol{b}$ and $\boldsymbol{c}$ represent different settings on the selector dials in Figure 4.21, and are related in the manner shown in Figure 4.22. The important point to note is that the angle θ between $\boldsymbol{a}$ and $\boldsymbol{b}$ is equal to the angle between $\boldsymbol{b}$ and $\boldsymbol{c}$, and that the value of $|C(\boldsymbol{b}, \boldsymbol{b}) + C(\boldsymbol{b}, \boldsymbol{c}) + C(\boldsymbol{a}, \boldsymbol{b}) - C(\boldsymbol{a}, \boldsymbol{c})|$ will depend on the value of the angle θ.

Bell was able to show that in any theory that combines locality and realism, so that there are no 'spooky' connections, and properties are possessed even in the absence of measurement, then the following inequality will always be satisfied:

$$|C(\boldsymbol{b}, \boldsymbol{b}) + C(\boldsymbol{b}, \boldsymbol{c}) + C(\boldsymbol{a}, \boldsymbol{b}) - C(\boldsymbol{a}, \boldsymbol{c})| \leq 2. \quad (4.12)$$

This is one version of a result known as **Bell's inequality**.

On the other hand, it is also easy to show that conventional quantum mechanics (which is neither local nor realist) predicts that $C(\boldsymbol{a}, \boldsymbol{b}) = -\cos\theta$, implying that

$$|C(\boldsymbol{b}, \boldsymbol{b}) + C(\boldsymbol{b}, \boldsymbol{c}) + C(\boldsymbol{a}, \boldsymbol{b}) - C(\boldsymbol{a}, \boldsymbol{c})| = |\cos(2\theta) - 2\cos(\theta) - 1|. \quad (4.13)$$

A graph showing how the value of this quantity varies with θ is shown in Figure 4.23. As you can see, there are values of θ (from 0 to $\pi/2$ rad for example) for which the quantum-mechanical prediction exceeds the upper limit allowed by local realism.

The conclusion to which we are driven by the above argument is known as **Bell's theorem**, and may be stated as follows:

> Any theory that has the properties of locality and realism cannot replicate all the predictions of quantum mechanics.

So, the EPR claim that physical theory must be characterized by locality and realism can, thanks to Bell's inequality (Equation 4.12), be tested experimentally.

Carrying out the required experiment is far from simple. Nonetheless, such experiments did become possible in the early 1980s. Some of the earliest were performed by the French physicist Alain Aspect and his collaborators. (These were

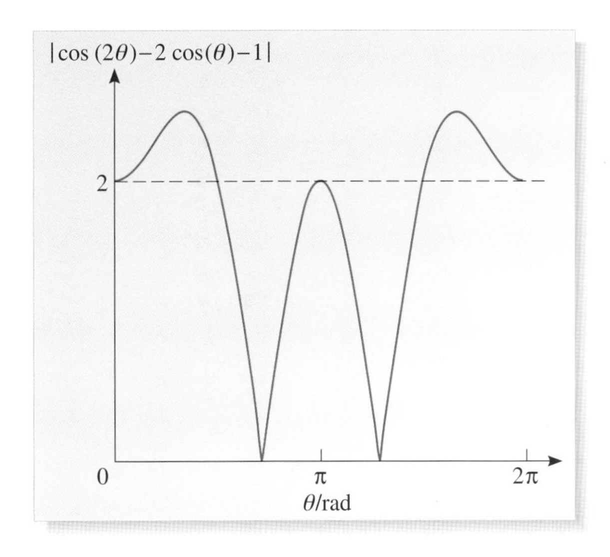

Figure 4.23 Graph of $|\cos(2\theta) - 2\cos(\theta) - 1|$ against θ.

actually variants of the idealized experiment we have been discussing, but they worked in the same sort of way and established the same general point.) Several such experiments have now been performed, and *they all support quantum mechanics rather than local realism.*

The experimental successes of quantum mechanics, even where its predictions conflict with those of local realism, show that Einstein, Podolsky and Rosen were wrong and that local realism is simply not acceptable as a general feature of physical theory. So, even if quantum mechanics is eventually replaced by some new theory of microscopic phenomena, Bell's theorem tells us that the days of EPR-style local realism are gone forever. Whatever its appeal, such a simple view of reality is inconsistent with what we now know about the physical world.

4.4 A third look at interpretation

The suggestion that quantum mechanics offers only an incomplete description of nature is an old one. Those who take this view have often suggested that the behaviour of electrons and other microscopic entities is actually fully deterministic (rather than inherently probabilistic as quantum mechanics asserts) and that a complete description of such entities involves certain **hidden variables** that we have not yet learned to determine.

Bell's theorem imposes serious limitations on the nature of any hidden variable theory that aims to be at least as successful as quantum mechanics, but it does not entirely rule out the possibility of formulating such a theory. There is one hidden variable theory in particular that provides a realist view of microphysics (that is, that electrons, etc. possess properties independent of measurements) and replicates the predictions of quantum mechanics. It achieves this by wholeheartedly embracing non-locality (as Bell's theorem implies that it must). Because this theory uses the formalism of quantum mechanics, it can actually be regarded as an interpretation of quantum mechanics, and as such it is referred to as the **ontological interpretation**. However, it is more widely know as **Bohm's theory**, in memory of its main proponent, David Bohm (1917–1992). A detailed account of this interpretation may be found in *The Undivided Universe* by D. Bohm and B. J. Hiley (Routledge, 1995).

Ontology is the branch of philosophy that concerns the nature of existence.

Bohm's theory echoes certain ideas originally presented in 1927 by Louis de Broglie. In essence, de Broglie suggested that the wavefunction of quantum mechanics describes a real wave that acts as a 'pilot wave' for particles, guiding them towards certain destinations. The wave and the particle, in this view, are not complementary ways of describing reality, but are both simultaneously real. Thus, in the case of two-slit electron diffraction for example, the pilot wave would be diffracted by its actual passage through two slits, and the electron, having passed through one slit or the other, would be acted upon by the pilot wave and guided towards those parts of the screen where quantum mechanics predicts that it will be found.

In Bohm's theory, a single particle always has a well defined position, which is the 'hidden variable' that conventional quantum mechanics does not recognize. But particles also have a wavefunction that satisfies Schrödinger's equation. The important new ingredient of Bohm's theory is a quantity called the *quantum potential* that may be determined from the mathematical form of the wavefunction. The quantum potential acts on the particle in much the same way that an ordinary potential would, and plays a role in determining the particle's trajectory. In the case of a two-slit diffraction experiment, for instance, the effect of the quantum potential is to cause electrons described by the same wavefunction Ψ to follow the trajectories shown in Figure 4.24. As you can see, this will result in what is conventionally described as an interference pattern when the electrons arrive at the screen, even though there is no 'interference' in this interpretation.

At first sight Bohm's theory looks rather like classical mechanics with the quantum potential as an added ingredient. However, the quantum potential has a number of features that would be considered very unusual in any conventional classical theory. Foremost among these is the fact that the quantum potential is non-local. One effect

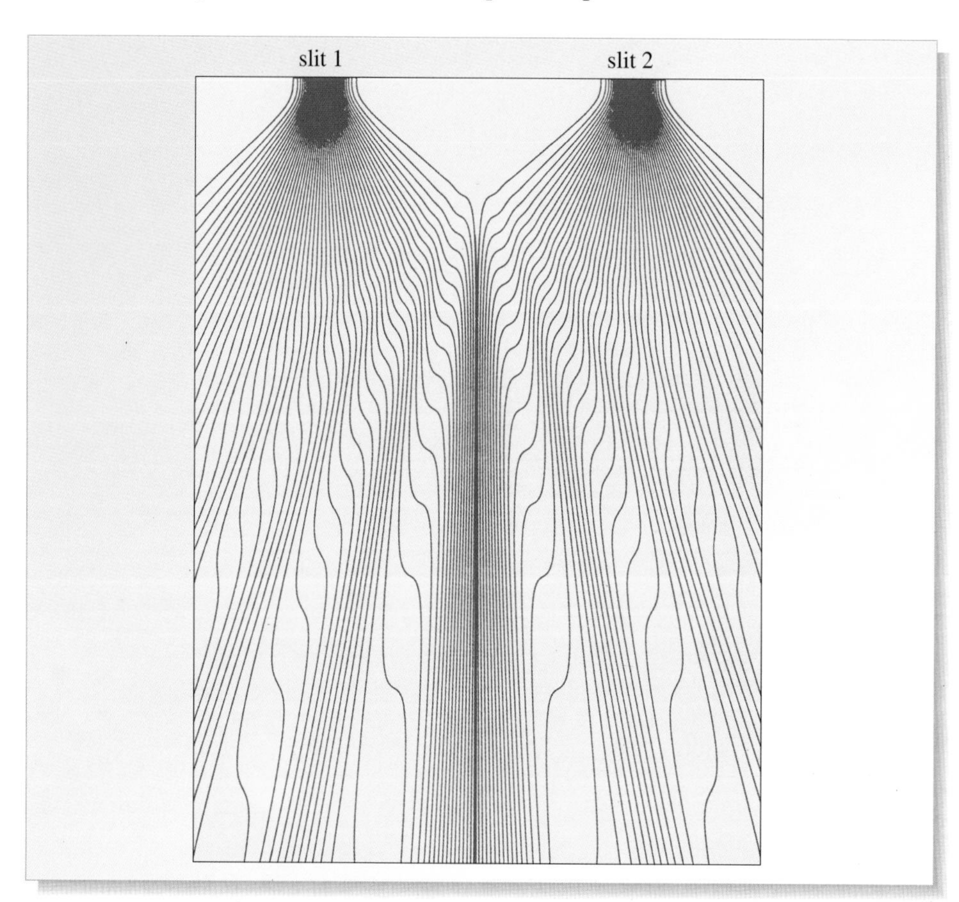

Figure 4.24 The (classical) trajectories of electrons passing through the slits of a two-slit diffraction apparatus according to Bohm's ontological interpretation of quantum mechanics.

of this is to make the trajectory of an electron passing through one of the slits in a two-slit diffraction apparatus depend on whether or not the other slit is open. Closing one of the slits alters the quantum potential everywhere and hence alters the trajectories of particles passing through the other slit. This alters the points at which the electrons arrive at the screen and causes the two-slit interference pattern to disappear. Another effect of the non-local quantum potential is to allow a realist theory, in which electrons may be regarded as possessing spin components even in the absence of measurements, to reproduce the correlations that are predicted by conventional quantum mechanics and observed in the experiments by Aspect and others.

In fact, any system that can be described using Schrödinger's quantum mechanics can also be described by Bohm's theory and there are no simple experimental grounds for preferring conventional quantum mechanics over Bohm's radical alternative. Nonetheless, there are other factors which have continued to ensure that Bohm's approach is viewed with suspicion by many physicists. For instance, serious difficulties are encountered when attempts are made to extend Bohm's interpretation so that it can be applied to forms of quantum theory that are consistent with Einstein's special theory of relativity. (Quantum mechanics is *non-relativistic* in this sense, so it can be subjected to Bohm's interpretation.) Also, in situations that involve several particles, the non-locality of Bohm's theory takes on a very awkward appearance which fails to appeal to most physicists. Despite these drawbacks, Bohm's theory does have its supporters even though they are certainly in a minority.

As you will appreciate, there is still much debate about the interpretation of quantum mechanics. The debate continues to be fuelled by our increasing ability to carry out experiments that the founders of quantum mechanics could only speak about, if they could conceive them at all. In such an atmosphere it is encouraging to think that the interpretation of quantum mechanics might yet provide the answer to some of the most ancient of philosophical problems, perhaps even that of Figure 4.25.

Figure 4.25 The Copenhagen chicken.

5 Closing items

5.1 Chapter summary

1 Quantum mechanics consists of a formalism and an interpretation.

2 The specification of a quantum system typically involves a number of parameters (such as m_e and $-e$) and a potential energy function $E_{pot}(x, y, z)$. These items are then used in the version of Schrödinger's time-dependent equation that is appropriate to the system being considered.

3 In quantum mechanics the state of a system is represented by its wavefunction Ψ. This is generally a function of time and may be determined, subject to appropriate conditions, by solving the appropriate version of Schrödinger's time-dependent equation.

4 The measurable properties of a system are called observables. By analysing the wavefunction that describes a particular state of a system, it is possible to predict the possible outcomes of any measurement of a given observable made while the system is in that state. It is also possible to predict the probability of each of those possible outcomes. This, however, is as far as quantum mechanics goes. It is not generally possible to predict the precise outcome of any particular measurement.

5 The possible outcomes of a measurement of a given observable are known as the eigenvalues of that observable. If the state of a system is such that a measurement of some selected observable has only one possible outcome, then that state is said to be an eigenstate of the selected observable, and may be said to correspond to the relevant eigenvalue. The time-independent wavefunctions that describe the essential features of stationary states, such as those of atoms or particles confined in potential wells, represent eigenstates of energy for those systems and therefore correspond to energy eigenvalues.

6 If $\Psi_1, \Psi_2, \Psi_3, \ldots \Psi_n$ are n solutions of the time-dependent Schrödinger equation for a given system, then that equation will also be satisfied by a linear superposition of the form

$$\Psi = c_1\Psi_1 + c_2\Psi_2 + c_3\Psi_3 + \ldots + c_n\Psi_n \tag{4.5}$$

where c_1, c_2, c_3, etc. are numbers (which may be complex).

7 Given a system in a state represented by the normalized wavefunction

$$\Psi = c_1\Psi_1 + c_2\Psi_2 + c_3\Psi_3 + \ldots + c_n\Psi_n$$

where the n normalized wavefunctions $\Psi_1, \Psi_2, \Psi_3, \ldots \Psi_n$ represent eigenstates of some observable A, and correspond, respectively, to eigenvalues a_1, a_2, a_3, $\ldots a_n$, then the only possible outcome of any measurement of that observable is one of those eigenvalues, and the probability that it will be a_i is given by

$$P(a_i) = |c_i|^2 \quad (\text{for } i = 1, 2, 3, \ldots n). \tag{4.8}$$

8 According to the conventional Copenhagen interpretation, a wavefunction provides the fullest possible description of the state of an individual quantum system. This implies that probabilities are an inherent part of the description of nature. It also implies that systems do not generally 'possess' values of observables, but that such values may be measured. The Copenhagen interpretation thereby gives great significance to the process of measurement, and calls into question naive views concerning the 'reality' of microscopic entities.

9 Measurement involves a physical interaction between a quantum system and a measuring device that results in the measuring device being left in some state that represents the outcome of the measurement. The conventional view is that when a measurement is made on a system, the system's state changes discontinuously and unpredictably to an eigenstate of the measured observable whose eigenvalue a is the actual result of the measurement. This is described by saying that the system's wavefunction has collapsed.

10 The example of Schrödinger's cat highlights the difficulty of interpreting macroscopic superposition states, particularly in the context of measurement. There are a number of proposals concerning the reason why such states are never observed.

11 According to the many worlds interpretation, all of the possibilities encompassed by quantum mechanics are realized, but the different possibilities occur in different universes.

12 The Einstein–Podolsky–Rosen argument shows that for a pair of electrons in an entangled spin state, a measurement performed at one location immediately determines which of various possible outcomes will be found at some other location. According to EPR, this shows that observable properties are possessed, and that conventional quantum mechanics is therefore incomplete.

13 Bell's theorem states that theories which combine realism and locality (in which properties are possessed independently of measurements, and there are no 'spooky' faster-than-light connections) cannot replicate all the predictions of quantum mechanics.

14 Experiments by Aspect and others support the predictions of quantum mechanics and conflict with locally realist theories of the kind advocated by EPR.

15 According to the ontological interpretation (Bohm's theory), microscopic entities do possess properties independent of measurements, but the predictions of conventional quantum mechanics are accounted for by the action of a non-local quantum potential.

5.2 Achievements

Having completed this chapter, you should be able to:

A1 Explain the meaning of all the newly defined (emboldened) terms introduced in this chapter.

A2 Summarize the essential features of quantum mechanics in terms of systems, states, observables, predicted measurement outcomes, indeterminacy and indeterminism.

A3 State the superposition principle, and, in the case of a wavefunction that is a superposition of normalized wavefunctions describing different eigenstates of some observable, relate the coefficients involved in that superposition to the probabilities that measurements of the related observable will yield the various eigenvalues of that observable.

A4 Outline the effect of a measurement on the wavefunction describing the state of a system, and discuss some of the issues surrounding the process of wavefunction collapse.

A5 Identify (in sufficiently simple cases) wavefunctions describing entangled states, and explain the significance of such states in terms of correlated measurement outcomes.

A6 Outline the EPR argument and explain the significance of Bell's inequality, Bell's theorem and the relevant experiments of Aspect and others.

A7 Describe in qualitative terms the key features of various interpretations of quantum mechanics; specifically the Copenhagen, many worlds and ontological interpretations.

5.3 End-of-chapter questions

Question 4.7 What is wrong with the following woefully inadequate summary of quantum mechanics?

'According to quantum mechanics, the Universe is all fuzzy. Everything suffers from indeterminacy, or, equivalently, indeterminism, and nothing is certain.'

Question 4.8 Suppose that $\Psi(\boldsymbol{r}_1, \boldsymbol{r}_2, t)$ is the time-dependent wavefunction describing a system of two particles passing through the double-slit diffraction apparatus that was described in Section 2. Given that the positions of the particles are indeterminate prior to any measurement of them, what significance can be attached to any given set of values for the six position coordinates (x_1, y_1, z_1) and (x_2, y_2, z_2)?

Question 4.9 The wavefunction describing the state of a certain atom is given by the linear superposition $\Psi = a(2\Psi_1 + 3\Psi_2)$, where Ψ_1 and Ψ_2 are normalized wavefunctions describing eigenstates of energy that correspond to the eigenvalues 1.00 eV and 2.00 eV, respectively. If Ψ is normalized, what is the value of the (real) constant a? What is the probability that a measurement of the atom's energy will have the result 1.00 eV? What is the probability that the outcome will be 1.50 eV?

Question 4.10 Summarize the key features of each of the following interpretations of quantum mechanics: the Copenhagen interpretation, the many worlds interpretation and the ontological interpretation.

Question 4.11 List some of the major challenges of quantum mechanics that seem to require an interpretive solution.

Question 4.12 Explain in general terms why the instant collapse of the wavefunction of a pair of electrons in an entangled state cannot be used to send a meaningful signal from one place to another at a speed greater than the speed of light. (This is a hard question.) ■

Chapter 5 Consolidation and skills development

1 Introduction

This final chapter will help you to consolidate what you have learned in this book. First, in Section 2, we will review the key ideas that have been introduced in Chapters 1 to 4 and clarify a little further the historical context in which the evolution of these ideas took place. Second, in Section 3, we will help you to develop your skill in obtaining information on physics topics from a range of sources.

Section 4 consists of a series of short questions which will help you revise and consolidate what you have learned in this book. Section 5 directs you to the interactive question package for *Quantum physics: an introduction*, and Section 6 invites you to try some of the longer questions contained in the *Physica* package.

2 Overview of Chapters 1 to 4

In Chapter 1 we began the story of the evolution of quantum physics by describing some of the problems which beset physicists round about 1900: phenomena for which the great theories of classical physics appeared to be unable to provide an adequate explanation. The eventual resolution of these (and other) problems required the most profound change in outlook since the foundation of modern physics in the days of Galileo and Newton. This new revolution was the quantum physics revolution.

The five problems which we discussed in Chapter 1 were:

1. *Understanding atoms* Why did they have the sizes and properties which they do have? Could these properties somehow be related to their structure?
2. *Atomic spectra* One specific aspect of atoms which needed explaining was their spectra. Why did they emit and absorb radiation only at certain wavelengths, that is, as spectral *lines*? In particular, the regularity in the wavelengths of the visible lines of the Balmer series of hydrogen seemed to require an explanation.
3. *Blackbody radiation* The shape of the blackbody spectrum was of particular concern to physicists. This spectrum should have had a quite straightforward explanation in terms of the classical concepts of equipartition and the number of radiation modes in a cavity, which were well known at the time. However, the classical prediction for the ultraviolet end of the spectrum was spectacularly wrong!
4. *The photoelectric effect* When a beam of light is incident on the surface of certain metals, electrons are ejected. But the details of such experiments cannot be explained using classical physics.
5. *The heat capacities of solids* Dulong and Petit's law, $C_{\rm m} = 3R$, for the heat capacities of solids, seems initially to be upheld by experimental observations. But there are some exceptions, such as diamond, and at low temperatures the heat capacities of all solids deviate considerably from this law.

The revolutionary development in physics which solved all these problems was quantum mechanics. Closed systems, whether atoms or cavities filled with radiation, share a radical new property: quantities such as energy no longer have a continuous range of values available to them, but are quantized. In particular, light of frequency f

is transferred only in packages of energy hf, that is, as light quanta. In effect, light has a particle aspect, the particles being photons each carrying energy hf. The constant h permeates every aspect of quantum theory. It is known as Planck's constant after the man who, in 1900, first glimpsed the quantum realm in the course of explaining the blackbody radiation curve. The constant h marks the divide between quantum and classical; all our experiences from birth onward belong to a world in which h is very small. But h is not insignificant in the world of atoms.

In 1905, Einstein applied the light quantum idea in what was in many ways his most radical work: his explanation of the photoelectric effect. Einstein's photoelectric equation, $\frac{1}{2}mv_{\text{max}}^2 = hf - \phi$, linked the work function ϕ (a property of the metal), Planck's constant h, the frequency of the light f and the measurable kinetic energy of the emitted electrons. This equation was verified in detail by Robert Millikan in 1916.

Before Millikan's work, the discovery which had really convinced physicists that the microscopic world was *different*, with quantum properties at its heart, was Einstein's 1907 explanation of the longstanding puzzle of the heat capacities of solids. Einstein found that the motion of atoms in solids was subject to quantum rules. It was this discovery which was to find its way into the consciousness of the young Niels Bohr in 1913. Rutherford had just made a momentous discovery: the nuclear atom. He had realized that this was the inescapable consequence of the α-particle scattering experiments of his younger colleagues Geiger and Marsden. It meant that most of the mass of an atom was concentrated within a nucleus, which had a radius of only $\sim 10^{-14}$ m and that this was the site of all the positive charge of the atom. This positive charge was balanced by the negative charge on the electrons, which occupied the rest of the volume of the atom, which had a typical radius of $\sim 10^{-10}$ m. Rutherford was very well aware that this picture had many problems, but it was his young visitor from Copenhagen, Niels Bohr, who showed how the nuclear atom could be made to work. Not only that, but Bohr's quantum version of the nuclear hydrogen atom solved a key part of the 'spectroscopy problem', namely, the Balmer series of hydrogen. The most crucial postulate of Bohr's model was that the angular momentum associated with the orbital motion of the electron is quantized, coming in integer multiples of $\hbar$.

Despite much effort by Bohr and others, the model was far less successful for heavier atoms. We can now see that Bohr's theory was a hybrid: a classical theory with certain quantum aspects bolted on, as it were. The full explanation of the properties of atoms, molecules, solids, nuclei, etc. had to await another and more profound revolution: the discovery of quantum mechanics itself. Nevertheless, Bohr's hydrogen atom was an essential step towards this discovery.

Chapter 1 concludes with a discussion of the 'wave–particle duality' dilemma. The idea that light transfers energy in quanta of size hf is now taken much further. Light, and, by implication, all electromagnetic radiation, has a particle aspect, the particles being *photons*. Not only do photons have energy $E = hf$, but also momentum (Compton's experiment) $p = E/c = hf/c = h/\lambda$. Yet electromagnetic radiation also displays wave-like properties. Many experiments with light reveal its wave-like aspect, but if, like Compton, we devise an experiment which asks about a particle-like aspect of light, like its momentum, we get the typical answer of a particle: 'momentum h/λ in such and such a direction.' Stated informally, wave–particle duality means: If an experiment poses a typical 'particle' question, we get a particle answer, but if we pose a 'wave' question we get a wave answer.

The revolutionary hypothesis made by Louis de Broglie was that electrons, and indeed all particles that we think of as matter, also exhibit wave–particle duality. He postulated that a particle with momentum of magnitude p has an associated wavelength, known as the de Broglie wavelength, $\lambda_{dB} = h/p$. Davisson and Germer showed that you do indeed get diffraction patterns when electrons pass through suitable gratings, such as metallic crystals.

Chapter 2 begins with an investigation of diffraction experiments with electrons which highlight the wave nature inherent in the behaviour of matter at the atomic level. Having established that matter on this scale behaves like waves, we might well ask 'where in quantum physics do particles get their 'particle-ness''? In particular, how do we get back our common sense idea that particles must, after all, be found within some particular volume? A volume which must, for example, be able to move down a cathode ray tube under the influence of electric and magnetic forces. This problem is partly resolved by the concept of *wave packets*. However, one aspect of wave packets was soon recognized by Heisenberg: if the wave packet representing a particle is confined to a smaller and smaller volume, then the spread of wavelengths required to describe that wave packet becomes larger and larger. This fact was elucidated by Fourier's analysis of waves, which had been developed at the beginning of the nineteenth century. But now we know from the de Broglie relationship, $\lambda_{dB} = h/p$, that a spread in λ must mean a spread in p. This leads straight to the first statement of Heisenberg's uncertainty principle which says that, as a wave packet defines the position of a particle more and more closely, the momentum of that particle becomes less and less well defined. In our 'big' (macroscopic) world where the constant h appears to be very small, this uncertainty is not apparent since the wave packet for a football, say, is many, many, orders of magnitude smaller than the size of the football itself. But for electrons, things are the other way round. The wave packet or wavefunction for an electron in a hydrogen atom is *much* more extended in space than the electron itself.

If electrons, as well as photons, exhibit wave–particle duality, is there a differential equation which describes 'electron waves' in the same way that Maxwell's equations describe electromagnetic waves? Yes! It is Schrödinger's equation. The wavefunctions, $\Psi(x, y, z, t)$, that are the solutions to Schrödinger's time-dependent equation for a particle in a particular situation, contain all the information that it is possible to have about that particle. For unbound particles the wavefunctions describe travelling waves and for bound particles they describe standing waves. In the latter case, it is possible to separate the time and space variation of the wave-functions and approach the problem with the simpler, time-independent form of Schrödinger's equation.

The reason why microscopic systems, like atoms, exist only in particular energy states becomes clear: just as the equations for sound in an organ pipe or on a guitar string give acceptable solutions only for certain wavelengths, so Schrödinger's equation for a confined particle has acceptable solutions $\psi(x, y, z)$ only for certain wavelengths, each of which correspond to a definite value of the energy.

The wavefunctions must be interpreted probabilistically: in one dimension, the probability of finding a particle in a small interval Δx is $|\psi(x)|^2 \Delta x$. In three dimensions, the probability of finding a particle in a small volume ΔV is $|\psi(x, y, z)|^2 \Delta V$.

Chapter 3 is devoted to the quantum mechanics of atoms, beginning with the simplest atom, hydrogen. The Cartesian coordinates (x, y, z) are not now very convenient: it is much simpler to write the distance dependence of the Coulomb potential as $1/r$ rather than $1/\sqrt{x^2 + y^2 + z^2}$. For this reason, it is more natural to solve Schrödinger's equation for the hydrogen atom using spherical polar coordinates (r, θ, ϕ) in which case the wavefunction is written $\psi(r, \theta, \phi)$.

It turns out that solutions of Schrödinger's equation for the hydrogen atom are labelled by various quantum numbers. The first of these is the *principal quantum number* n, which ranges from 1 upward through all integers. The energy of a state of the hydrogen atom is determined by n and is $(-13.6/n^2)$ eV. The negative energy means that at least 13.6 eV must be *added* to a hydrogen atom in its ground state ($n = 1$) to allow the electron to become unbound, in other words, to ionize the hydrogen atom. The second quantum number is the *orbital angular momentum quantum number* l, where, in a given energy level, l can take integer values from 0 to $n - 1$. Since the magnitude L of the orbital angular momentum is given by $\sqrt{l(l+1)}\,\hbar$, an orbital angular momentum of zero is allowed, unlike in Bohr's model where the lowest allowed magnitude of angular momentum was $\hbar$.

To further specify the state of an electron in the hydrogen atom, we must also give the *orbital magnetic quantum number* m_l which specifies the z-component, L_z, of the electron's orbital angular momentum as $L_z = m_l\hbar$. This new quantum number takes values $-l, -l + 1, -l + 2, \ldots, 0, \ldots, l - 2, l - 1, l$; that is, $2l + 1$ values in all.

There are thus, in general, a number of states for any given n all having energy $(-13.6/n^2)$ eV, but they are nevertheless distinct, with spatial probability densities $|\psi|^2$ depending on l and m_l (as well as on n).

In order to fully specify the state of the electron we must also take account of its spin. All electrons have a spin of $\frac{1}{2}$ ($s = \frac{1}{2}$). The z-component of the electron's spin, S_z, is given by $S_z = m_s\hbar$ where $m_s = \pm\frac{1}{2}$. Counting up all the possible combinations, we find that the total number of states with principal quantum number n, and therefore with an energy of $(-13.6/n^2)$ eV, is $2n^2$. This will prove to be vital for understanding the Periodic Table of the elements.

To understand atoms of elements heavier than hydrogen, one must solve Schrödinger's equation for an electron in an atom containing Z electrons, each with electric charge $-e$, and with a nucleus carrying a charge $+Ze$, that is, the charge of Z protons. The atomic number Z characterizes each chemical element. Writing down Schrödinger's equation for an atom with Z electrons is relatively straightforward, solving it is not. However, we can model the potential energy function for each electron in the atom by assuming that it moves in an electric field which is a superposition of the attractive potential due to the nucleus and the repulsion due to the other $Z - 1$ electrons. The wavefunctions ψ will therefore not be identical to those in a hydrogen atom, but they *can* still be labelled by the quantum numbers n, l, m_l, m_s. In particular, there are still $2n^2$ states for each value of principal quantum number n. The sequence of values of $2n^2$ is 2, 8, 18, 32, … — numbers of great significance to chemistry.

A new consideration enters here: in hydrogen, an electron in an excited state spontaneously loses energy by emitting photons until it reaches the ground state in which $E = -13.6$ eV. But *Pauli's exclusion principle* tells us that only one electron in an atom can have a particular set of quantum numbers n, l, m_l, m_s. Hence, an electron cannot lose energy to fall into a lower energy state if that state is occupied. This is the key idea underlying the way the structure of atoms builds up with increasing atomic number Z. The Pauli exclusion principle is also of crucial importance in the physics of atomic nuclei, solids, white dwarf stars … all matter, in fact.

Chapter 4 addresses the question of how quantum mechanics is to be understood. Some important new concepts are required for this discussion.

One of the most fundamental characteristics of quantum mechanics is the principle of superposition. This means that if Ψ_1, Ψ_2, … Ψ_n are solutions of the time-dependent Schrödinger equation, then so is any superposition $\Psi = c_1\Psi_1 + c_2\Psi_2 + \ldots + c_n\Psi_n$, where $c_1, c_2, \ldots, c_n$ are numbers. Thus the interference pattern which appears when an electron goes through a double slit and strikes a screen is proportional to $|\Psi_1 + \Psi_2|^2$ where Ψ_1 and Ψ_2 are the wavefunctions that describe electrons going through slit 1 and slit 2, respectively. In quantum mechanics, the probability of finding some particular outcome is always of the form of the modulus squared of an amplitude. For example, for a state $\Psi = c_1\Psi_1 + c_2\Psi_2$, the probability of somehow finding the system in state 1 is proportional to $|c_1|^2$ and the probability of finding the system in state 2 is proportional to $|c_2|^2$. This is related to what you saw in an earlier chapter: the fact that the probability of finding an electron, which is described by the wavefunction $\Psi(x, t)$, within an interval Δx, is also proportional to the modulus squared of an amplitude, namely $|\Psi(x, t)|^2 \Delta x$.

The concept of measurement plays a key role in the interpretation of quantum mechanics. It is important to note that, like 'force', 'energy', 'work', etc., 'measurement' is a technical term which does not have its everyday meaning. In particular, it is not necessarily an action by a person. What is it then? Answering this question involves a number of other ideas. The first of these is *observable*, meaning any measurable property of the physical system, such as the position of an electron, or its spin along some given direction. The second is *eigenstate*. The state of a system is an eigenstate of an observable when a measurement of that observable is certain to give some particular value for that observable; that value is the *eigenvalue* of the observable characterizing that state. For example, the ground state of the hydrogen atom is an eigenstate of energy (eigenvalue −13.6 eV). However, it is certainly not an eigenstate of position since any experiment which asked where the electron was could yield any position (x, y, z) for which $|\psi(x, y, z)|^2$ was greater than zero. The essence of a measurement is that the system, after a measurement, collapses into the eigenstate characterized by the observed eigenvalue.

Another characteristic of a measurement is that it involves something irreversible happening in the macroscopic realm. For example, a spot might appear on a film or a detector might click, recording, in either case, the arrival of an electron. Such an electron cannot be said to *have* a position until its position is measured. This is true whether the electron is described by the time-independent hydrogen atom wavefunction or by the time-dependent wavefunction which accounts for its interaction with diffraction slits. The same is true for all observables for which the state of the system is not an eigenstate. A system in a state described by $\Psi = c_1\Psi_1 + c_2\Psi_2$ is not in a state described by either Ψ_1 or Ψ_2 until it has been measured to be so. After such a measurement, Ψ will collapse to state Ψ_i with probability $|c_i|^2$. This is why the two-slit interference pattern disappears if you make a measurement determining which slit the electron passes through.

In fact, systems do not have values of observables until the observables are measured. This is the clear implication of experiments carried out since the 1980s and prompted by the work of Bell in the 1960s. Bell in turn was responding to the profound critique of quantum mechanics in 1935 by Einstein, Podolsky and Rosen (known as the EPR argument).

It is for reasons such as this that Niels Bohr once said: 'If you're not shocked by quantum mechanics, you haven't understood it'. Quantum mechanics directly contradicts some of the deeply ingrained habits of mind which we have acquired as a matter of survival since we were born. Survival, that is, in the world in which the constant h appears to be very small. Quantum mechanics has proven inexhaustibly successful in explaining the properties of this world in terms of what is going on in the realm where h is significant. As a mathematical formalism for predicting the results of experiments, quantum mechanics is unchallenged as *the* fundamental theory in terms of which all our models of the world on the microscopic level must be formulated. And yet ... what does it all mean? Because the picture of the world presented by quantum mechanics is so irreducibly different from our everyday world, we will probably never understand quantum mechanics *if, by understand it, we mean construct mental models based on what happens in our everyday world.* Chapter 4 discusses what can legitimately be said about how the formalism of quantum mechanics can be interpreted. You should be aware that this is a subject where it can seem that there are almost as many views as there are physicists. However, most of them adhere to some version of the Copenhagen interpretation hammered out by Bohr, Heisenberg, Pauli, Born and others.

The advent of quantum mechanics led to a vast extension in the *scope* of physics. This book has been your introduction to the basic concepts of quantum mechanics and in it, and particularly in the next book in the series (*Quantum physics of matter*), we give you a glimpse of how, indeed, it does allow us to understand the structure of matter on a small scale.

3 How to find out more physics

A vast amount of interesting, inspiring and useful physics lies beyond the scope of the present course. There is also a huge amount of information about physics 'out there' in both print and electronic media and, in this section, we give you some guidance about accessing this information. Perhaps you simply wish to satisfy your curiosity about some aspect which has grabbed your imagination, or maybe you want to include information from sources beyond the course in an essay you are writing; possibly you need some specific information related to physics in connection with your job. One solution, naturally, is to take further physics courses, but there are other ways to get the information. Here, we simply make some remarks which we hope will increase your skill at finding physics information. The points we make apply whether or not this is the last formal physics course that you take.

Before we go on to discuss the different sources of physics information, we give a word of warning about differences in notation, etc. which you will undoubtedly encounter.

3.1 Different notations and conventions

In the series of books which comprises *The Physical World* we have tried very hard to keep a consistent style and stick with our chosen conventions. As you consult a variety of books and journals, you will find that not all authors and publishers have made the same decisions that we have. This is not a serious problem as long as you are aware of the different possibilities. Here are a number of points of which you should be aware.

- *Vector notation* Where we write, for example, a position vector as $\boldsymbol{r}$, you might see $\mathbf{r}$, $\vec{r}$ and other variants. Some books, unfortunately, are not very careful or consistent over the use of vector notation.
- *Systems of units* We have used SI units in *The Physical World*. However, you may still find books that use cgs (centimetre, gram, second) units instead of SI units, which are based on the rationalized mks (metre, kilogram, second) system. This is by no means a trivial difference. The most commonly encountered form of cgs (Gaussian) results in the equations of electromagnetism looking quite

different. For example, ε_0 and μ_0 don't appear, and many equations, including the Lorentz force law, include the speed of light c. (Recall that in SI units $c = 1/\sqrt{\varepsilon_0\mu_0}$.)

A conspicuous example: in SI units, the magnitude of the force between two point charges, q_1 and q_2, a distance r apart is:
$F = \dfrac{1}{4\pi\varepsilon_0}\dfrac{q_1 q_2}{r^2}$, but in Gaussian cgs units there is no $\dfrac{1}{4\pi\varepsilon_0}$ term. If you see references to units such as ergs, dynes, gauss, maxwells, oersteds, statamperes, statvolts, esus, emus, etc., then you are reading a text which uses cgs units. We give just two conversion factors which may be useful: 1 joule = 10^7 ergs; 1 newton = 10^5 dynes. For definitions of cgs electromagnetic units, you should look for a suitable appendix in the back of a good electromagnetism textbook.

- *Customary units* There are various units which are often used by physicists which fall outside the strict SI system. These are often convenient for particular situations. Among these is the electronvolt (eV) a convenient energy unit which you have already met. In a nuclear context, you will often encounter MeV and GeV (megaelectronvolt, 10^6 eV and gigaelectronvolt, 10^9 eV, respectively) reflecting the larger energies involved in nuclear processes. A convenient unit still much used for expressing the wavelength of light and atomic sizes is the angstrom (Å) which is 10^{-10} m or 0.1 nm. You might well also encounter the traditional unit for pressure, the torr. This is still used for describing a vacuum. One torr is the pressure due to one mm of mercury, i.e ~130 Pa.
- *Calculus* You should be aware that derivatives with respect to time are often represented by a dot, and a double dot for a second derivative. Thus,

$$\dot{x} \text{ means } \frac{dx}{dt}, \text{ and } \ddot{x} \text{ means } \frac{d^2x}{dt^2}.$$

- *Other mathematics* Sometimes (including in *The Physical World*) you will see e^x written as $\exp x$. This is especially helpful when x is something like $hc/\lambda kT$. Another common variation is the use of $\wedge$ instead of $\times$ to denote vector product: thus,

$$\boldsymbol{F} = q(\mathscr{E} + \boldsymbol{v} \wedge \boldsymbol{B}) \text{ is an alternative form of } \boldsymbol{F} = q(\mathscr{E} + \boldsymbol{v} \times \boldsymbol{B}).$$

You should be aware that more advanced physics is plagued by differing conventions which have often led to error and confusion.

- *Symbols* Different texts do not necessarily employ the same symbols for the same quantities. A few examples are:

the electric field appears variously as $\mathscr{E}$, $\boldsymbol{E}$, $\mathbf{E}$, E or others;

frequency as f or ν;

the normal reaction force might be $\boldsymbol{N}$ or $\boldsymbol{R}$;

kinetic energy might be E_{kin} or K and potential energy E_{pot} or U;

current can be i or I, and so on.

3.2 Printed media

There have never before been so many good 'popular' scientific books available and there is also a huge range of reputable textbooks ranging from elementary to monographs at the highest level. How you can access these, as well as reference books and scientific journals, will be discussed in Section 3.3. Here, we first make some general remarks on the kinds of different print items you might encounter as you seek scientific information.

There is a huge range of what might be called periodical literature. At one end of the market are the scientific journals themselves. These constitute the primary means by which physicists make their results known to their colleagues (but see the comment on electronic preprints in Section 3.4). All papers published in these journals are subject to peer review or refereeing. That is to say, they are carefully studied by other experts who send a report to the journal editor who will forward it anonymously to the author of the paper. This report is the basis for the decision as to whether the paper is accepted as it is, or maybe with revisions by the author. Like any human institution, this is not a perfect system and can lead to heated debates between authors and referees. Nevertheless, it does ensure that, by and large, clearly erroneous, trivial or unoriginal papers are rejected. It also filters out the many crank papers that all journal editors receive.

Crank papers and articles do, however, get into print even though some of the popular style magazines do take great care, through careful editing, to publish authoritative articles. You will notice, for example, that the more serious popular magazines, such as *Scientific American*, publish the institutional affiliations of their contributors, together with references to original work in refereed journals. But there is a continuum of quality that you should be aware of, leading all the way down to the Sunday newspaper which infamously reproduced the photo of the Second World War bomber spotted on the Moon! Clearly, it is not at this extreme end of the market where you should feel the need to exercise careful judgement, but in the places where you might expect better. In particular, it is regrettable that the standard of scientific editing in even the most reputable daily papers sometimes allows outrageous scientific errors to appear in news stories.

The system of scientific publications has evolved greatly over the last 300 years, and will probably change rapidly as electronic media develop. Always, however, it has been central in maintaining the ideals of science.

3.3 Libraries

Libraries remain the greatest and most reliable source of information. Even, probably, more than the Internet (see Section 3.4) since the Internet itself is likely to be accessible from a good library, certainly a university library.

The key to getting the best out of libraries is to remember that the library staff make their living out of helping library users find things out. Libraries can probably do things for you that you didn't even know could be done.

The books in the library represent a wonderful resource whether you use the catalogue (certainly computerized) or just browse the shelves. (For many of us, there is a particular pleasure in the serendipitous discovery of something interesting in an unexpected place.) We shall not explain here the different catalogue systems which go under such names as 'Dewey' or 'Library of Congress'. Library users get used to both systems and the librarian can almost certainly give you a leaflet or other guide explaining the system used locally.

Apart from the ordinary book stock on the shelves, a good library can provide many or all of the following sources of physics information:

1 An extensive reference library comprising reference books that remain in the library. Particular reference books include *Handbook of Mathematical Functions* by M. Abramowitz and I. A. Stegun (which gives all the standard integrals, etc. anyone is likely to need); the 16 volume Dictionary of Scientific Biography — all you need to know about famous scientists; *Table of Isotopes* — complete data about radioactive nuclei and their decay properties; etc.

2 Periodicals, ranging from the professional journals to the better popular magazines. Larger and older libraries have 'journal runs' of the main serious journals going back many years. If you are using newer or smaller libraries, you may need to avail yourself of facilities mentioned next if you wish to study a particular article.
3 Facilities for borrowing books from certain very large libraries (e.g. the British Library) which hold much larger stocks than the smaller or more recently established libraries can hope to have. Many libraries, particularly university libraries, have a similar facility for arranging for photocopies of particular journal articles not held locally to be sent from the large reference libraries. Such provision might incur a fee.
4 Bibliographic facilities which enable you to find works (books or journal articles) by particular authors or writings on particular subjects. Of more specialized use is the *Citation Index* which allows you to track who has cited (referred to) particular papers by a named author.
5 Large information databases, often held on CD-ROM.
6 Access to the Internet.

3.4 The Internet

The World Wide Web is an immensely valuable source of physics information and has rapidly become an indispensable part of the life of every working physicist and physics student. If, for example, you want to know the value of the electron charge or proton mass, or any such quantity, there are many places on the Web that not only tell you, but give the most up-to-date value with experimental errors. Similarly, you can find information on the conversion between systems of units.

Very good Web sites are maintained by the large Physical Societies, and these keep frequently updated pages describing recent significant discoveries, usually with links to the original published papers. These might be a good starting point if you wish to explore the large volume of authoritative information about the lives of scientists and the history of science. It is not usually too hard to distinguish the genuinely scholarly sites from those presenting such things as high school projects or nationalistic propaganda (yes, it exists).

Some of the Physical Society sites (e.g. Institute of Physics in the UK www.iop.org, www.physicsweb.org, or the American Physical Society www.aps.org in the US) are good places to find listings of physics departments, national laboratories and other physics institutions around the world. The physics departments of major universities are often good sources of information concerning the research that they do, and also often have good lists of 'useful links' (e.g. the Open University Physics and Astronomy Department site http://physics.open.ac.uk).

Large national or international institutions (e.g. CERN, www.cern.ch or NASA, www.nasa.gov) also maintain large sources of data relating to their work. Certain large news organizations (e.g. the BBC) also give popularized accounts of recent scientific discoveries, especially stories involving astronomy. These are often updated daily. Here we need to warn that some cautious scepticism about the authoritativeness of what you read is necessary. Worse still: all those crank papers that never see the light of day in refereed journals are *so easily* placed on the Web. Watch out! There's a lot of it about.

Physicists also make great use of the electronic preprint archives on the Web. For many years physicists have circulated copies of their research papers to colleagues in the form of preprints. Their colleagues understand that these have often not yet been through the refereeing procedure, and such papers are not included in a researcher's list of publications. Nevertheless, such preprints do give advance notice of research and help to establish priority. But now, the circulation of paper preprints has largely been superseded by the use of electronic preprint archives from which anyone can extract a copy of a paper with an interesting title. Such a preprint might have originated from any country in the world.

3.5 What sources do you believe? The authority problem

Physicists, along with all natural scientists, proudly claim that Nature herself is the final authority. There are stories from the life of Galileo (1564–1642), one of the true founders of physics, which illustrate this very well. For example, it was handed down for centuries, as part of Aristotelian dogma, that ice was heavier than water. Why then, you may ask, does ice float on water? Aristotelian dogma had an answer to this: it was because ice on the surface of water had trouble breaking through the surface. What did Galileo do? He simply pushed a piece of floating ice under the water. Of course, it came back up to the top in defiance of Aristotelian authority.

The frontier of physics has moved far beyond questions of whether ice floats on water. In this book, we have been asking you to believe very strange things concerning Nature at the microscopic level. None of us can personally check everything; checking is done at the level of the scientific community. When new results are published, the details of how these are found are presented so that others can verify the findings ... or not, as the case may be. This involves trust, and judgement concerning whom to trust is part of the education of a physicist.

These remarks are made partly as a warning. There are many people out there who make claims at odds with what the community as a whole has come to believe. This has, perhaps, become particularly conspicuous with the advent of the World Wide Web where anyone can say anything they fancy — and they do. In a large population you will always find some who believe: that the Earth is flat; that they have constructed a source of perpetual free energy; or even that relativity and/or quantum mechanics are nonsensical and only maintained by a conspiratorial scientific establishment. Some of these people even claim scientific training. If you surf the Web, you will certainly find them, and they use print as well.

So what do you do? You can't do all the experiments necessary to test everything and you probably don't have the specialist knowledge required to assess all the arguments. But when you encounter some of the claims mentioned above, stand back and consider what bearing the physics you know might have on them. Physicists would remind you that relativity is important for the global positioning system that gets your next flight to the right place, and quantum mechanics is needed to understand transistors and the laser in your CD-ROM drive. The question of who one believes on issues like nuclear power, global warming, etc. is a difficult one. Probably the best advice is to be moderately sceptical and to seek alternative points of view. A good scientific training is of immense help to you in detecting fallacious arguments or errors of logic. With this training, it will be easier for you to decide whether a set of premises actually does imply the claimed conclusion or whether a certain series of observations necessarily leads to a proposed hypothesis.

However, it is worth remembering that none of us is immune to the temptation to view very complex situations from an angle that focuses on arguments that will, in the end, confirm what we wish to be true. We must remember this both with respect to ourselves as well as to those who might wish to persuade us. Science presents objectivity as a goal, not always easily achieved, but the difficulty in achieving it is not an excuse to indulge our wishful thinking.

4 Basic skills and knowledge test

You should be able to answer the following questions using what you have learned in Chapters 1 to 4.

Leave your answers in terms of π, $\sqrt{2}$, h, etc. where appropriate.

Question 5.1 Using Balmer's formula (Equation 1.1) for the wavelengths of the visible spectral lines of hydrogen, calculate the wavelength of the fifth line in the Balmer series of hydrogen. Why is *this* line *not* visible to the naked eye? What would be the limiting wavelength in the Balmer series as n becomes very large?

Question 5.2 Was it at high frequencies or at low frequencies that the classical expression for blackbody radiation went catastrophically wrong?

Question 5.3 Concerning the photoelectric effect: (a) does the kinetic energy of the ejected electrons depend upon the *intensity* or the *wavelength* of the incident radiation? (b) Is the threshold value of the wavelength, λ_t, an upper limit, above which no electrons are ejected, or a lower limit, below which no electrons are ejected?

Question 5.4 In what way do the molar heat capacities of solids depart from the Dulong–Petit law?

Question 5.5 A diode laser of the kind used in medium-speed laser printers emits 10 mW of infrared radiation of wavelength 780 nm. How many photons per second does this correspond to?

Question 5.6 From the graph in Figure 5.1, showing the maximum kinetic energy of electrons emitted from a certain metal against frequency of the incident light, estimate the work function of that metal. Give your answer in electronvolts.

Question 5.7 Why, according to classical theory, would atoms containing moving electrons be unstable?

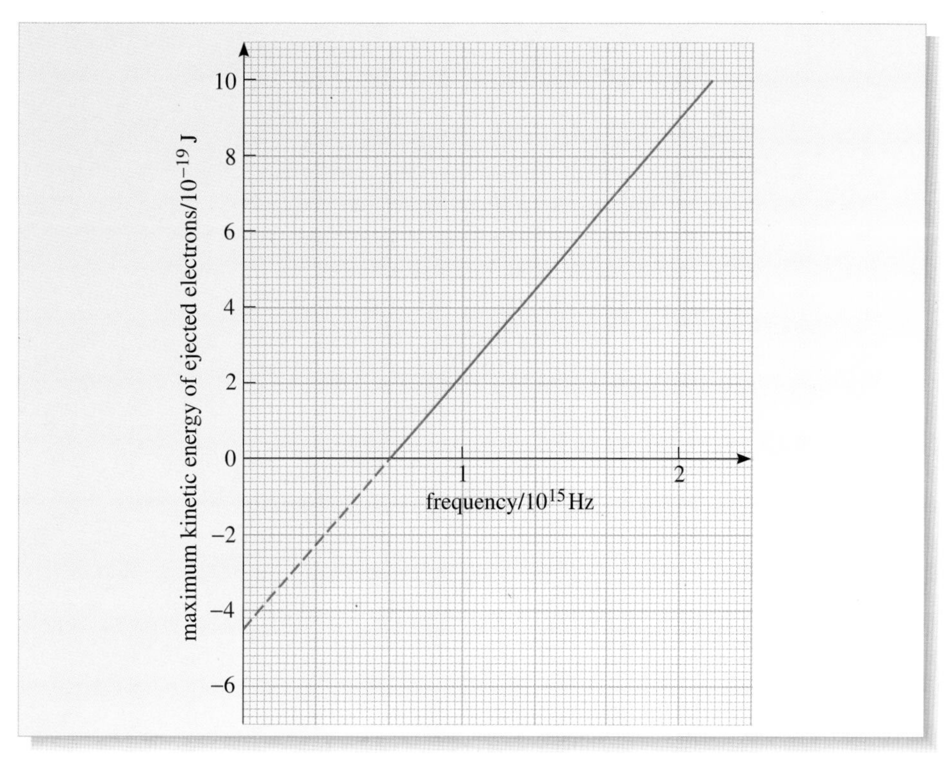

Figure 5.1 For use with Question 4.6.

Question 5.8 At what distance from a uranium nucleus ($Z = 92$) is the magnitude of the force on an α-particle equal to the magnitude of the force on an α-particle due to a vanadium nucleus ($Z = 23$), at a distance of 10^{-14} m?

Question 5.9 Consider the statement: 'The effect on a hydrogen atom in its ground state of absorbing a photon of energy 13.5 eV will be quite different from the effect of absorbing a photon of energy 13.7 eV.' Explain.

Question 5.10 (a) The momentum of a proton is doubled; what will this do to its de Broglie wavelength? (b) Its kinetic energy is doubled; what will be the effect on its de Broglie wavelength? (Assume that $v_{\text{proton}} \ll c$ in each case.)

Question 5.11 If Davisson and Germer had used higher energy electrons in their experiment, would the diffraction pattern have been more spread out or less spread out in angle?

Question 5.12 What happens if you try to determine which slit the electron passed through in a two-slit diffraction experiment?

Question 5.13 Two protons, A and B, are both described as having kinetic energy 1 MeV. Proton A has twice as wide a spread of de Broglie wavelength components in its wave packet as proton B. Which proton has a more precisely defined position?

Question 5.14 A particle is bound in an infinite square well. What can you say about the wavefunction outside the well?

Question 5.15 Consider a particle bound in an infinite square well. At how many points (within the well, but excluding the walls) can you be sure that the particle would never be found if it was in: (a) the ground state, (b) the first excited state, (c) the state with quantum number n?

Question 5.16 (a) If the width of an infinite square well is halved, what effect does this have on the energy levels? (b) How does this relate to Heisenberg's uncertainty principle?

Question 5.17 Is the following statement true or false? 'A particle bound in a finite square well can have a substantial probability of being found outside the walls of the well.'

Question 5.18 What does it mean when it is said that the energy levels of a particle in a three-dimensional infinite cubical well are 'degenerate'?

Question 5.19 Does the concept of 'degeneracy' have any relevance to the energy levels of a hydrogen atom?

Question 5.20 What is the magnitude of the angular momentum of the ground state of the hydrogen atom in (a) the Bohr model, and (b) the Schrödinger model?

Question 5.21 Give the *principal* and *orbital angular momentum* quantum numbers for the following states specified in spectroscopic notation: (a) 3p, (b) 4s, (c) 5f.

Question 5.22 Which of the following states cannot exist for the electron in a hydrogen atom and why: 3f, 4f, 1p, 2p?

Question 5.23 Are the following statements true or false?

(a) 'There is a spherical region at a finite distance from the nucleus where the probability of finding a 2s electron in a hydrogen atom is zero.'

(b) 'There is a spherical region at a finite distance from the nucleus where the probability of finding a 2p electron in a hydrogen atom is zero.'

Question 5.24 Why are *two* different states available to an electron of a particular set of quantum numbers n, l, m_l?

Question 5.25 How many states in the hydrogen atom are there with principal quantum number $n = 3$?

Question 5.26 Which parts of the following statement are true and which are false? (a) The ground state of fluorine ($Z = 9$) can be written in spectroscopic notation as $1s^2\,2s^2\,2p^5$, (b) that of neon ($Z = 10$) can be written as $1s^2\,2s^2\,2p^6$, and (c) that of sodium ($Z = 11$) as $1s^2\,2s^2\,2p^7$.

Question 5.27 If, in a two-slit diffraction experiment with electrons, the normalized wavefunction corresponding to the experiment with slit 2 closed is Ψ_1 and that corresponding to slit 1 closed is Ψ_2, what is the expression for $|\,\Psi\,|^2$ which will give the intensity distribution for the electrons when *both* slits are open?

Question 5.28 What equation must be solved in order to obtain the wavefunctions $\Psi(x, y, z, t)$ that describe the behaviour of a particle?

Question 5.29 When an observable is measured for a quantum state, (a) what are the possible outcomes? (b) does the state 'have' these values prior to measurement?

Question 5.30 Explain in a few words what is meant by indeterminacy, and indeterminism.

5 Interactive questions

Open University students should leave the text at this point and use the interactive question package for *Quantum physics: an introduction*. When you have completed the questions you should return to this text. You should not spend more than 2 hours on this package.

The interactive question package includes a random number feature that alters the values used in many of the questions each time those questions are accessed. This means that if you try the questions again, as part of your end of course revision for instance, you will find that many of them will have changed, at least in their numerical content.

6 *Physica* problems

Open University students should leave the text at this point and tackle the *Physica* problems that relate to *Quantum physics: an introduction*. You should not spend more than 2 hours on this package.

Answers and comments

Q1.1 Using Balmer's formula

$$\lambda = 364.56\left\{\frac{n^2}{n^2-4}\right\}\text{nm},$$

the wavelengths of the four spectral lines corresponding to $n = 3, 4, 5$ and 6 are 656.21 nm, 486.08 nm, 434.00 nm and 410.13 nm, respectively. These agree with the measured wavelengths of the visible hydrogen lines to within 0.01 nm. The wavelength of the line corresponding to $n = 7$ is 396.97 nm which lies in the ultraviolet region of the spectrum.

Q1.2 The number of modes, Δn, in the cavity of volume V, between the wavelengths λ and $\lambda + \Delta\lambda$ is given by

$$\begin{aligned}\Delta n &= \frac{8\pi V\Delta\lambda}{\lambda^4}\\ &= \frac{8\pi(5\times10^{-2}\,\text{m})^3\times2\times10^{-9}\,\text{m}}{(450\times10^{-9}\,\text{m})^4}\\ &= 1.53\times10^{14}.\end{aligned}$$

The total energy in the cavity in this wavelength range is given by

$$\begin{aligned}\Delta nkT &= (1.53\times10^{14}\times1.38\times10^{-23}\times2000)\,\text{J}\\ &= 4.22\times10^{-6}\,\text{J}.\end{aligned}$$

Q1.3 In this case, to find the total energy in the cavity in the given wavelength range, we multiply the number of modes Δn (from Q1.2 above) by Planck's expression for the average energy per mode. Thus the total energy is

$$\frac{\Delta nhc}{\lambda}\times\frac{1}{\exp\left(\dfrac{hc}{\lambda kT}\right)-1}$$

$$= \frac{1.53\times10^{14}\times6.63\times10^{-34}\,\text{J s}\times3\times10^{8}\,\text{m s}^{-1}}{450\times10^{-9}\,\text{m}}$$

$$\times\frac{1}{\exp\left(\dfrac{6.63\times10^{-34}\,\text{J s}\times3\times10^{8}\,\text{m s}^{-1}}{450\times10^{-9}\,\text{m}\times1.38\times10^{-23}\,\text{J K}^{-1}\times2000\,\text{K}}\right)-1}$$

$$= 7.50\times10^{-12}\,\text{J}.$$

As you can see this is considerably smaller than the corresponding classical value (given in the answer to Q1.2).

Q1.4 (a) Between 6×10^{14} Hz and 2×10^{15} Hz the line rises from zero to 9.2×10^{-19} J. Planck's constant is given by the gradient, and is therefore

$$h \approx \frac{9.2\times10^{-19}}{1.4\times10^{15}}\,\text{J s} = 6.6\times10^{-34}\,\text{J s}.$$

(b) The work function of the metal can be simply read off as the magnitude of the intercept on the vertical scale: 4×10^{-19} J.

Q1.5 (a) The radiation is emitted equally to all parts of a sphere of area 4π m^2. The target area presented by a sodium atom at the surface of the metal is $\pi\times(0.5\times10^{-10})^2\,\text{m}^2 = 2.5\times10^{-21}\times\pi\,\text{m}^2$. Thus the rate at which energy is incident on a single atom is

$$8\,\text{J s}^{-1}\times\frac{2.5\times10^{-21}\pi\,\text{m}^2}{4\pi\,\text{m}^2} = 5\times10^{-21}\,\text{J s}^{-1}.$$

(b) The time required to accumulate 3.4×10^{-19} J at the rate given in part (a) is

$$\frac{3.4\times10^{-19}}{5\times10^{-21}}\,\text{s} = 68\,\text{s}.$$

(c) The wave theory of light is unable to account for the very small ($<10^{-9}$ s) measured time lag in the photoelectric effect. The mystery is cleared up by Einstein's idea that each electron is ejected by a single quantum of light. If we assume that the light quanta have just enough energy to eject electrons, then the calculation in part (b) implies that each individual atom will have to wait an *average* of 68 seconds to be hit by a light quantum. However, some atoms will be hit and will emit electrons immediately the first light quanta arrive at the target.

Q1.6 Using Einstein's photoelectric equation,

$$\begin{aligned}\tfrac{1}{2}m_e v_{max}^2 &= hf - \phi\\ &= \frac{hc}{\lambda} - \phi.\end{aligned}$$

Substituting in the values, and remembering to convert the work function into joules, we find

$$\begin{aligned}\tfrac{1}{2}m_e v_{max}^2 &= \frac{6.63\times10^{-34}\times3.00\times10^{8}}{450\times10^{-9}}\,\text{J}\\ &\quad -2.13\times1.6\times10^{-19}\,\text{J}\\ &= 4.42\times10^{-19}\,\text{J} - 3.41\times10^{-19}\,\text{J}\\ &= 1.01\times10^{-19}\,\text{J}.\end{aligned}$$

Thus, $v_{\max} = \sqrt{\dfrac{2.02\times10^{-19}}{9.11\times10^{-31}}}\ \mathrm{m\,s^{-1}}$

$= 4.7\times10^5\ \mathrm{m\,s^{-1}}$.

Q1.7 We use Einstein's photoelectric equation,

$$\tfrac{1}{2}m_e v_{\max}^2 = hf - \phi$$

where ϕ is the work function of sodium.

The photoelectric threshold occurs when the electrons are *just* unable to escape, so it corresponds to a kinetic energy of zero, i.e.

$$\phi = hf_t = \frac{hc}{\lambda_t}.$$

Hence $\phi = \dfrac{6.63\times10^{-34}\times3.00\times10^8}{542\times10^{-9}}\ \mathrm{J}$

$= 3.67\times10^{-19}\ \mathrm{J}$.

Q1.8 In the limit of high temperature, i.e. when $\theta_E/T \ll 1$, we can replace $\exp(\theta_E/T)$ by $(1 + \theta_E/T)$ in Equation 1.8 which then becomes

$$C_m(T) = 3R\left(\frac{\theta_E}{T}\right)^2\left(1+\frac{\theta_E}{T}\right)\left(1+\frac{\theta_E}{T}-1\right)^{-2}$$

$$= 3R\left(1+\frac{\theta_E}{T}\right)$$

$$\approx 3R.$$

Q1.9 (a) In *Static fields and potentials* it is shown that the electrostatic potential energy of a point charge q_1 in the field of another point charge q_2 is $\dfrac{q_1q_2}{4\pi\varepsilon_0 r}$, where r is the distance between the two point charges. In this problem, the α-particle carries a charge $2e$ and the nucleus a charge Ze, so that the potential energy is given by

$$\frac{2Ze^2}{4\pi\varepsilon_0 r} = \frac{Ze^2}{2\pi\varepsilon_0 r}.$$

(b) When the α-particle is far from the nucleus, its potential energy is zero, so its total energy is E_{kin}. When it reaches the point of closest approach it stops, so its kinetic energy is zero, and its potential energy and total energy are both $\dfrac{Ze^2}{2\pi\varepsilon_0 r_{\text{cl}}}$. Thus conservation of total energy requires

$\dfrac{Ze^2}{2\pi\varepsilon_0 r_{\text{cl}}} = E_{\text{kin}}$, from which $r_{\text{cl}} = \dfrac{Ze^2}{2\pi\varepsilon_0 E_{\text{kin}}}$.

Q1.10 (a) If the speed of the α-particle is doubled then its kinetic energy is increased by a factor of 4. Thus, its distance of closest approach will be *reduced* by a factor of 4 to 1.7×10^{-14} m.

(b) The expression for r_{cl} shows that if the charge on the nucleus (or equivalently the value of Z) is doubled then so is r_{cl}, which becomes 1.4×10^{-13} m.

(c) The mass of the target nucleus does not affect r_{cl} (as long as it is heavy compared to the mass of the α-particle). In this case, the distance of closest approach would again be 6.8×10^{-14} m.

Q1.11 Squaring Equation 1.11 gives

$$r^2 = \frac{n^2\hbar^2}{m_e^2 v^2}$$

and from Equation 1.10 we have

$$v^2 = \frac{e^2}{4\pi\varepsilon_0 m_e r}.$$

Substituting the expression for v^2 into the expression for r^2 and cancelling an r from each side gives

$$r = \frac{n^2\hbar^2\,4\pi\varepsilon_0}{m_e e^2}.$$

Putting $a_0 = \dfrac{4\pi\varepsilon_0\hbar^2}{m_e e^2}$, this can be written as $r = n^2 a_0$.

Q1.12 From Equation 1.10 we have

$$m_e v^2 = \frac{e^2}{4\pi\varepsilon_0 r}$$

and from Equation 1.11 we have

$$r = \frac{n\hbar}{m_e v}.$$

Substituting the expression for r into Equation 1.10 and cancelling a v and m_e from each side gives

$$v = \frac{e^2}{4\pi\varepsilon_0 n\hbar}.$$

The speed of the electron in the first ($n = 1$) Bohr orbit is then

$v_1 = \dfrac{(1.6\times10^{-19})^2\times9.0\times10^9}{1.05\times10^{-34}}\ \mathrm{m\,s^{-1}}$

$= 2.2\times10^6\ \mathrm{m\,s^{-1}}$.

(Where $1/4\pi\varepsilon_0 = 9.0 \times 10^9\,\text{N m}^2\,\text{C}^{-2}$.) This is less than 1% of the speed of light, so that relativistic effects should be very small. (These tiny effects *are* detected experimentally, however.)

Q1.13 (a) No. An unbound electron has a continuum of possible positive energy states, i.e. the total energy of an unbound electron can, in principle, have any positive value. The electron's total energy is quantized (i.e. restricted to certain definite values) only when it is bound in the atom.

(b) From Equation 1.13 an electron in the $n = 2$ energy level has a total energy of $-3.4\,\text{eV}$, so the electron will be ejected from the atom and will escape with a residual kinetic energy of $4.0\,\text{eV} - 3.4\,\text{eV} = +0.6\,\text{eV}$.

Q1.14 From Figure 1.34, the Brackett series arises from transitions terminating at the $n = 4$ orbit, which means we should put $q = 4$ in Equation 1.14. This leads to the required result

$$\lambda_{n\to q} = 1458\left\{\frac{n^2}{n^2 - 16}\right\}\text{nm}.$$

For $\lambda_{n\to q} = 1736\,\text{nm}$, we have $\left\{\dfrac{n^2}{n^2 - 16}\right\} = \dfrac{1736}{1458} = 1.191.$

Thus, $n^2 = 1.191n^2 - 19.06$, i.e. $n^2 = 99.8$. Clearly then, $n = 10$ and so the transition is from $n = 10$ to $n = 4$.

Q1.15 In each case, it is necessary to use the de Broglie formula, $\lambda_{\text{dB}} = h/p$ (Equation 1.15) with $h = 6.6 \times 10^{-34}\,\text{J s}$.

(a) The billiard ball has mass 0.1 kg and travels at $2\,\text{m s}^{-1}$, so the magnitude of its momentum is $0.2\,\text{kg m s}^{-1}$. Its de Broglie wavelength is therefore $\lambda_{\text{dB}} = 3.3 \times 10^{-33}\,\text{m}$.

Comment *This wavelength is approximately 10^{23} times smaller than the diameter of an atom! There is no known physical system that could give rise to observable diffraction effects for wavelengths of this size.*

(b) The dust particle has mass $10^{-6}\,\text{kg}$ and a speed of $0.1\,\text{m s}^{-1}$, so λ_{dB} is $6.6 \times 10^{-27}\,\text{m}$.

Comment *Although this is about 10^6 times larger than the wavelength you calculated in (a) it is again much too small to produce observable diffraction effects.*

(c) The mass of the electron is $m_{\text{e}} = 9.1 \times 10^{-31}\,\text{kg}$ and 10 eV is equal to $10 \times 1.6 \times 10^{-19}\,\text{J} = 1.6 \times 10^{-18}\,\text{J}$. Using the relation $p = \sqrt{2m_{\text{e}}E_{\text{kin}}}$, the magnitude of the electron's momentum is $1.7 \times 10^{-24}\,\text{kg m s}^{-1}$ and $\lambda_{\text{dB}} = 3.9 \times 10^{-10}\,\text{m}$.

Q1.16 Figure 1.7 shows how the spectrum of blackbody radiation varies with temperature. As you can see, the peak in the spectrum moves to shorter wavelengths with increasing temperature. Since a peak at 2.3 μm ($2.3 \times 10^{-6}\,\text{m}$) is about half-way between the peaks of the curves at temperatures of 1500 K and 1250 K, a reasonable estimate of the furnace temperature is about 1350 K.

Q1.17 The threshold wavelength is $\lambda_{\text{t}} = 542\,\text{nm}$, so the work function is

$$\phi = \frac{hc}{\lambda_{\text{t}}} = \frac{6.63 \times 10^{-34} \times 3.0 \times 10^8}{542 \times 10^{-9} \times 1.6 \times 10^{-19}}\,\text{eV} = 2.29\,\text{eV}.$$

A quantum of the incident radiation has energy

$$\frac{hc}{\lambda} = \frac{6.63 \times 10^{-34} \times 3.0 \times 10^8}{425 \times 10^{-9} \times 1.6 \times 10^{-19}}\,\text{eV} = 2.93\,\text{eV},$$

and the most energetic electron ejected therefore has energy $2.93\,\text{eV} - 2.29\,\text{eV} = 0.64\,\text{eV}$.

Q1.18 The Einstein temperature is given by $\theta_{\text{E}} = hf_{\text{v}}/k$. Therefore the vibrational frequency of the atoms is

$$f_{\text{v}} = \frac{\theta_{\text{E}}k}{h} = \frac{1300\,\text{K} \times 1.38 \times 10^{-23}\,\text{J K}^{-1}}{6.63 \times 10^{-34}\,\text{J s}} = 2.7 \times 10^{13}\,\text{s}^{-1}.$$

Q1.19 (i) Characteristic line spectra of the elements.

(ii) The shape of the blackbody spectrum, in particular the 'ultraviolet catastrophe'.

(iii) The existence of the threshold frequency and the fact that the electrons' maximum kinetic energy is proportional to frequency not intensity of the light (Einstein's photoelectric equation).

(iv) The temperature dependence of the molar heat capacity, i.e. its decrease with decreasing temperature.

Q1.20 The middle of the visible spectrum corresponds to a wavelength of about 580 nm, and the corresponding light quantum energy is thus

$$\frac{hc}{\lambda} = \frac{6.63 \times 10^{-34} \times 3 \times 10^8}{580 \times 10^{-9}}\,\text{J} = 3.4 \times 10^{-19}\,\text{J}.$$

Thus $10^{-18}\,\text{J}$ would be supplied by three quanta.

Q1.21 (a) In classical physics the average energy of an oscillator of natural frequency f in a large assembly of such oscillators is given by kT.

(b) Using Planck's quantum hypothesis this average energy becomes $\dfrac{hf}{(e^{hf/kT}-1)}$.

Q1.22 (a) The mysterious nature of the positive sphere or the fact that the atoms can only be stable for a maximum of a few thousand years due to radiation by accelerating electrons. Experimental evidence against Thomson's model is the fact that α-particles bombarding a thin foil are observed to be deflected through angles up to 150°.

(b) Electrons orbiting outside the nucleus lose energy due to emission of electromagnetic radiation and should collapse into the nucleus. The model sheds no light on the Periodic Table of the elements.

(c) Neither model can explain atomic line spectra or why atomic diameters are of the order of 10^{-10} m.

Q1.23 The distance of closest approach is given by

$$\begin{aligned} r_{\text{cl}} &= \frac{2Ze^2}{4\pi\varepsilon_0 E_{\text{kin}}} \\ &= \frac{2\times 78\times(1.6\times10^{-19})^2\times 9.0\times10^9}{1.3\times10^{-12}}\,\text{m} \\ &= 2.76\times10^{-14}\,\text{m}. \end{aligned}$$

If the results of scattering experiments are in agreement with theory (using Coulomb's law for the interaction between the α-particle and the nucleus) then this provides an experimental upper limit on the radius of a platinum nucleus.

Q1.24 The magnitude of the momentum of the ball is $0.06\times 60\,\text{kg m s}^{-1} = 3.6\,\text{kg m s}^{-1}$, and using Equation 1.15, its de Broglie wavelength is

$$\frac{6.63\times10^{-34}}{3.6}\,\text{m} = 1.8\times10^{-34}\,\text{m}.$$

This is a minuscule wavelength so it is probably not worth setting up an experiment to observe ball-diffraction at Wimbledon!

Q2.1 Laplace would probably not have made this statement because the results of the two-slit experiment demonstrate that the statement is incorrect. Even if he had *all* the data on an electron projected towards the slit, he could predict *only* the *probability* that it will be detected in a certain region of the screen. In this sense, he could not predict the future with *certainty*. (Quantum mechanics normally allows us to predict the *possible* results of a measurement and the *probabilities* with which these possible results will occur.)

Q2.2 (a) At the slit, the position of each electron is determined to within w in the y-direction. (In fact, this is a considerable overestimate: a proper statistical analysis would suggest something like $w/3$.) The minimum uncertainty this introduces in the y-component of momentum is given by Heisenberg's uncertainty principle as

$$\Delta p_y = \frac{\hbar}{2w}.$$

The angular spread of the beam after the slit, θ, is given approximately for small angles, by

$$\theta \approx \frac{\Delta y}{x} = \frac{\Delta v_y}{v_x} = \frac{\Delta p_y}{p_x}.$$

Now p_x is related to the wavelength of the beam by the de Broglie formula $p_x = h/\lambda$, so substituting for Δp_y and p_x gives

$$\theta \approx \frac{\hbar\,\lambda}{2wh}$$

and hence $\theta \approx \dfrac{\lambda}{4\pi w}$

which is within about an order of magnitude of λ/w.

(b) The de Broglie wavelength of the bullets is given by $\lambda = h/mv$, and the angular spread of the beam after the slit is given by

$$\begin{aligned} \theta &\approx \frac{\lambda}{w} = \frac{h}{mvw} \\ &= \frac{6.6\times10^{-34}\,\text{J s}}{0.1\,\text{kg}\times 1500\,\text{m s}^{-1}\times 0.01\,\text{m}} \\ &\approx 4\times10^{-34}\,\text{rad}. \end{aligned}$$

A very small number!

Q2.3 From *Static fields and potentials*

$$E_{\text{pot}} = \frac{-e^2}{4\pi\varepsilon_0 r}.$$

Also, $E_{\text{kin}} = \dfrac{p^2}{2m_{\text{e}}}$.

Therefore, $E_{\text{tot}} = \dfrac{-e^2}{4\pi\varepsilon_0 r} + \dfrac{p^2}{2m_{\text{e}}}$.

Q2.4

$$E_{\text{tot}} = \frac{-e^2}{4\pi\varepsilon_0 r} + \frac{p_x^2 + p_y^2 + p_z^2}{2m_\text{e}}$$
$$\approx \frac{-e^2}{4\pi\varepsilon_0 r} + \frac{(\Delta p_x)^2 + (\Delta p_y)^2 + (\Delta p_z)^2}{2m_\text{e}}.$$

But $(\Delta p_x)^2 + (\Delta p_y)^2 + (\Delta p_z)^2 \approx 3(\hbar/2r)^2$.

So $$E_{\text{tot}} \approx \frac{-e^2}{4\pi\varepsilon_0 r} + \frac{3\hbar^2}{8m_\text{e}r^2}.$$

Q2.5 To find the minimum value of E_{tot}, we find the point on the graph of E_{tot} against r where the gradient is zero. At this point, $\text{d}E_{\text{tot}}/\text{d}r$ is zero. So taking the approximate expression for E_{tot} from Q2.4 and differentiating it with respect to r,

$$\frac{\text{d}E_{\text{tot}}}{\text{d}r} \approx \frac{e^2}{4\pi\varepsilon_0 r^2} - \frac{3\hbar^2}{4m_\text{e}r^3}.$$

At the minimum of energy, the gradient is equal to zero, so

$$\frac{e^2}{4\pi\varepsilon_0 r_\text{H}^2} = \frac{3\hbar^2}{4m_\text{e}r_\text{H}^3},$$

i.e.

$$r_\text{H} = \frac{3\pi\varepsilon_0\hbar^2}{m_\text{e}e^2}$$
$$= \frac{3\pi \times (8.85 \times 10^{-12}\,\text{C}^2\,\text{N}^{-1}\,\text{m}^{-2}) \times (1.06 \times 10^{-34}\,\text{J s})^2}{(9.11 \times 10^{-31}\,\text{kg}) \times (1.60 \times 10^{-19}\,\text{C})^2}$$
$$\approx 0.4 \times 10^{-10}\,\text{m}.$$

Thus to a first approximation the hydrogen atom has a radius of the order of 10^{-10} m.

Q2.6 (a) Equation 2.7a, $\Delta x\,\Delta p_x \geq \hbar/2$, gives limits to our simultaneous knowledge of the electron's position coordinate x and momentum component p_x. Similarly, Equation 2.8, $\Delta E\,\Delta t \geq \hbar/2$ gives limits to our knowledge of the electron's energy E if it is measured in a finite time interval Δt. Equations 2.7 and 2.8 are true no matter how accurately the experimenter tries to take his measurements — the equations give *fundamental* limits to our knowledge of x, p_x, E and t.

(b) Yes. The uncertainty principle applies only to *simultaneous* knowledge of position and momentum. It does *not* limit the accuracy to which these two quantities can be known at *different* times.

Q2.7 Using Equation 2.8, $\Delta E\,\Delta t \geq \hbar/2$, since $\Delta t \approx 10^{-8}$ s, $\Delta E \gtrsim 5.5 \times 10^{-27}$ J. This short, but finite, time implies an indeterminacy in the energy of the excited atomic states. Such an indeterminacy in energy will result in a corresponding indeterminacy in the frequency and wavelength of spectral lines, because the frequency of light emitted when the atom falls from a state of energy E_a to one with E_b is $f = (E_\text{a} - E_\text{b})/h$ (Chapter 1).

Comment *The intrinsic width, or spread, $\Delta\lambda$, of the wavelength of spectral lines due to this effect, is called the 'natural width' of the spectral lines.*

Q2.8 The spread of 20 electron masses in the mass of the new elementary particle is not due to any experimental limitation — it is ten times larger than the imprecision due to measurement errors! The spread arises from the fundamental limits set by Heisenberg's uncertainty principle. Since $m_\text{e} = 9.1 \times 10^{-31}$ kg, the uncertainty in the measurement of the total relativistic energy of the particle is (when it is stationary),

$$\Delta E = 20 \times (9.1 \times 10^{-31}\,\text{kg}) \times (3 \times 10^8\,\text{m s}^{-1})^2$$
$$= 1.64 \times 10^{-12}\,\text{J}.$$

Equation 2.8, $\Delta E\Delta t \geq \hbar/2$, therefore implies that

$$\Delta t \geq \frac{1.06 \times 10^{-34}\,\text{J s}}{2 \times 1.64 \times 10^{-12}\,\text{J}} \approx 3.2 \times 10^{-23}\,\text{s},$$

i.e. the measurement of the particle's energy must have been made over a *minimum* time interval of 3.2×10^{-23} s. This is associated with the *short lifetime* of the particle. (Many elementary particles have a very short lifetime — such particles are rapidly annihilated, or they rapidly decay into other particles.)

Q2.9 As it stands the statement is wrong. Quantum mechanics *does* describe such systems, but there is far more to quantum mechanics than a discussion of quantized energy. A better formulation of the statement might be something along the following lines:

'Schrödinger's approach to quantum mechanics is based on the use of wavefunctions that satisfy a differential equation known as the time-dependent Schrödinger equation. This equation involves a potential energy function that allows the equation to take on a form appropriate to specific problems. In the case of a particle confined within a potential well, the Schrödinger equation implies that the measured energy of the particle will generally be quantized into discrete energy levels. However, in other situations, such as that of a free particle, the measured value of energy may be any of a continuous range of possible outcomes.'

Q2.10 (a) Since the total energy of the confined particle is $9h^2/(8m_eD^2)$, the wavefunction describing its behaviour must be characterized by the number $n = 3$ (Equation 2.18). Hence, there are *three* half-wavelengths of the probability wave between the confining walls (Figure 2.36a).

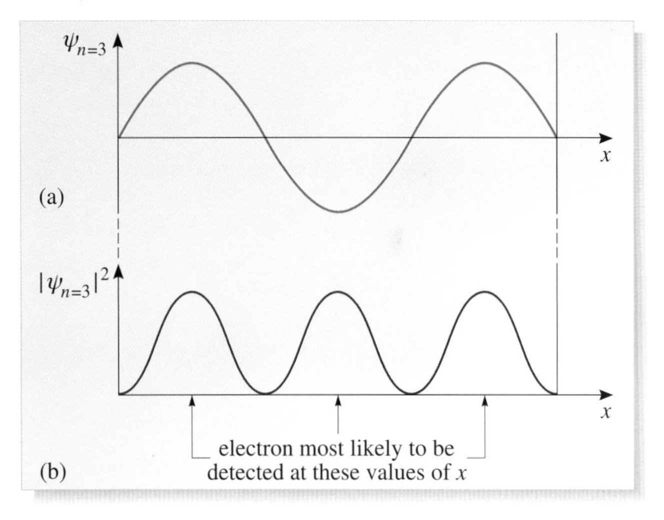

Figure 2.36 Q2.10.

(b) According to Equation 2.21, the electron is most likely to be detected where $|\psi_{n=3}|^2$ is a maximum (Figure 2.36b). There are three such maxima.

(c) A single photon would be ejected with energy

$$E = \frac{9h^2}{8m_eD^2} - \frac{h^2}{8m_eD^2} = \frac{h^2}{m_eD^2}.$$

Remember from Chapter 1 that when an electron makes a transition between two energy levels, a single photon is ejected with an energy equal to the spacing of the two levels.

Q2.11 The particle does not lose any energy in escaping from the well, so after its escape its energy will still be E_{tot}. Its *kinetic* energy will, of course, change. Inside the well $E_{kin} = E_{tot}$; outside the well, $E_{kin} = E_{tot} - W$, accounting for the longer wavelength outside the well.

Q2.12 Preparation Figure 2.37 shows the shape of the well in the two cases. For the infinite well, the energy levels of the confined electron are given by

$$E_{tot} = \frac{n^2h^2}{8m_eD^2} \quad \text{where } n = 1, 2, 3, \text{etc.}$$

The known quantity is $D = 0.5\,\text{nm} = 5 \times 10^{-10}\,\text{m}$.

A useful conversion factor is $1\,\text{eV} = 1.60 \times 10^{-19}\,\text{J}$.

The angular wavenumber is defined by $k = 2\pi/\lambda$, where λ is the wavelength.

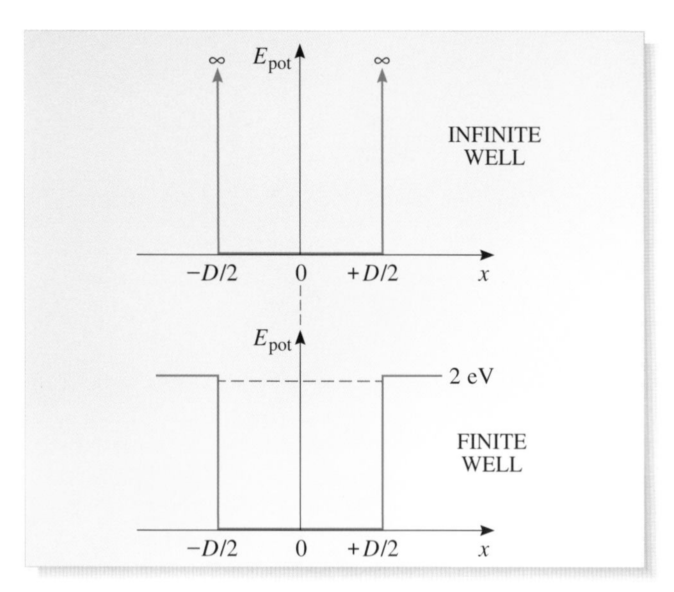

Figure 2.37 Preparation stage of Q2.12.

Working In the case of the *infinite well*, the lowest energy level is that for which $n = 1$, so

$$E_1 = \frac{h^2}{8m_eD^2} = \frac{(6.6 \times 10^{-34}\,\text{J s})^2}{8 \times 9.1 \times 10^{-31}\,\text{kg} \times (5 \times 10^{-10}\,\text{m})^2}$$

$$= 2.4 \times 10^{-19}\,\text{J}$$

$$= \frac{2.4 \times 10^{-19}\,\text{J}}{1.6 \times 10^{-19}\,\text{J eV}^{-1}} = 1.5\,\text{eV}.$$

The wavefunction of the electron will have the shape shown in the top diagram in Figure 2.38 and is zero at the boundary of the well.

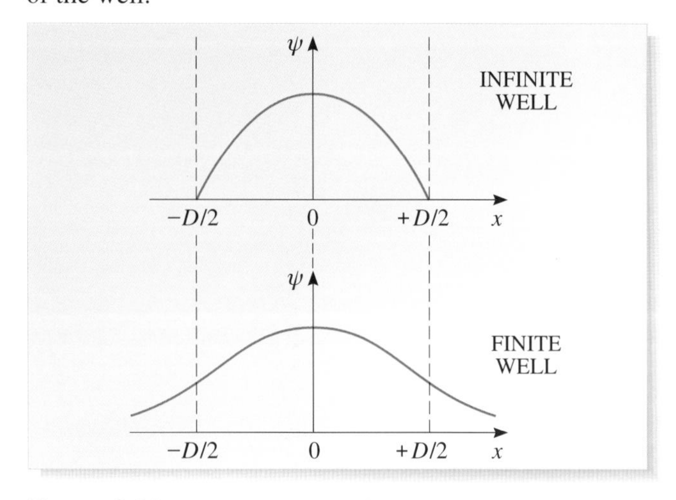

Figure 2.38 Wavefunctions (for Q2.12).

In the case of the *finite well*, the height of the potential well is still greater than the total energy of the particle, so the particle will still be confined within the well but the wavefunction will penetrate to some distance into the classically forbidden region as sketched in the bottom diagram in Figure 2.38. (This wavefunction has the form $\psi = \psi_0 \cos kx$ in the region inside the well.)

The wavefunction now clearly has a longer wavelength, since it has to match an exponentially decaying value at the boundary. Hence it has a smaller value of the angular wavenumber k than does the wavefunction in the case of the infinite well. The value of k is related to the energy according to $k = \sqrt{2mE_{tot}}/\hbar$ so a smaller value of k corresponds to a smaller value of E_{tot}.

Thus the energy of the particle in its lowest energy state is smaller when it is confined in the finite well than when it is confined in the infinite well.

Checking Is this result physically reasonable? According to Heisenberg's uncertainty principle, it is. In an infinite well (especially a very narrow one), we are fairly certain of the particle's position because it is completely trapped with no possibility of being found outside the confines of the well. Hence the uncertainty in its position (Δx in one dimension) is small. The corresponding uncertainty in its momentum (Δp_x) is therefore high. If there is a large spread of possible momenta for the particle, on average its momentum must be high and its average kinetic energy must also be high. So the lowest energy state of the particle is of greater energy the deeper (and narrower) the well.

Comment *It is a general feature that the more tightly a particle is confined the greater its energy. You will be able to see this clearly in the multimedia package 'Stepping through Schrödinger's equation' that accompanies this book.*

Q2.13 For a particle confined in an infinite square well, the allowed energies are given by Equation 2.18

$$E_{tot} = \frac{n^2h^2}{8mD^2}, \quad \text{where } n = 1 \text{ or } 2 \text{ or } 3\text{, etc.}$$

The minimum energy of the bead occurs when $n = 1$. With $D = 0.1$ m,

$$E_1 = \frac{(6.6 \times 10^{-34}\,\text{J s})^2}{8 \times 10^{-3}\,\text{kg} \times 1 \times 10^{-2}\,\text{m}^2}$$
$$= 5.4 \times 10^{-63}\,\text{J}.$$

Classically, if we interpret this energy as a kinetic energy, then the bead must have speed v such that $E_1 = \frac{1}{2}mv^2$, so

$$v = \sqrt{\frac{2E_1}{m}} = \sqrt{\frac{2 \times 5.4 \times 10^{-63}\,\text{J}}{1 \times 10^{-3}\,\text{kg}}}$$
$$= 3.3 \times 10^{-30}\,\text{m s}^{-1}.$$

The time t taken to cross the box at this speed is

$$t = \frac{0.1\,\text{m}}{3.3 \times 10^{-30}\,\text{m s}^{-1}} = 3 \times 10^{28}\,\text{s},$$

i.e. about 10^{21} years!

Q2.14 The bead has a mass of 1×10^{-3} kg and speed $0.2\,\text{m s}^{-1}$ (= 0.1 m/0.5 s), so its kinetic energy is given by $\frac{1}{2}mv^2 = 2 \times 10^{-5}$ J. The allowed energy levels of the bead are given by Equation 2.18.

$$E_{tot} = \frac{h^2}{8mD^2} \times n^2 = \frac{(6.6 \times 10^{-34}\,\text{J s})^2}{8 \times 10^{-3}\,\text{kg} \times 10^{-2}\,\text{m}^2} \times n^2$$
$$= 5.4 \times 10^{-63} \times n^2\,\text{J}.$$

Equating this quantity with the total energy of the glass bead (2×10^{-5} J) gives

$$n^2 = \frac{2 \times 10^{-5}}{5.4 \times 10^{-63}} = 3.7 \times 10^{57}$$

so $\quad n = 6 \times 10^{28}$.

This extremely large value for n implies that the standing wave representing the bead has a correspondingly large number of nodes between the walls of the box. In this case, the distance between nodes of the wave is too short to be measurable.

Q2.15 (a) $E_{tot} = \dfrac{h^2}{8m_eD^2}(n_1^2 + n_2^2 + n_3^2)$

where n_1, n_2 and n_3 = 1 or 2 or 3, etc., and where $D = 3 \times 10^{-10}$ m.

(b) The lowest energy level is characterized by $n_1 = n_2 = n_3 = 1$, and so $E_{tot} = 3h^2/(8m_eD^2)$.

The second lowest energy level is characterized by ($n_1 = 1$, $n_2 = 1$ and $n_3 = 2$) or ($n_1 = 1$, $n_2 = 2$, $n_3 = 1$) or ($n_1 = 2$, $n_2 = 1$ and $n_3 = 1$): these three states are degenerate since they each describe the electron when it has total energy of $E_{tot} = 6h^2/(8m_eD^2)$.

According to this model, the spacing ΔE of these two energy levels is

$$\Delta E = \frac{6h^2}{8m_eD^2} - \frac{3h^2}{8m_eD^2} = \frac{3h^2}{8m_eD^2},$$

i.e. $$\Delta E = \frac{3 \times (6.6 \times 10^{-34}\,\text{J s})^2}{8 \times (9.1 \times 10^{-31}\,\text{kg}) \times (3 \times 10^{-10}\,\text{m})^2}$$
$$= 2 \times 10^{-18}\,\text{J}.$$

This prediction agrees very well with the *experimentally measured* value of the spacing 1.6×10^{-18} J (Chapter 1)!

(c) The model does not take account of the fact that the electrostatic attractive force on the electron varies with its distance from the centre of the nucleus.

Comment *A more careful investigation of the model shows that its prediction for the spacings of the electron's higher energy levels is in gross disagreement with experiment.*

Q2.16 The energy levels of the particle are given by

$$E_{\text{tot}} = \frac{h^2}{8m_e D^2}(n_1^2 + n_2^2)$$

where D is the length of each side of the square. The particle's three lowest energy levels are characterized by:

(i) ($n_1 = 1$, $n_2 = 1$): there is no degeneracy here.

(ii) ($n_1 = 2$, $n_2 = 1$) and ($n_1 = 1$, $n_2 = 2$): these two states are degenerate since they *both* describe the particle when it has a total energy of $5h^2/(8mD^2)$.

(iii) ($n_1 = 2$, $n_2 = 2$): there is no degeneracy here.

Q2.17 Rearranging Equation 2.9 and the using the relation $k = 2\pi/\lambda$ we obtain:

$$E_{\text{kin}} = \frac{\hbar^2 k^2}{2m} = \frac{h^2}{2m\lambda^2}.$$

So the kinetic energy of the particle is

$$E_{\text{kin}} = \frac{(6.6\times10^{-34})^2}{2\times1.00\times10^{-30}\times(7.00\times10^{-11})^2}\,\text{J}$$
$$= 4.4\times10^{-17}\,\text{J}.$$

Q2.18 The mass of the α-particle is $4m_p$ where m_p is the mass of the proton. Using Equation 2.9 and the relation $k = 2\pi/\lambda$, and remembering that we must convert the kinetic energy from MeV into joules, we find

$$\lambda = \frac{h}{\sqrt{2mE_{\text{kin}}}}$$
$$= \frac{6.6\times10^{-34}}{\sqrt{2\times4\times1.67\times10^{-27}\times6.00\times10^{6}\times1.6\times10^{-19}}}\,\text{m}.$$
$$= 5.8\times10^{-15}\,\text{m}.$$

Interestingly, this value is very similar to the radius of the radium nucleus.

Q2.19 Using the energy form of the uncertainty principle, $\Delta E\Delta t \geq \hbar/2$ we find a minimum uncertainty in the energy of the excited state of

$$\Delta E = \hbar/2\Delta t = (1.1\times10^{-34}/2\times10^{-9})\,\text{J}$$
$$= 5.5\times10^{-26}\,\text{J}$$
$$= 3.4\times10^{-7}\,\text{eV}.$$

Q2.20 (a) The total energy levels of the confined particle are given by Equation 2.18.

$$E_{\text{tot}} = \frac{h^2 n^2}{8mD^2}, \quad \text{where } n = 1 \text{ or } 2 \text{ or } 3, \text{ etc.}$$

Using the information given in the question,

$$2.69\times10^{-36}\,\text{J} = \frac{(6.6\times10^{-34}\,\text{J s})^2\times n^2}{8\times(1\times10^{-10}\,\text{kg})\times(1\times10^{-10}\,\text{m})^2}.$$

From this equation, you should easily be able to show that $n^2 = 49.40$, so the value n of the quantum number that characterizes the particle's wavefunction (which must be an integer) is 7.

(b) When the particle's behaviour is characterized by the $n = 7$ wavefunction (Figure 2.39a), the relative probabilities of detecting the particle in different regions are proportional to $|\psi_{n=7}|^2$ (Figure 2.39b). The particle is most likely to be detected at the values of x at which $|\psi_{n=7}|^2$ has a maximum value.

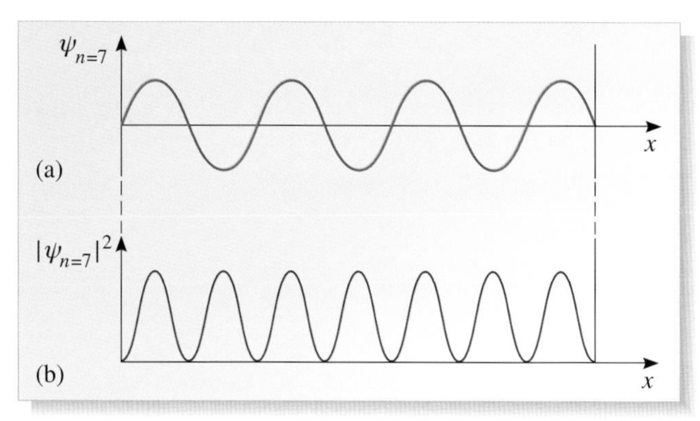

Figure 2.39 Q2.20b.

Q2.21 For $x > D/2$, $E_{\text{pot}} = W$, so the time-independent Schrödinger equation appropriate to this region is

$$\frac{d^2\psi}{dx^2} + \frac{2m}{\hbar^2}(E_{\text{tot}} - W)\psi = 0.$$

We have $\psi = Ae^{-\kappa x}$. Differentiating this twice gives

$$\frac{d^2\psi}{dx^2} = \kappa^2 Ae^{-\kappa x}.$$

Substituting into the Schrödinger equation

$$\kappa^2 Ae^{-\kappa x} + \frac{2m}{\hbar^2}(E_{\text{tot}} - W)Ae^{-\kappa x} = 0.$$

So ψ is a solution to the Schrödinger equation with

$$\kappa^2 = -\frac{2m}{\hbar^2}(E_{\text{tot}} - W) = \frac{2m}{\hbar^2}(W - E_{\text{tot}}),$$

(here κ^2 is a positive number since $W > E_{\text{tot}}$).

Q2.22 The energy of the allowed states in a three-dimensional cubical box is given by Equation 2.23

$$E_{\text{tot}} = \frac{h^2}{8mD^2}(n_1^2 + n_2^2 + n_3^2).$$

Since we are given in this case $E_{\text{tot}} = 19h^2/4mD^2$, we see that $(n_1^2 + n_2^2 + n_3^2) = 38$. There are two different sets of quantum numbers which will give this value, i.e. 6, 1, 1, and 5, 3, 2. The first set can be assigned in three different ways: $(n_1 = 6, n_2 = 1, n_3 = 1)$; $(n_1 = 1, n_2 = 6, n_3 = 1)$; $(n_1 = 1, n_2 = 1, n_3 = 6)$. The second set can be assigned in six different ways (5, 3, 2); (5, 2, 3); (3, 5, 2); (3, 2, 5); (2, 5, 3); (2, 3, 5). Hence the energy level is nine-fold degenerate.

Q3.1 The coordinates of the four points in the Cartesian and spherical polar coordinate systems are shown in Figure 3.43.

(a) The origin in both coordinate systems is denoted by (0, 0, 0) and appropriate units. In the spherical polar coordinate system the origin is at $r = 0$ m, $\theta = 0°$ and $\phi = 0°$.

(b) The distance r of the point from the origin is 2 m. Since the point is on the x-axis, $\theta = 90°$ and $\phi = 0°$. Thus, the spherical polar coordinates of the point are (2 m, 90°, 0°).

(c) The r-coordinate is given by Pythagoras' theorem

$$r = \sqrt{x^2 + y^2} = \sqrt{1^2 + 1^2}\ \text{m} = \sqrt{2}\ \text{m}.$$

Since the point is in the xy-plane, $\theta = 90°$, and since it is equidistant from the x- and y-axes, $\phi = 45°$. The spherical polar coordinates of the point are therefore $(\sqrt{2}\ \text{m}, 90°, 45°)$.

(d) In this case, $r = \sqrt{1^2 + 1^2 + 1^2}\ \text{m} = \sqrt{3}\ \text{m}$. The θ-coordinate is given by $\cos\theta = 1/\sqrt{3}$, i.e. $\theta = 55°$. The ϕ-coordinate is given by $\cos\phi = 1/\sqrt{2}$, i.e. $\phi = 45°$. Hence, the spherical polar coordinates of this point are $(\sqrt{3}\ \text{m}, 55°, 45°)$.

Q3.2 According to Equation 3.5, the possible values of the electron's orbital angular momentum quantum number are $l = 0$, $l = 1$ and $l = 2$. Equation 3.4 then tells us that the possible values of the magnitude of the electron's orbital angular momentum are $L = 0$, $L = \sqrt{2}\,\hbar$ and $L = \sqrt{6}\,\hbar$.

Q3.3 (a) $L = \sqrt{l(l+1)}\,\hbar$ (Equation 3.4) and $L_z = m_l\hbar$ (Equation 3.6).

(b) When $l = 2$, $L = \sqrt{2(2+1)}\,\hbar = \sqrt{6}\,\hbar$ (using Equation 3.4). The possible values of m_l are −2, −1, 0, +1, +2, using Equation 3.7a. Hence the possible values of L_z are $-2\hbar$, $-\hbar$, 0, $+\hbar$, $+2\hbar$, using Equation 3.6.

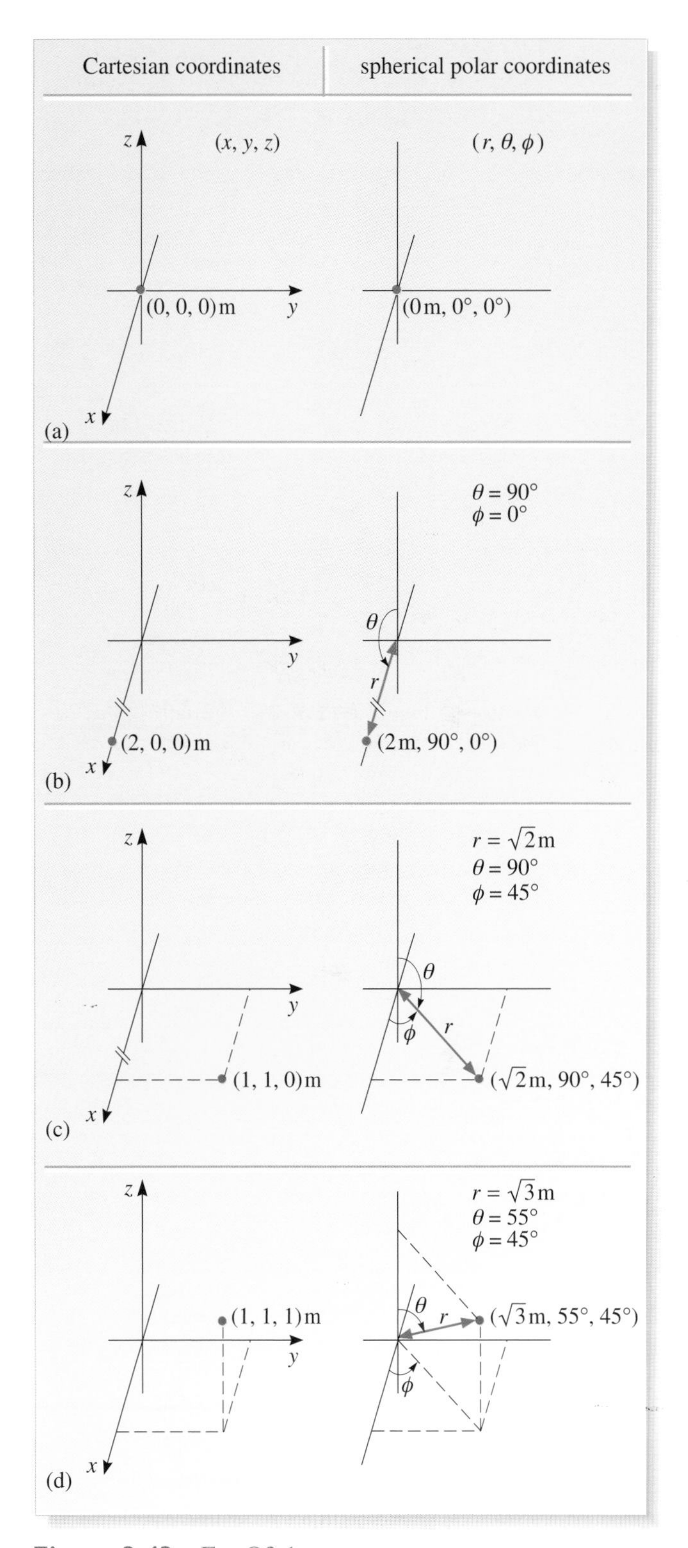

Figure 3.43 For Q3.1.

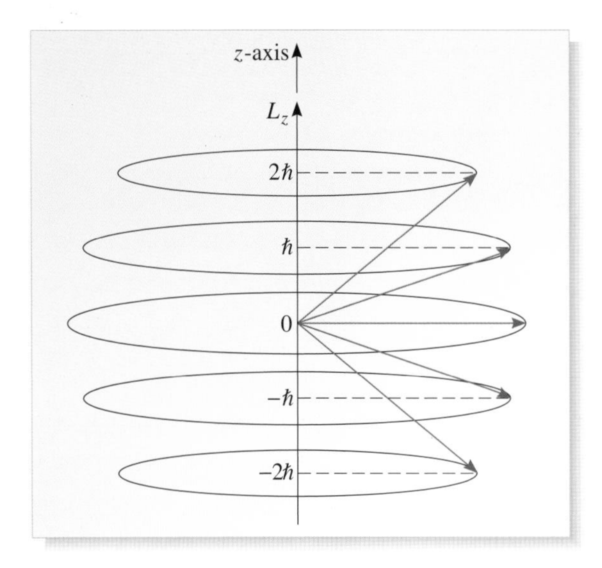

Figure 3.44 Answer to Q3.3(c).

(c) See Figure 3.44.

Q3.4 (a) Since $L = \sqrt{l(l+1)}\,\hbar$ (Equation 3.4) and $L_z = m_l\hbar$ (Equation 3.6), and since the magnitude of m_l is always less than or equal to l (Equation 3.7), the maximum magnitude of L_z is *always* less than L. (If you find this difficult to see algebraically, try to prove the truth of the statement for a few different values of l. For example, if $l = 2$, $L = \sqrt{6}\,\hbar$ and the maximum magnitude of L_z is $2\hbar$ the maximum value of L_z is less than L in this case.)

If $\boldsymbol{L}$ were parallel or antiparallel to the z-axis, the component of the electron's angular momentum vector perpendicular to the z-axis would be zero, i.e. L_z would be equal to L in this hypothetical case. However, we have just shown that L_z is always less than L, so the vector $\boldsymbol{L}$ cannot be aligned along the z-axis.

(b) If the angular momentum quantum number l of the electron is very large, there is a correspondingly large number of possible orientations of its orbital angular momentum vector $\boldsymbol{L}$ with respect to an arbitrary z-axis (Figure 3.45). Eventually, when l is sufficiently large, there are so many allowed values of L_z that it is impossible to observe this quantization. This is the so-called classical limit. (Remember, according to classical mechanics, the angular momentum of the electron should not be quantized!)

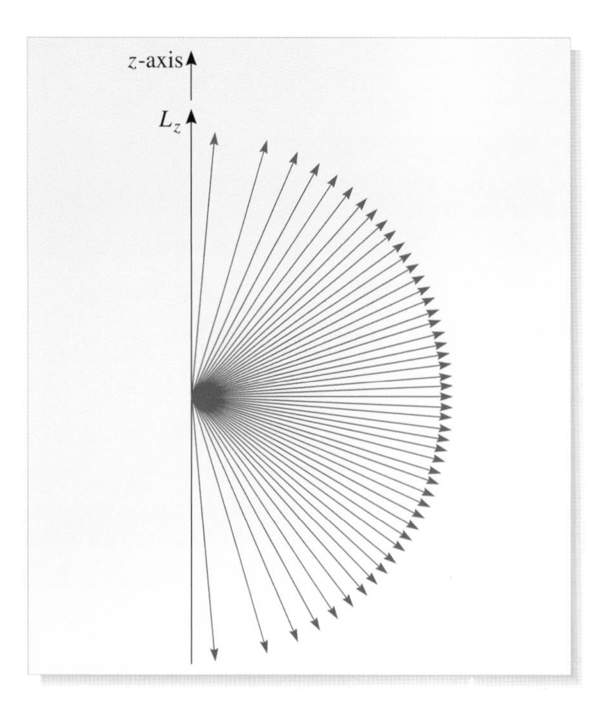

Figure 3.45 For use with Q3.4.

Q3.5 When the hydrogen atom is subject to a magnetic field, the electron wavefunctions characterized by the *same* values of n and l but by *different* values of m_l will correspond to different energy levels. When $l = 1$, $m_l = -1$, 0 or +1 (Equation 3.7a), i.e. there are three possible values of m_l. This implies that when the hydrogen atom is subject to a net magnetic field, the $n = 2$, $l = 1$ energy level will split into three energy levels. Similarly, the $n = 3$, $l = 2$ energy level will split into five. These splittings are shown in Figure 3.46. Notice from this figure that (a) there is no energy shift if $m_l = 0$; (b) the spacings between adjacent energy levels are the same; and (c) the energy increases as the value of m_l increases. These properties all follow from Equation 3.8a. For clarity, the splittings are much exaggerated compared with the difference between the $n = 2$ and $n = 3$ levels.

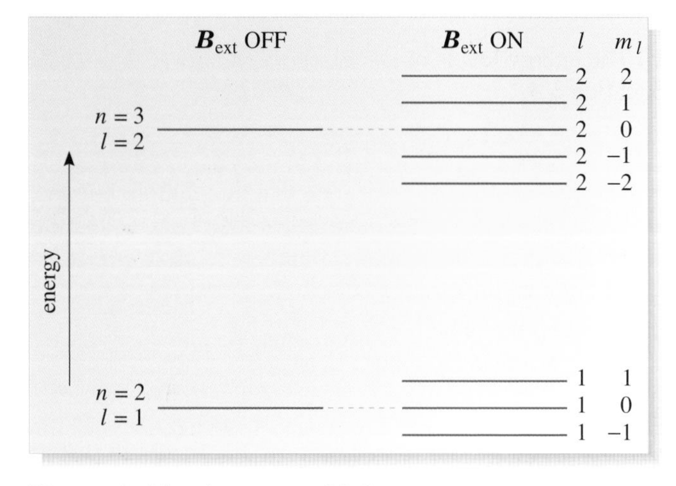

Figure 3.46 Answer to Q3.5.

Q3.6 (a) Table 3.8 shows the completed Table 3.2.

Table 3.8 Quantum states of the electron in the hydrogen atom. (See Q3.6.)

	$l = 0$	$l = 1$	$l = 2$	$l = 3$	$l = 4$
$n = 1$	1s				
$n = 2$	2s	2p			
$n = 3$	3s	3p	3d		
$n = 4$	4s	4p	4d	4f	
$n = 5$	5s	5p	5d	5f	5g

(b) Each quantum state has its own unique set of the quantum numbers n, l and m_l. Corresponding to an orbital angular momentum quantum number l, there are $2l + 1$ possible values of m_l. So for an s state ($l = 0$) there is just one possible value of m_l, and for an f state ($l = 3$) there are 7 possible values of m_l. Combining this with the results in Table 3.8, we can deduce that

for $n = 1$, there is 1 quantum state,

for $n = 2$, there are $1 + 3 = 4$ quantum states,

for $n = 3$, there are $1 + 3 + 5 = 9$ quantum states,

for $n = 4$, there are $1 + 3 + 5 + 7 = 16$ quantum states,

for $n = 5$, there are $1 + 3 + 5 + 7 + 9 = 25$ quantum states.

This illustrates the general result that there are n^2 different quantum states associated with the principal quantum number, n, according to Schrödinger's theory.

Q3.7 (a) The extra piece of information that must be inserted in the arrow-shaped box (top right-hand corner of the figure) is the potential energy function $E_{el} = -e^2/4\pi\varepsilon_0 r$; this is Equation 3.1.

(b) The energy levels of the confined electron are given by Equation 3.3:

$$E_{tot} = -\frac{13.6}{n^2}\,\text{eV}.$$

This equation tells us that the second lowest energy level is -3.40 eV ($n = 2$).

(c) Your completed version of Table 3.2 (Table 3.8, Q3.6) should show that when the electron is in the $n = 1$ energy level ($E_{tot} = -13.6$ eV), it can occupy only a 1s ($n = 1$, $l = 0$) quantum state. Also, Table 3.8 should show that when the electron is in the $n = 2$ energy level, it can occupy a 2s or a 2p quantum state, and that when it is in the $n = 3$ energy level, it can occupy a 3s or a 3p or a 3d quantum state.

Hence, the wavefunctions that describe the confined electron with $E_{tot} = -3.40$ eV ($n = 2$) are ψ_{2s} and ψ_{2p}, and the wavefunctions that describe the confined electron with $E_{tot} = -1.51$ eV ($n = 3$) are ψ_{3s}, ψ_{3p} and ψ_{3d}.

Q3.8 (a) Since 2p quantum states are characterized by the principal quantum number $n = 2$, we apply Equation 3.3 to get

$$E_{tot} = -(13.6\,\text{eV})/2^2 = -3.40\,\text{eV}.$$

(b) No, the electron could be in a 2s quantum state (Table 3.8 and Figure 3.10).

(c) The electron is most likely to be detected at a distance of about 0.5×10^{-10} m from the centre of the nucleus (Figure 3.14).

(d) According to the Bohr model, it should be possible to specify the position and momentum of the orbiting electron simultaneously, contrary to the Heisenberg uncertainty principle. However, according to the Schrödinger equation, the behaviour of the electron is described by its wavefunction, which tells us the *probability* that the electron will be detected in different regions of the hydrogen atom. This quantum-mechanical picture asserts that when the electron is in the ground state, it is *most likely* to be detected at the distance of about 0.5×10^{-10} m from the centre of the nucleus (Figure 3.14). The Bohr model incorrectly predicts that the electron in its ground state will *always* be detected at the Bohr radius $\sim 0.5 \times 10^{-10}$ m (Chapter 1).

Bohr assumed that the angular momentum of the electron was quantized in units of $\hbar$, so that in the ground state the electron would have angular momentum of magnitude $\hbar$ (i.e. $h/2\pi$). The correct quantum-mechanical description, on the other hand, predicts that the electron will have zero angular momentum in the ground state.

Q3.9 In working out the allowed radiative transitions, the following conditions must be applied.

1. For the emission of radiation, an electron must make a transition to a state with *lower* energy. Thus all radiative transitions must correspond to a downward shift between energy levels.
2. The l quantum number can change by one unit only, i.e. $\Delta l = \pm 1$ (Equation 3.10). This means that transitions can occur only to one of the immediate neighbouring columns in Figure 3.19. The allowed transitions are shown in Figure 3.47 (overleaf).

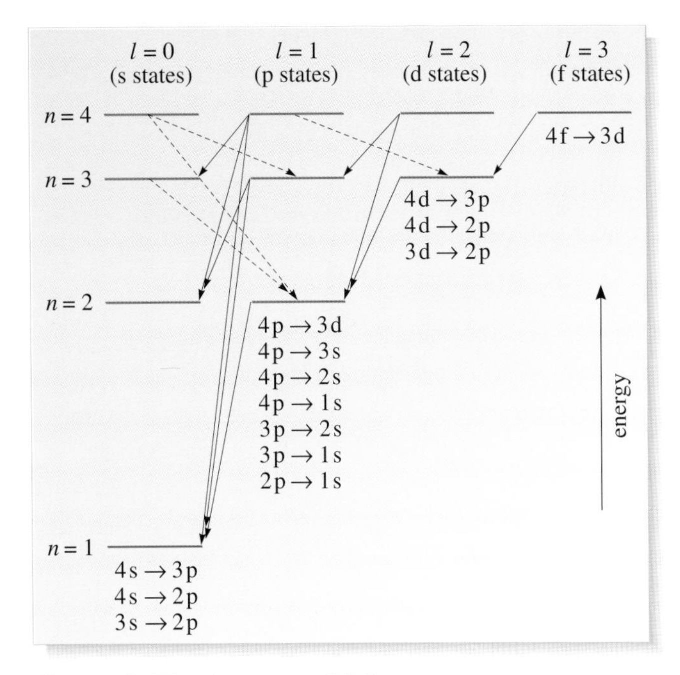

Figure 3.47 Answer to Q3.9.

Q3.10 (a) In Question 3.5, you showed that the 3d ($n = 3$, $l = 2$) level is split into *five* energy levels in an external magnetic field, and the 2p ($n = 2$, $l = 1$) state is split into *three* levels (Figure 3.46). In order to work out the allowed transitions, we must apply the selection rules $\Delta l = \pm 1$, $\Delta m_l = 0$ or ± 1 (Equations 3.10 and 3.11). These show that there are nine possible transitions, and these are indicated by the arrows on Figure 3.48.

Comment *However, because of the equal spacing between the split levels (Q3.5), transitions with the same value of Δm_l correspond to the same energy difference and therefore to the same frequency. Thus, although Figure 3.48 shows that nine different transitions take place, only* three *spectral lines will be observed experimentally.*

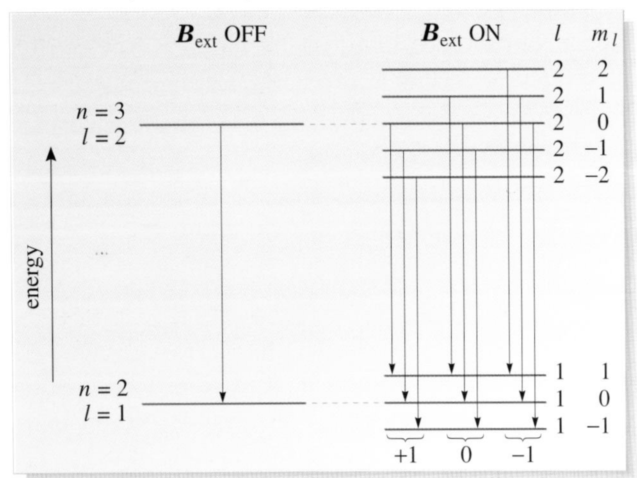

Figure 3.48 Answer to Q3.10.

(b) If the hydrogen atom were not subject to an external magnetic field, neither the 3d nor the 2p level would be split, so only *one* emission line would be produced by transitions between these energy levels.

Q3.11 See Table 3.9, opposite (the completed Table 3.3).

Q3.12 (a) According to Equation 3.12, $S = \sqrt{s(s+1)}\,\hbar$. Since s is always equal to $\frac{1}{2}$, S is always equal to $\sqrt{3}\,\hbar/2$, $(= 0.866\hbar)$.

According to Equation 3.13, $S_z = m_s\hbar$. Since $m_s = +\frac{1}{2}$ or $-\frac{1}{2}$, $S_z = +\hbar/2$ or $-\hbar/2$.

(b) See Figure 3.49.

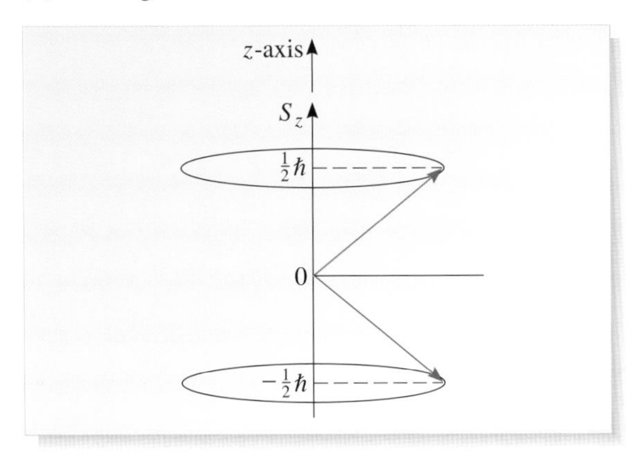

Figure 3.49 Answer to Q3.12(b).

Q3.13 (a) True. The number of orientations to an arbitrarily chosen z-axis is given by $(2s + 1)$, which, with $s = \frac{1}{2}$, is equal to 2 (Box 3.3). The two allowed values of m_s are $\pm\frac{1}{2}$.

(b) False. The magnitude of the spin angular momentum is given by $S = \sqrt{s(s+1)}\,\hbar$. It does not depend on m_s, and has only one value $S = \sqrt{3}\,\hbar/2$.

(c) True. The spin of the electron has to be taken into account whether or not the electron is in a hydrogen atom.

Q3.14 (a) Ten, since its atomic number $Z = 10$.

(b) Eleven. Since its mass number $A = 21$, the nucleus contains 21 particles of which ten are protons, from part (a) of the question — so it must contain 11 neutrons.

(c) Ten. It must contain the same number of electrons as protons, since the atom is neutral.

(d) $+1.6 \times 10^{-18}$ C, since the total charge of the ten protons is $10 \times (+1.6 \times 10^{-19}$ C).

(e) Yes, since all isotopes of neon have atomic number $Z = 10$.

Table 3.9 The Bohr model of the hydrogen atom compared with the Schrödinger model (Q3.11).

	Bohr's model	Schrödinger's model
1 What are the energy levels of the electron when the hydrogen atom is *not* subject to a net magnetic field?	$E_{tot} = -\frac{13.6}{n^2}$ eV where n is Bohr's quantum number; $n = 1$ or 2 or 3, etc.	$E_{tot} = -\frac{13.6}{n^2}$ eV where n is the principal quantum number of the electron; $n = 1$ or 2 or 3, etc.
2 What are the possible values of the magnitude of the electron's orbital angular momentum?	$L = n\hbar$, $n = 1$ or 2 or 3, etc. (Section 5 of Chapter 1).	$L = \sqrt{l(l+1)}\,\hbar$ (Eqn 3.4) where l is the electron's orbital angular momentum quantum number; $l = 0, 1, \ldots, n-1$.
3 Can the electron have zero orbital angular momentum?	No, the minimum value of L is $\hbar$ (see equation in box above).	Yes (see equation in box above).
4 What quantum numbers specify each possible quantum state of the electron?	Bohr quantum number n.	n, l and m_l (Section 2.4) (In Section 4, you will see that the Schrödinger model is incomplete and a fourth quantum number is needed.)
5 What is the role of the electron's quantum number n?	It specifies the magnitude of the electron's angular momentum *and* the electron's energy levels.	It specifies the electron's energy levels (Eqn 3.3).
6 Does the theory correctly predict the intensities of atomic hydrogen's spectral lines?	No.	Yes (Section 3.2).

Q3.15 (a) In standard notation: Be $1s^2\,2s^2$. In box notation:

Be	↑↓	↑↓
	1s	2s

(b) The two electrons in the 1s quantum states have the quantum numbers $n = 1$, $l = 0$, $m_l = 0$, $m_s = +\frac{1}{2}$ and $n = 1$, $l = 0$, $m_l = 0$, $m_s = -\frac{1}{2}$. The two electrons in the 2s quantum states have the quantum numbers $n = 2$, $l = 0$, $m_l = 0$, $m_s = +\frac{1}{2}$ and $n = 2$, $l = 0$, $m_l = 0$, $m_s = -\frac{1}{2}$.

Q3.16 In box notation, the electronic structure of a nitrogen atom in its ground state is:

N	↑↓	↑↓	↑	↑	↑
	1s	2s		2p	

Notice that each of the three electrons in 2p quantum states is unpaired. In this way, Hund's rule is satisfied — the total spin of the electrons in the atom has its maximum possible value.

Q3.17 (a) In a set of d quantum states, the electrons can have five different values of m_l: $-2, -1, 0, +1, +2$. In each of these states, the spin of the electron can be 'up' or 'down', that is $m_s = +\frac{1}{2}$ or $m_s = -\frac{1}{2}$. Hence, the total number of available quantum states is $5 \times 2 = 10$.

(b) Since the electrons can have *seven* different values of m_l ($-3, -2, -1, 0, 1, 2, 3$) when $l = 3$, the total number of quantum states available is $7 \times 2 = 14$, using the same reasoning as in part (a).

(Don't forget to enter your answers to this question in Table 3.6.)

Q3.18 According to Figure 3.29, the order in which the quantum states will be filled is 1s → 2s → 2p → 3s → 3p → 4s → 3d, and so on. Using the data given in Table 3.6, which gives the number of quantum states associated with s ($l = 0$), p ($l = 1$) and d ($l = 2$) states, you should have found that the electronic structure of a rubidium atom in its ground state is, in standard notation:

Rb $1s^2\,2s^2\,2p^6\,3s^2\,3p^6\,4s^2\,3d^{10}\,4p^6\,5s^1$.

Q3.19 (a) $ns^2\,np^6$, where n is the principal quantum number of the outer shell.

(b) $4s^2\,4p^6$. Since krypton is in Period 4 of the Periodic Table, the principal quantum number of each electron in its outer shell is $n = 4$.

(c) The outer shell of the bromine atom will have the structure $4s^2\,4p^5$, since the atom contains one fewer electron than the krypton atom. Bromine is in Group VII, the one next to the group of noble gas elements (Group 0).

It is, like krypton, in Period 4 since in *both* cases the principal quantum number of the outer shell is $n = 4$.

Q3.20 My estimated measurements from Figure 3.33 are $E(3p) = -3.0\,\text{eV}$ and $E(3s) = -5.2\,\text{eV}$. Hence

$$\Delta E = (-3.0 - (-5.2))\,\text{eV} = 2.2\,\text{eV} = 3.5 \times 10^{-19}\,\text{J}.$$

The wavelength is therefore

$$\lambda = \frac{ch}{\Delta E} = \frac{(3.0\times10^{8}\,\text{m}\,\text{s}^{-1})\times(6.6\times10^{-34}\,\text{J}\,\text{s})}{(3.5\times10^{-19}\,\text{J})} = 570\,\text{nm},$$

which is the required value, within the ~5% uncertainties of my measurements.

Q3.21 In the s states the active electron has zero orbital angular momentum. Therefore the spin angular momentum has no orbital angular momentum with which to interact.

Q3.22 In the triplets, both electrons have the same value of m_s, so, since they obey the exclusion principle, they cannot also have the same values of n, l and m_l.

Q3.23 The angular spread of the beam is given by

$$\theta = \lambda/w$$

so the size of the spot after 100 m is $100\theta\,\text{m} = (100 \times 633 \times 10^{-9}/10^{-3})\,\text{m} = 0.063\,\text{m}$ or 6.3 cm.

Q3.24 Amplification can only occur if there are more atoms in the upper level than in the lower level of the laser transition. In the scheme shown in Figure 3.37, this requires most of the atoms to be pumped out of the ground state, which is quite a difficult task. In the scheme shown in Figure 3.39, it only necessary to have more atoms in level E_2 than in level E_1. This is much easier to achieve because not many atoms occupy the excited levels to begin with, and pumping can easily achieve the necessary preponderance of atoms in level E_2 over those in E_1. The scheme based on Figure 3.39 is therefore likely to be more successful.

Q3.25 The emitted light has frequency $f = c/\lambda$ and its photons have energy

$$E = hf = \frac{hc}{\lambda} = \frac{6.6\times10^{-34}\,\text{J}\,\text{s}\times3.00\times10^{8}\,\text{m}\,\text{s}^{-1}}{1.1\times10^{-6}\,\text{m}} = 1.8\times10^{-19}\,\text{J}.$$

The number of photons emitted in the pulse is

$$(200 \times 10^{3}\,\text{J})/(1.8 \times 10^{-19}\,\text{J}) = 1.1 \times 10^{24}.$$

This is equal to the number of atoms that have undergone stimulated emission to form the pulse.

Q3.26 (a) Four quantum numbers are required to completely specify the wavefunction. They are n, l, m_l and m_s.

(b) There are $2n^2 = 32$ states at this energy. They are

$n = 4, l = 0, m_l = 0$ $\quad m_s = \pm\frac{1}{2}$ (2 states)
$n = 4, l = 1, m_l = 0, \pm1$ $\quad m_s = \pm\frac{1}{2}$ (6 states)
$n = 4, l = 2, m_l = 0, \pm1, \pm2$ $\quad m_s = \pm\frac{1}{2}$ (10 states)
$n = 4, l = 3, m_l = 0, \pm1, \pm2, \pm3$ $\quad m_s = \pm\frac{1}{2}$ (14 states).

Q3.27 The energy of the $m_l = 0$ state is unaltered by the magnetic field, but the energies of the $m_l = \pm1$ states are shifted by

$$E_{\text{mag}} = \pm\frac{e\hbar B_{\text{ext}}}{2m_{\text{e}}}. \qquad \text{(Eqn 3.8c)}$$

So $$E_{\text{mag}} = \pm\left(\frac{1.6\times10^{-19}\times1.1\times10^{-34}\times1\times10^{-2}}{2\times9.1\times10^{-31}}\right)\text{J} = \pm9.7\times10^{-26}\,\text{J} = \pm6.0\times10^{-7}\,\text{eV}.$$

Q3.28 There are eight allowed transitions. They are shown on Figure 3.50. The transitions can go in either direction.

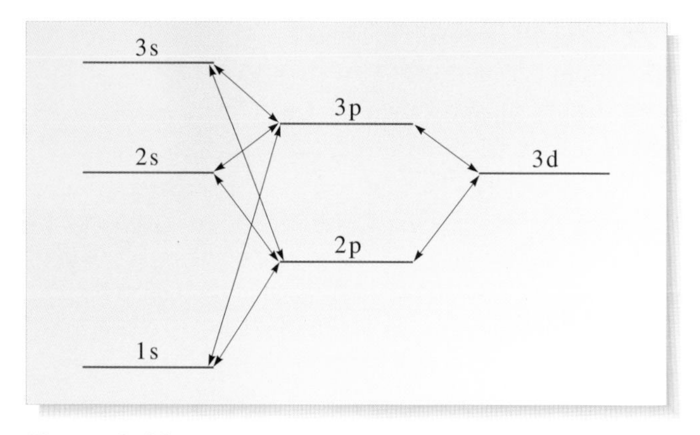

Figure 3.50 For answer to Q3.28.

Q3.29 In box notation the electronic structure of the ground state of oxygen is written as follows:

O [↑↓] [↑↓] [↑↓ | ↑ | ↑]
1s 2s 2p

In standard notation:

O $1s^2\,2s^2\,2p^4$.

Q3.30 Suppose the laser transition involves an upper energy level of E_2 and a lower energy level of E_1, and that there are more atoms in the lower level than the upper level. Then, a photon of energy $E_2 - E_1$ is more likely to encounter an atom in the lower level than in the upper level, so it is more likely to cause absorption than stimulated emission. The population of level E_2 will therefore increase, but it can never exceed the population of level E_1 because, when the two populations become equal, just as many atoms undergo stimulated emission as undergo absorption and equilibrium is reached. More realistically, the number of atoms in the lower level will always be more than the number in the upper level, because spontaneous emission continuously occurs and photons are lost from the system. With more atoms in the lower level than in the upper level, absorption always outweighs stimulated emission and no amplification can occur. To achieve amplification, pumping must be carried out to achieve a greater population of atoms in the upper level than in the lower level (a population inversion).

Q4.1 Although the results of individual measurements performed on a quantum system in a given state are not generally predictable, the possible results of that measurement and the probability of each of those results can be predicted. Hence the words 'different behaviour' might be interpreted as meaning 'predicting different possible outcomes and/or different probabilities for each of those possible outcomes'. In fact, even that interpretation is still not sufficient to ensure validity of the given statement due to the phenomenon of degeneracy. However, we shall not pursue that point here.

Q4.2 From Chapter 3, the energy levels of the hydrogen atom are given by $E_{\text{tot}} = (-13.6\,\text{eV})/n^2$, where n is the principal quantum number of the state. Thus: (a) $E(1\text{s}) = -13.6\,\text{eV}$, and (b) $E(2\text{p}) = -3.40\,\text{eV}$.

Q4.3 The fact that the given state is normalized is guaranteed by (i) the fact that each of the eigenstates Ψ_k is normalized, *and* (ii) the sum of the squared moduli of all of the coefficients in the linear superposition is equal to 1.

In this case those coefficients are $1/\sqrt{10}$, $2/\sqrt{10}$, $-2/\sqrt{10}$, $1/\sqrt{10}$; and

$$\left|\frac{1}{\sqrt{10}}\right|^2 + \left|\frac{2}{\sqrt{10}}\right|^2 + \left|\frac{-2}{\sqrt{10}}\right|^2 + \left|\frac{1}{\sqrt{10}}\right|^2 = \frac{1}{10} + \frac{4}{10} + \frac{4}{10} + \frac{1}{10} = 1.$$

The probability of each eigenvalue is given by the square of the modulus of the coefficient of the corresponding eigenstate. There is no eigenstate corresponding to an eigenvalue of $5\hbar$, so that is not a possible result and therefore has zero probability. For the other results, the probabilities are

$$P(\hbar) = 1/10 \quad \text{and} \quad P(3\hbar) = 4/10.$$

Q4.4 Immediately after the measurement the wavefunction of the atom will be Ψ_1.

Q4.5 The four paired terms in the wavefunction Ψ correspond to the following possible measurement outcomes

$$S_z(\text{A}) = +\hbar/2,\ S_z(\text{B}) = -\hbar/2$$
$$S_z(\text{A}) = +\hbar/2,\ S_z(\text{B}) = +\hbar/2$$
$$S_z(\text{A}) = -\hbar/2,\ S_z(\text{B}) = +\hbar/2$$
$$S_z(\text{A}) = -\hbar/2,\ S_z(\text{B}) = -\hbar/2.$$

Regardless of which electron has its spin component measured first, the result for either electron, may turn out to be either $+\hbar/2$ or $-\hbar/2$, so all possible outcomes may arise. The given wavefunction therefore describes a state in which, prior to any measurement, the spin component is indeterminate. Once the spin component of an electron has been measured, the wavefunction of the pair will collapse and it must do so in such a way that it reflects the measured result. So, if $S_z(\text{A})$ is found to be $+\hbar/2$, the wavefunction will collapse to $\Psi_+(\text{A})[\Psi_+(\text{B}) + \Psi_-(\text{B})]/\sqrt{2}$, indicating that when $S_z(\text{B})$ is measured the result is equally likely to be $+\hbar/2$ or $-\hbar/2$. Similar conclusions apply whatever the result of the first measurement, so there is no correlation between the result of the first and second measurements.

Q4.6

$$\Psi = \tfrac{1}{2} \times [\Psi_+(\text{A})\Psi_-(\text{B}) + \Psi_+(\text{A})\Psi_+(\text{B}) + \Psi_-(\text{A})\Psi_+(\text{B}) + \Psi_-(\text{A})\Psi_-(\text{B})].$$

can be written as the product $\Psi = \Psi(\text{A})\,\Psi(\text{B})$, provided we make the identifications:

$$\Psi(\text{A}) = \tfrac{1}{\sqrt{2}}[\Psi_+(\text{A}) + \Psi_-(\text{A})]$$

and $$\Psi(\text{B}) = \tfrac{1}{\sqrt{2}}[\Psi_+(\text{B}) + \Psi_-(\text{B})].$$

This can be confirmed by simply multiplying out the brackets.

Q4.7 'According to quantum mechanics, the Universe is all fuzzy. Everything suffers from indeterminacy, or, equivalently, indeterminism, and nothing is certain.'

Obviously the first sentence of the above statement is so vague as to be meaningless. It is true that indeterminacy and indeterminism are features of quantum mechanics, but it is too sweeping to say that 'everything' suffers from them. (They are not features of classical mechanics for instance.) Also, it is not true that indeterminacy and indeterminism are equivalent as is implied by the statement. Finally, it is just not true that nothing is certain. Even in quantum mechanics there are measurement outcomes that can be predicted with probability 1, i.e. with certainty. (Any system that is in a state represented by an eigenstate of the measured observable immediately prior to the measurement will furnish such a prediction.)

Q4.8 The six coordinates (x_1, y_1, z_1) and (x_2, y_2, z_2), should not be thought of as the position coordinates of the two particles, since those positions are indeterminate. Rather, each combination of six values should be thought of as a possible configuration of the system when it is measured. The time-dependent wavefunction determines the probability density that is to be associated with each of these possible configurations of the two particle system at time t.

Q4.9 Given that Ψ_1 and Ψ_2 are both normalized, the normalization of $\Psi = a(2\Psi_1 + 3\Psi_2)$ requires that $4a^2 + 9a^2 = 1$. Since a is real, this implies that $a = 1/\sqrt{13}$. The probability of finding that the atom's energy is 1.00 eV is $4a^2 = 4/13$. Since 1.50 eV is not an eigenvalue that corresponds to either of the eigenstates in the superposition, the probability of obtaining 1.50 eV as the outcome of an energy measurement on the given state is zero.

Q4.10 *The Copenhagen interpretation.* Long regarded as the conventional interpretation of quantum mechanics, this interpretation was developed by Bohr and others in the early days of quantum mechanics. It regards the standard formalism as providing the most complete possible account of an individual system. Accepting the indeterminacy and indeterminism of quantum mechanics and Bohr's principle of complementarity, it regards probabilities as an inherent and unavoidable part of the theory, and treats measurement as an unanalysable interaction between a (logically) classical measuring apparatus and a quantum system. There is no experiment that conflicts with this interpretation, which therefore remains the working hypothesis of most physicists.

The many worlds interpretation. An interpretation of quantum mechanics which, in one version at least, asserts the existence of a large number of parallel universes, each of which realizes one of the possibilities encompassed by the branching wavefunction describing the state of a system and the relative state of the rest of the Universe.

The ontological interpretation (Bohm's theory). A realist interpretation of quantum mechanics that replicates all the predictions of quantum mechanics, but does so at the price of introducing a highly non-local 'quantum potential'. It is related to de Broglie's theory, in which the wavefunction was supposed to act as a 'pilot wave', influencing the (classical) motion of particles and guiding them towards the destinations that quantum mechanics predicted.

Q4.11 Most of the outstanding interpretative issues surround the measurement process, its role and its significance. Different authors would list different concerns, but amongst those raised in this chapter are:

- The emergence of classical, everyday experience from a supposedly underlying quantum physics. Of the many possibilities described by an evolving wavefunction, why do we experience only one?
- Related to this, the evolution of the wavefunction and its collapse. Is there any form of time evolution other than that described by the Schrödinger equation? Does the wavefunction ever really collapse?
- Also related is the issue of macroscopic superpositions, such as Schrödinger's partly alive and partly dead cat. What do such states signify? Why do we not observe them?
- Does the measurement process really require us to regard the measuring apparatus as classical, and not amenable to quantum analysis?
- What is the role of probability in physics? Is it inherent and unavoidable, or is any such use of probability merely a sign that we do not know everything we could about a system?
- Do electrons and other such microphysical entities really exist? If so, in what sense do we mean that they exist? Are they there when nobody looks?

Q4.12 Although the collapse is supposed to be communicated instantly, the information that it conveys is not under the control of the experimenter. The result could have gone either way, so there is no way of using the non-local connection to send a message.

Q5.1 The fifth line in the Balmer series would have $n = 7$ in Equation 1.1, i.e.

$$\lambda = 364.56\frac{49}{45}\,\text{nm (in fact this is} = 396.97\,\text{nm). This}$$

corresponds to light in the 'near ultraviolet' and hence it has too short a wavelength to be visible. As n becomes very large, the -4 in the denominator of Equation 1.1 becomes negligible compared to n^2, so λ approaches 364.56 nm as n tends to infinity.

Q5.2 The classical prediction for blackbody radiation was wrong for high frequencies (short wavelength λ), hence the name 'ultraviolet catastrophe', ultraviolet being off the short wavelength (violet) end of the visible spectrum.

It is for high frequencies that the energy quantum hf becomes very large, thus making it more and more difficult to supply the energy to excite these modes. Thus, at high frequencies, the average energy per mode is much lower than the classical prediction.

Q5.3 (a) The kinetic energy of the electrons depends on the wavelength of the incident radiation. The wavelength determines the energy of each photon whereas the intensity relates to the total energy of many photons — one can have a high intensity with a large number of low-energy photons. It is the energy of each photon which is the key property here.

(b) λ_t is an upper limit on the wavelength. Short wavelength corresponds to high frequency ($f\lambda = c$) which corresponds to high energy ($E = hf$). This last bit of reasoning should become automatic for you.

Q5.4 The molar heat capacities of solids become very small at very low temperatures. (In the case of diamond however, the molar heat capacity is considerably lower than $3R$ even at room temperature.)

Q5.5 10 mW is $10^{-2}\,\mathrm{J\,s^{-1}}$. Each photon has an energy of $hf = hc/\lambda$. The number of photons per second is just the number of joules per second divided by the number of joules per photon. (*Note*: $1\,\mathrm{W} = 1\,\mathrm{J\,s^{-1}}$.) The energy per photon is

$$\frac{hc}{\lambda} = \frac{6.63\times10^{-34}\,\mathrm{J\,s}\times3\times10^{8}\,\mathrm{m\,s^{-1}}}{780\times10^{-9}\,\mathrm{m}} \approx 2.6\times10^{-19}\,\mathrm{J}.$$

So the number of photons per second is (roughly)

$$\frac{1\times10^{-2}\,\mathrm{J\,s^{-1}}}{2.6\times10^{-19}\,\mathrm{J}} \approx 4\times10^{16}\,\mathrm{s^{-1}}.$$

Q5.6 The intercept on the vertical axis occurs at about -4.5×10^{-19} J. Indicating a work function of

$$4.5\times10^{-19}\,\mathrm{J} = \frac{4.5\times10^{-19}}{1.6\times10^{-19}}\,\mathrm{eV} = 2.8\,\mathrm{eV}.$$

Q5.7 Electrons moving in, for example, circular orbits in an atom would be constantly accelerating. According to classical theory, accelerating charges radiate electromagnetic radiation and thus lose energy. The electrons should therefore eventually fall into the nucleus as they lose their energy. (Just as a satellite moving in an orbit low enough to lose energy to atmospheric friction will fall to Earth.)

Q5.8 To obtain the same magnitude of the force, Z_1Z_2/r^2 must be the same in each case. Since $Z_1 = 2$ is fixed (the α-particle), if we increase Z_2 by a factor of four, we must increase r^2 by a factor of four, so r, the α-particle–uranium separation, must be doubled to 2×10^{-14} m.

Q5.9 A photon having energy >13.6 eV can remove the electron from the atom completely, i.e. the atom will be ionized and the electron will have a positive energy and can move to infinity. A photon with energy <13.6 eV cannot ionize a hydrogen atom in its ground state.

When we say that the energy of the ground state of the hydrogen atom is −13.6 eV, this is with respect to a zero of energy defined as that which the proton and electron would have if the electron was at rest (no kinetic energy) and at an infinite distance from the proton (the potential energy $-e^2/(4\pi\varepsilon_0 r)$ is effectively zero at $r = \infty$, i.e. 'very far away').

Q5.10 (a) Since $\lambda = h/p$, doubling the momentum will halve its de Broglie wavelength.

(b) Since kinetic energy E_{kin} (non-relativistically) is

$\frac{1}{2}mv^2 = \frac{p^2}{2m}$, we have momentum $p = \sqrt{2mE_{\mathrm{kin}}}$.

Hence doubling E_{kin} multiplies p by $\sqrt{2}$, and thus its de Broglie wavelength is decreased, being multiplied by $1/\sqrt{2}$.

Comment *The statement $v \ll c$ is a signal to use non-relativistic equations.*

Q5.11 Higher energy means shorter wavelength, as in the previous question. Shorter wavelength means a less spread out diffraction pattern. For a single slit, the angular width of the central maximum is $\theta \approx \lambda/w$.

(It is a general fact about diffraction that the shorter the wavelength, the smaller the angle through which the waves are diffracted. You can remember this if you realize that we don't generally notice the diffraction of light since the wavelength is much smaller than the sizes of everyday objects. For the case of electrons, the expression $n\lambda = d\sin\theta_n$ shows that smaller λ means smaller $\sin\theta_n$ and hence the smaller θ_n will be for any given order n. You should be familiar with the fact that, unless $\theta > \pi/2$, larger θ implies larger $\sin\theta$.)

Q5.12 If you determine which of the slits the electron goes through, the two-slit diffraction pattern will be destroyed. (This fact points to some of the most profoundly mysterious aspects of the quantum world.)

Q5.13 Proton A could be more closely defined in position. This is an immediate consequence of the uncertainty principle, since B has the more closely defined momentum (since it has the more closely defined de Broglie wavelength).

Comment *Behind this is the purely mathematical fact, originating in Fourier theory, that it requires more frequency components (or equivalently, wavelength components) to describe a wave packet which is confined to a smaller region than are required for a more extended wave packet.*

Q5.14 The wavefunction for a particle bound in an infinite square well must be *zero* outside the well. This means that the particle has zero probability of being found outside the well.

Q5.15 (a) None. (b) One. (c) $n-1$.

The nth wavefunction has $n-1$ *nodes* between the walls, i.e. $n-1$ places where it has zero amplitude; this can be seen in Figure 5.2. It is a general property of solutions of Schrödinger's equation that the shorter the wavelength, the higher the energy.

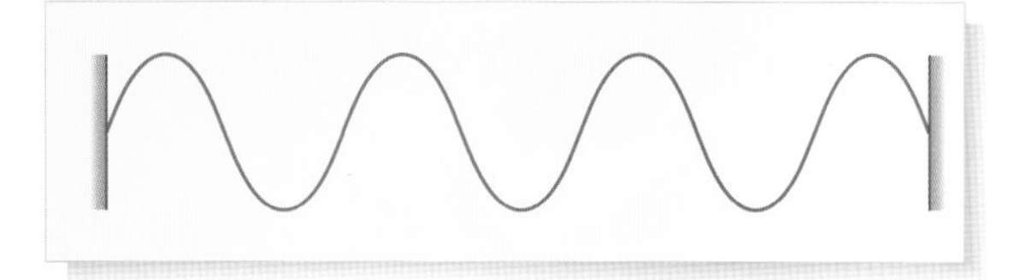

Figure 5.2 The $n = 7$ wavefunction for a particle in an infinite square well. There are six places between the walls where $\psi = 0$. For Q5.15.

Q5.16 (a) All the energies will be multiplied by a factor of four. This is because the energy of the nth state of an infinite square well of width D is $n^2h^2/(8mD^2)$.

(b) If the particle is confined to a narrower well, then its position is better defined simply because it must be located within the well. The uncertainty principle suggests that its momentum must then be less well defined. This is indeed the case since the magnitude of the momentum is increased as the wavelength is shortened (to fit waves of given n into the well). Since the bound state is a standing wave involving a superposition of motions in each direction across the well, the uncertainty in momentum is increased since the two possible equal and opposite momenta are both increased.

Q5.17 True. This can be seen by examining, for example, the wavefunctions in Figure 5.3. (D is the width of the well.)

Q5.18 An energy level is degenerate when there are several different quantum states with that same energy. For example, the energy levels of a particle confined in a three-dimensional infinite cubical well of side D are given by

$$E_{\text{tot}} = \frac{h^2}{8mD^2}(n_1^2 + n_2^2 + n_3^2)$$

where you will see that $n_1^2 + n_2^2 + n_3^2$ can take the value 6 for three different combinations of n_1, n_2 and n_3.

Q5.19 Yes, the concept of degeneracy is very relevant to the hydrogen atom. Within the model presented in this course, there are $2n^2$ states with the same energy for principal quantum number n, the factor 2 being due to spin.

Comment *In fact, relativistic effects slightly 'lift the degeneracy' leading to 'fine structure' in the spectrum revealed by spectroscopic instruments more powerful than simple prism or grating spectrometers.*

Q5.20 (a) $\hbar$, (b) 0.

Comment *Bohr, and therefore the world, were very lucky that the Coulomb potential gives the correct energies even though the states in the Bohr model have the wrong orbital angular momenta.*

Q5.21 (a) $n = 3$, $l = 1$; (b) $n = 4$, $l = 0$; (c) $n = 5$, $l = 3$.

Q5.22 3f and 1p are not allowed since these have $l = n$, breaking the rule that the maximum value of l is $n - 1$.

Q5.23 (a) True: s states are spherically symmetric and the wavefunction passes through zero at $n - 1$ places on the r-axis (see Figure 3.15).

(b) False: The electron cloud picture for a 2p electron is shown in Figure 3.16. There is no value of r for which $|\psi|^2 = 0$ at all angles.

Q5.24 Because there are two possible values of the z-component, S_z, of the electron spin angular momentum and so *only* two electrons can have particular n, l, m_l because the Pauli exclusion principle states that only one electron can have a particular set of four quantum numbers n, l, m_l and m_s.

Q5.25 $2n^2 = 18$.

Comment *If you have some knowledge of chemistry you may recognize the series of numbers given by $2n^2$, i.e. 2, 8, 18, 32 …*

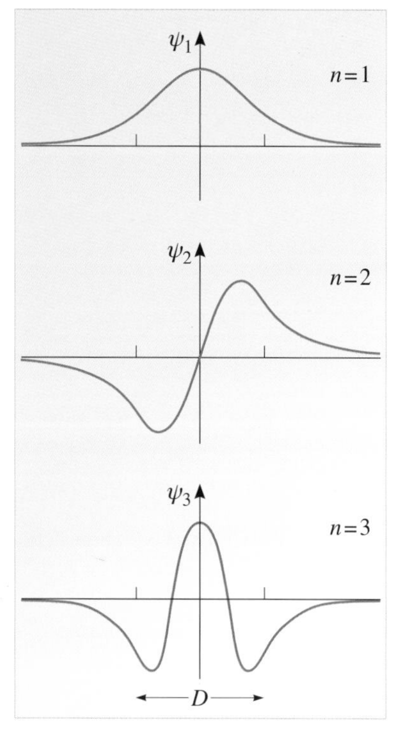

Figure 5.3 The first three (lowest energy) wavefunctions for a particle in a finite square well. For Q5.17.

Q5.26 The notation given for (a) fluorine and (b) neon is correct, whereas that for (c) sodium is incorrect. The Pauli exclusion principle does not permit seven electrons in the 2p subshell. The seventh electron in the ground state of sodium goes into a 3s state. So the correct notation for sodium is $1s^2\,2s^2\,2p^6\,3s^1$.

Q5.27 It is $\frac{1}{2}|\Psi_1 + \Psi_2|^2$. The occurrence of interference effects requires you to add before you square.

You can get some idea by comparing $\cos^2\theta + \sin^2\theta = 1$, with $(\cos\theta + \sin\theta)^2 = \cos^2\theta + \sin^2\theta + 2\sin\theta\cos\theta = 1 + \cos 2\theta$.

The first quantity (square then add) is constant while the second (add then square) is oscillatory.

Q5.28 The time-dependent Schrödinger equation.

Q5.29 (a) The possible values are the eigenvalues of the observable. (b) The values are not possessed prior to the measurement, as Bell's theorem shows.

Q5.30 *Indeterminacy* refers to the fact that, in general, observables cannot be simultaneously measured to give unique values for a system. A good example is that of position and momentum which do not have simultaneously well-defined values for a state.

Indeterminism refers to the fact that, in quantum mechanics, identical states with specific measured values do not lead to identical future measured values.

Comment *This superficially contradicts one of the basic traditional tenets of science, namely, that the same situations must have the same outcomes. To apply this tenet to quantum physics we must acknowledge the fact that 'outcomes' may mean a probability distribution rather than a single result.*

Suggestions for further reading

P. C. W. Davies and J. R. Brown (eds.) (1986), *The Ghost in the Atom*, Cambridge.

Roland Omnès (1999), *Understanding Quantum Mechanics*, Princeton.

David Deutsch (1997), *The Fabric of Reality*, Penguin Books.

D. Bohm and B. J. Hiley (1995), *The Undivided Universe*, Routledge.

Acknowledgements

Grateful acknowledgement is made to the following sources for permission to reproduce material in this book:

Front cover – High power, frequency doubled, titanium–sapphire laser pumped by a copper vapour laser. Courtesy of Dr D. W. Coutts, Oxford Institute for Laser Science, Clarendon Laboratory, University of Oxford;

Fig. 1.1, 1.16 American Institute of Physics; *Fig. 1.7 Thermal radiation and the origin of quantum theory,* Eisberg, R. M., Fundamentals of Modern Physics, Chapter 2, Copyright © (1961). Reprinted by permission of John Wiley & Sons, Inc; *Fig. 1.10* Basle University; *Fig. 1.14, 1.21, 1.30* Science and Society Picture Library/Science Museum; *Fig. 1.19* Caltech; *Fig. 1.27* Cavendish Laboratory University of Cambridge; *Fig. 1.36* Ullstein Bilderdienst; *Fig. 1.39* Reprinted with permission from *Nature* 122: 279–282, Copyright (1928) Macmillan Magazines Limited;

Fig. 2.1 Institut Solvay; *p. 48* Electron two-slit interference pattern, Jonsson, Professor C, Kalmbach Publishing Co. Milwaukee, WI; *Fig. 2.11* Bildersienst Suddent Scher Kerlaos; *Fig. 2.14* Corbis/Bettmann; *Fig. 2.15* Taken from Walter Moore, *Schrödinger Life and Thought,* (1989) p. 193/Cambridge University Press; *Fig. 2.24* Camera Press;

Fig. 3.1 Doug Martin/Science Photo Library; *Fig. 3.2* Courtesy of Sebastian de Echaniz; *Figs. 3.12, 3.15, 3.16, 3.17* Electron cloud pictures. Thanks are due to Robert Hasson; *Fig. 3.24* Science and Society Picture Library/Science Museum; *Fig. 3.38* University of California, Berkeley; *Fig. 3.41a* Philippe Plailly/Eurelios/Science Photo Library; *Fig. 3.42b* NASA/Science Photo Library;

Poem p.165 Oxymoron Humour Archive, http://paul.merton.ox.ac.uk *Fig. 4.16* Photo obtained from University of St. Andrews Library; *Fig. 4.20* CERN; *Fig. 4.24* Bohm, D. and Hiley, B. J. (1993) *The Undivided Universe,* Routledge; *Fig. 4.25* Nick D. Kim.

Every effort has been made to trace all the copyright owners, but if any has been inadvertently overlooked, the publishers will be pleased to make the necessary arrangements at the first opportunity.

Index

Entries and page numbers in **bold type** refer to key words which are printed in **bold** in the text and which are defined in the Glossary.